ADVANCES IN ELECTRONICS AND ELECTRON PHYSICS

VOLUME 44

CONTRIBUTORS TO THIS VOLUME

A. A. Barybin
David J. Bates
Richard I. Knight
Lloyd Kusak
Steven T. Manson
Aris Silzars
Salvatore Spinella
J. L. Teszner
S. Teszner
Gernot M. R. Winkler

Advances in Electronics and Electron Physics

EDITED BY
L. MARTON

Smithsonian Institution, Washington, D.C.

Assistant Editor
CLAIRE MARTON

EDITORIAL BOARD

T. E. Allibone E. R. Piore
H. B. G. Casimir M. Ponte
W. G. Dow A. Rose
A. O. C. Nier L. P. Smith
F. K. Willenbrock

VOLUME 44

1977

ACADEMIC PRESS   New York San Francisco London
A Subsidiary of Harcourt Brace Jovanovich, Publishers

COPYRIGHT © 1977, BY ACADEMIC PRESS, INC.
ALL RIGHTS RESERVED.
NO PART OF THIS PUBLICATION MAY BE REPRODUCED OR
TRANSMITTED IN ANY FORM OR BY ANY MEANS, ELECTRONIC
OR MECHANICAL, INCLUDING PHOTOCOPY, RECORDING, OR ANY
INFORMATION STORAGE AND RETRIEVAL SYSTEM, WITHOUT
PERMISSION IN WRITING FROM THE PUBLISHER.

ACADEMIC PRESS, INC.
111 Fifth Avenue, New York, New York 10003

United Kingdom Edition published by
ACADEMIC PRESS, INC. (LONDON) LTD.
24/28 Oval Road, London NW1

LIBRARY OF CONGRESS CATALOG CARD NUMBER: 49–7504

ISBN 0–12–014644–4

PRINTED IN THE UNITED STATES OF AMERICA

CONTENTS

CONTRIBUTORS TO VOLUME 44	vii
FOREWORD	ix

Atomic Photoelectron Spectroscopy. II
STEVEN T. MANSON

I. Photoionization Cross Sections	1
II. Photoelectron Angular Distributions	15
III. Summary and Concluding Remarks	29
References	30

Timekeeping and Its Applications
GERNOT M. R. WINKLER

I. Introduction	34
II. Clocks	43
III. Time Determination	73
IV. Remote Time Measurements—Synchronization	77
V. Applications	88
References	90

Electrodynamic Concepts of Wave Interactions in Thin-Film Semiconductor Structures. I
A. A. BARYBIN

I. Introduction	99
II. Searches for Methods of Design and Analysis of Solid-State Traveling-Wave Amplifiers	100
III. Generalized Theory of Normal-Mode Excitation in Active Polarized-Medium Waveguides by External Sources	110
References	135

Microwave Power Semiconductor Devices. II
Critical Review
S. TESZNER AND J. L. TESZNER

Three-Terminal Devices

I. Bipolar Transistors	141
References for Section I	172
II. Field-Effect Transistors	174
References for Section II	213

Present Trends and Future Development Prospects for Microwave Power Semiconductor Devices

I. Present Trends	216
II. Future Development Prospects	217

Electron-Bombarded Semiconductor Devices
DAVID J. BATES, RICHARD I. KNIGHT,
SALVATORE SPINELLA, AND ARIS SILZARS

I. Introduction	221
II. Analytic Foundation	232
III. Physical Limitations to EBS Capabilities	238
IV. Target Assembly	246
V. Some Device Configurations and Performance	260
VI. Life and Reliability	276
VII. Future Potential	277
References	280

Basic Concepts of Minicomputers
LLOYD KUSAK

I. Historical Development	283
II. Characteristics of Minicomputer CPUs	285
III. Instruction Repertoire	287
IV. Data Checking	293
V. Bus Architecture	295
VI. Power-Fail Protection	297
VII. Memory Protect	298
VIII. Memory Interleaving	298
IX. Cache Memory	299
X. Interrupt Structure	300
XI. Direct Memory Access	305
XII. Microprogrammable Systems	306
XIII. Nonminicomputer Systems	308
XIV. Peripheral Devices	310
XV. Computer Consoles	310
XVI. Printing Devices	312
XVII. Data Entry I/O	313
XVIII. Magnetic Tapes	314
XIX. Tape Cassettes	315
XX. Disk Systems	315
XXI. Floppy Disks	316
XXII. Process Control Peripherals	317
XXIII. IEEE Standard Digital Interface	319
XXIV. Data Set Interfaces or Modems	320
XXV. Software	321
XXVI. Disk Operating Systems	325
XXVII. Growth of Minicomputer Applications	328
Glossary	330
References	344

AUTHOR INDEX	347
SUBJECT INDEX	359

CONTRIBUTORS TO VOLUME 44

Numbers in parentheses indicate the pages on which the authors' contributions begin.

A. A. BARYBIN, Department of Electron-Ion Processing of Solids, V. I. ULYANOV (Lenin) Electrical Engineering Institute, Leningrad, U.S.S.R. (99)

DAVID J. BATES, Watkins-Johnson Company, Palo Alto, California (221)

RICHARD I. KNIGHT, Watkins-Johnson Company, Palo Alto, California (221)

LLOYD KUSAK, Hewlett-Packard Company, Rolling Meadows, Illinois (283)

STEVEN T. MANSON, Department of Physics, Georgia State University, Atlanta, Georgia (1)

ARIS SILZARS, Tektronix, Inc., Beaverton, Oregon (221)

SALVATORE SPINELLA, Watkins-Johnson Company, Palo Alto, California (221)

J. L. TESZNER, Direction des Recherches et Moyens d'Essais, France (141)

S. TESZNER, Centre National d'Etudes des Telecommunications, France (141)

GERNOT M. R. WINKLER, U.S. Naval Observatory, Washington, D.C. (33)

FOREWORD

The second part of S. T. Manson's review on "Atomic Photoelectron Spectroscopy" (the first part appeared in Volume 41 of these Advances) introduces our present volume. While the first part concentrated on the theoretical treatment of photoionization and of photoelectron angular distributions, in this second part the emphasis is on a comparison of theory with experiment, on the predictive power of theory, and on areas in which work is called for.

G. M. R. Winkler discusses the very important and interesting subject of "Timekeeping and Its Applications." Present-day physics and technology require an increasingly precise knowledge of time, both elapsed and "absolute." The philosophical and historical discussion of the subject in the introductory part of the review puts the whole problem in its proper perspective, so that the actual methods used can be succinctly presented. The review closes with a short survey of the principal applications of modern timekeeping methods.

In his review of "Electrodynamic Concepts of Wave Interactions in Thin Film Semiconductor Structures" A. A. Barybin shows how investigations of wave propagation and instabilities in solid state plasmas led to the conception of solid state traveling-wave amplifiers. After a discussion of methods of design and analysis, he investigates in this first part of the review the general theory of normal mode excitation of such waveguides by external sources.

In our Volume 39 (1975) we published the first part of a review on "Microwave Power Semiconductor Devices" by S. Teszner and J. L. Teszner. Whereas the first part was devoted to a discussion of two-terminal devices, the second part considers in detail two three-terminal devices: the bipolar transistor and the field effect transistor. For both classes of devices the general theory is followed by an examination of the technologies necessary for the production of these devices, as well as a discussion of their electrical characteristics. The review ends with an extrapolation to future development prospects for microwave power semiconductor devices.

D. J. Bates, R. I. Knight, S. Spinella, and A. Silzars present a review of "Electron Bombarded Semiconductor Devices." The development of these devices went through a long period of neglect, and it is only recently that their usefulness has been recognized. As the authors point out, the EBS is a hybrid vacuum tube-semiconductor device, in which both technologies are of essentially equal importance. A short survey of the principle of operation is followed by a discussion of the general characteristics and an analysis of the relevant devices. An examination of the physical limitations to EBS capabilities is followed by a review of the technologies required for the production of the devices, as well as a survey of some device configurations and their performance.

The last review in this volume, by L. Kusak, is on "Basic Concepts of Minicomputers." A historical survey is followed by a description of the characteristics of the central processing unit of the minicomputer, of its memory, of the instruction repertoire, of the data checking facilities, of the protection against power failures, and of memory interactions. After a discussion of memory access, the author examines in detail many of the related technologies and applications.

The following is a list of reviews we expect to publish in forthcoming volumes:

In Situ Electron Microscopy of Thin Films	A. Barna, P. B. Barna, J. P. Pocza and I. Pozsgai
High Injection in a Two-Dimensional Transistor	W. L. Engl
Physics of Ion Beams from a Discharge Source	G. Gautherin and C. Lejeune
Physics of Ion Source Discharges	G. Gautherin and C. Lejeune
Terminology and Classification of Particle Beams	B. W. Schumacher
On Teaching of Electronics	H. E. Bergeson and G. Cassidy
Wave Propagation and Instability in Thin Film Semiconductor Structures. II	A. A. Barybin
The Gunn–Hilson Effect	M. P. Shaw
A Review of Applications of Superconductivity	W. B. Fowler
Minicomputer Technology	C. W. Rose
Digital Filters	S. A. White
Physical Electronics and Modeling of MOS Devices	J. N. Churchill, T. W. Collins, and F. E. Holmstrom
Thin Film Electronics Technology	T. P. Brody
Characterization of MOSFETs Operating in Weak Inversion	R. J. Van Overstraeten
Electron Impact Processes	S. Chung
Sonar	F. N. Spiess
Microchannel Electron Multipliers	R. F. Potter
Electron Attachment and Detachment	R. S. Berry
Noise in Solid State Devices	E. R. Chenette and A. van der Ziel
Radar Signal Processing	Merrill I. Skolnik
Electron Beam Controlled Lasers	Charles Cason
Amorphous Semiconductors	H. Scher and G. Pfister
Electron Beams in Microfabrication. I and II	P. R. Thornton
Photoacoustic Spectroscopy	A. Rosencwaig
Design Automation of Digital Systems. I and II	W. G. Magnuson and Robert J. Smith
Wire Antennas	P. A. Ramsdale

FOREWORD

Electron Microdiffraction	J. M. Cowley
Ion Beam Technology Applied to Electron Microscopy	J. Franks
Microprocessors in Physics	A. J. Davies
Time-Resolved Laser Fluorescence Spectroscopy	J. Delpech
The Edelweiss System	J. Arsac
A Computational Critique of an Algorithm for the Enhancement of Bright Field Electron Microscopy	T. A. Welton
Magnetic Liquid Fluid Dynamics	R. E. Rosensweig
Fundamental Analysis of Electron–Atom Collisions Processes	H. Kleinpoppen
Auger Spectroscopy	N. J. Taylor
Electronic Clocks and Watches	A. Gnädinger
Review of Hydromagnetic Shocks and Waves	A. Jaumotte and Hirsch
Beam Waveguides and Guided Propagation	L. Ronchi
Recent Developments in Electron Beam Deflection Systems	E. F. Ritz, Jr.
Seeing with Sound	A. F. Brown

Supplementary Volumes:

Image Transmission Systems	W. K. Pratt
Computer Techniques for Image Processing in Electron Microscopy	W. G. Saxton
High-Voltage and High-Power Applications of Thyristors	G. Karady

Our thanks go again to the many friends who helped us organize this and all other volumes of *Advances in Electronics and Electron Physics*. We are also grateful for the suggestions which have reached us and hope that many more will follow in the future.

L. MARTON
C. MARTON

Atomic Photoelectron Spectroscopy. II*

STEVEN T. MANSON

Department of Physics
Georgia State University
Atlanta, Georgia

I. Photoionization Cross Sections 1
II. Photoelectron Angular Distributions 15
III. Summary and Concluding Remarks 29
   References 30

I. Photoionization Cross Sections

In this section, the subject of partial photoionization cross sections, i.e., subshell photoelectron cross sections, will be discussed. The discussion will focus on three aspects of photoelectron cross sections: the variety of phenomena that are observed, the physical explanation of these phenomena, and the comparison of experimental results with the predictions of the various theoretical formulations described in Section II,A in Part I of this review (Manson, 1976), hereafter referred to as I.

In connection with the comparison of theory and experiment, note that in an energy region where the cross section for a particular subshell dominates the total photoionization cross section, measurements of photoabsorption will provide substantially the same information as photoelectron spectroscopy of that subshell. Thus, in these cases, we shall compare theory with the results of photoabsorption experiments when no photoelectron results are available.

Figure 1 shows the photoionization cross section of free atoms of sodium in the threshold region. The 3s is the only energetically accessible subshell in this energy range, so the cross section is entirely due to the 3s and is thus entirely equivalent to photoelectron spectroscopy results. The experimental cross section (Hudson and Carter, 1967a,b) shows a zero minimum just above threshold, and this behavior is reproduced quite well by the simple central field HS calculation (Manson and Cooper, 1968). This minimum,

---

* This work was supported by the National Science Foundation and the U.S. Army Research Office.

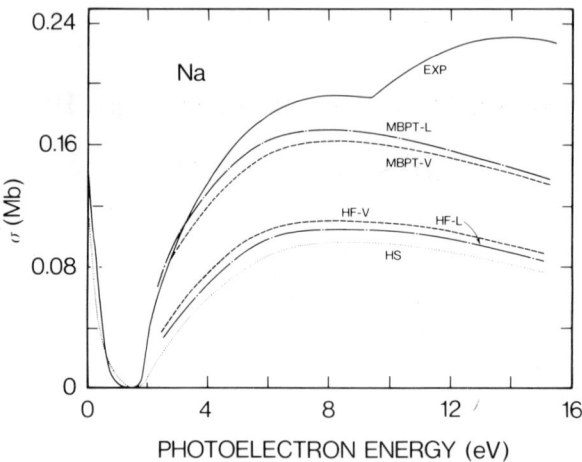

FIG. 1. Photoionization cross section of sodium. The experimental results are from Hudson and Carter (1967a,b), the HS results from Manson and Cooper (1968), and the HF and MBPT results from Chang (1975).

which is quite pronounced in sodium, is known as a Cooper (1964) minimum and was first explained by Seaton (1951). It is due to the $3s \to \varepsilon p$ dipole matrix element going through a zero because of the positive and negative contributions just cancelling. If we use the convention that all wave functions have positive slope at the origin, the dipole matrix element is negative at threshold and becomes positive above the Cooper (1964) minimum. Using this convention, it can be shown that *all* $\langle nl|r|\varepsilon l'\rangle$ dipole matrix elements are positive at high enough energy (Fano and Cooper, 1968). Thus, if the dipole matrix element for a photoionization process is negative at threshold, it is certain that a Cooper minimum will appear at some energy.

Also shown in Fig. 1 are the theoretical HF and MBPT results (Chang, 1975), but only above the minimum since at lower energies they are significantly the same as the HS cross section. At the higher energies, it is seen that the HS and HF results are significantly the same; the MBPT results are much closer to the experimental values but still fail to reproduce the "structure" seen just above 9 eV. Since the MBPT calculation includes the effects of ground state correlation, final state correlation, *and* interchannel coupling, it is difficult to see what could be omitted that would cause the structure seen. Thus, as pointed out by Chang (1975), "a review of the experimental situation would be desirable."

Cooper minima are found quite often in the photoionization of atoms in the *ground* state and some general rules about their behavior over the periodic system can be enunciated owing to extensive calculations (Cooper, 1962, 1964; Manson and Cooper, 1968; Fano and Cooper, 1968; Combet

Farnoux, 1969, 1971, 1972; Kennedy and Manson, 1972; Manson, 1973). First it is found that they appear only in $l \to l + 1$ transitions, although one case of a minimum appearing in an $l \to l - 1$ photoionizing transition has been found through photoelectron polarization measurements (Heinzmann et al., 1976). Cooper minima appear only for outer and near outer subshells and only for subshells whose wave functions have nodes, i.e., they do not appear for photoionization of 1s, 2p, 3d, or 4f electrons. In addition, the Cooper minimum appears for a given (noded) subshell as soon as it becomes bound in the ground state and, with increasing Z, moves toward threshold and into the discrete spectrum, although not necessarily monotonically. Finally, the Cooper minima are generally not zero minima because for np, nd, and nf subshells, even though the $l \to l + 1$ dipole matrix element vanishes, the $l \to l - 1$ does not; in fact, the minimum can be overwhelmed entirely by the $l \to l - 1$ channel and not show up experimentally as a minimum at all.

Some examples of calculated Cooper minima (Manson and Cooper, 1968) are shown in Fig. 2 for the $3p \to \varepsilon d$ photoionizing transition. From these central field HS results it is seen that the Cooper minimum for argon ($Z = 18$) has moved in slightly by copper ($Z = 29$) and is just below threshold for germanium ($Z = 32$).

Note that the above discussion refers to ground state atoms. Some very interesting effects have recently been uncovered in theoretical calculations of photoionization excited atomic states (Msezane and Manson, 1975). In particular, for the 5d excited state of cesium, shown in Fig. 3, the photoionization cross section has two minima. One, at higher energies (shown in the insert), is the ordinary Cooper minimum. Just above threshold, however, a very dramatic minimum in the cross section is found. This minimum is also due to the vanishing of the $5d \to \varepsilon f$ dipole matrix element and this channel has a double minimum, a phenomenon not found for ground states. In addition, a minimum is found in the $5d \to \varepsilon p$ channel and this minimum is quite close to the first $5d \to \varepsilon f$ minimum, thus giving rise to the dramatic situation shown. The calculations show that this is not an isolated case (Msezane and Manson, 1977). The $5d \to \varepsilon f$ double minimum is found to appear for $Z = 34$–55, and the $5d \to \varepsilon p$ minimum for $Z = 9$–55. In addition, the calculations show minima in the $4d \to \varepsilon p$ channel (when 4d is an excited state) over a wide range of Z's, and even for excited 3d states a $3d \to \varepsilon p$ minimum is found in a number of elements, despite the fact that the 3d wave function is nodeless. These minima are also found for higher excited d states and, although no experimental confirmation of these effects has yet been reported, they are expected soon.

The detailed reasons for these excited d state effects are presently under scrutiny. They illustrate, however, a fundamental physical difference between ground and excited atomic states. Ground states have a spatial extent

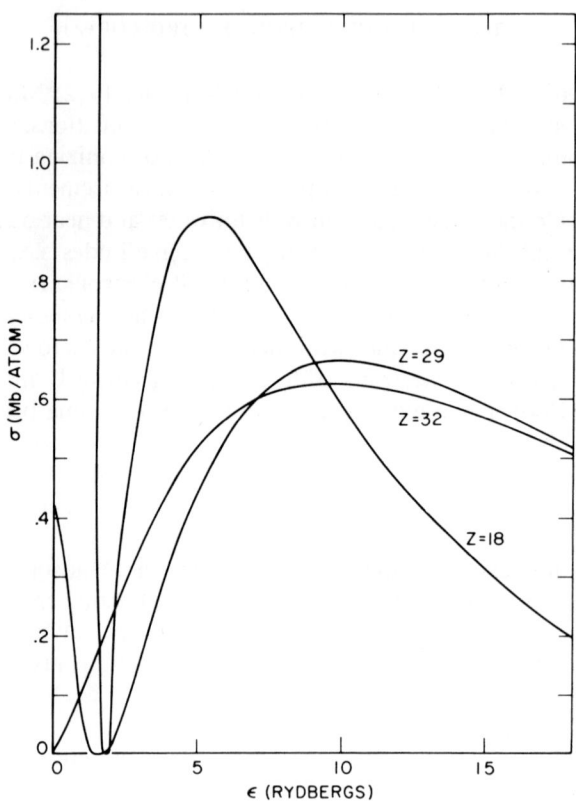

FIG. 2. Theoretical HS $3p \to \varepsilon d$ photoionization cross sections for $Z = 18$, 29, and 32 (Manson and Cooper, 1968).

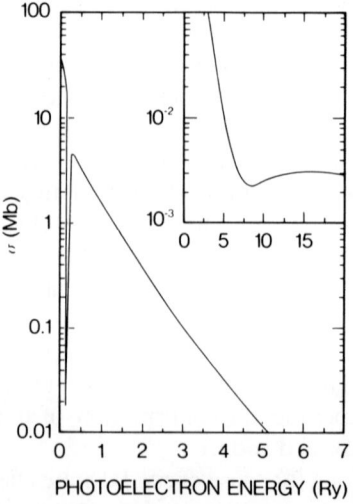

FIG. 3. Theoretical HS cross section for photoionization of cesium in the excited 5d state (Msezane and Manson, 1975).

of $\sim a_0$ from the atomic nucleus, while excited states are much larger. Thus the ground state wave function interacts with the continuum wave function over just a small region of space, while the excited state interaction extends over a much larger region where the character of the continuum wave function can be considerably different from its character near the nucleus. From another point of view, it seems that photoelectron spectroscopy of excited atomic states can provide valuable and basic information on continuum wave functions in regions of space inaccessible from the ground state.

Another phenomenon that is found throughout the periodic system is illustrated in Fig. 4. Here the $3d \to \varepsilon f$ photoionization cross section, cal-

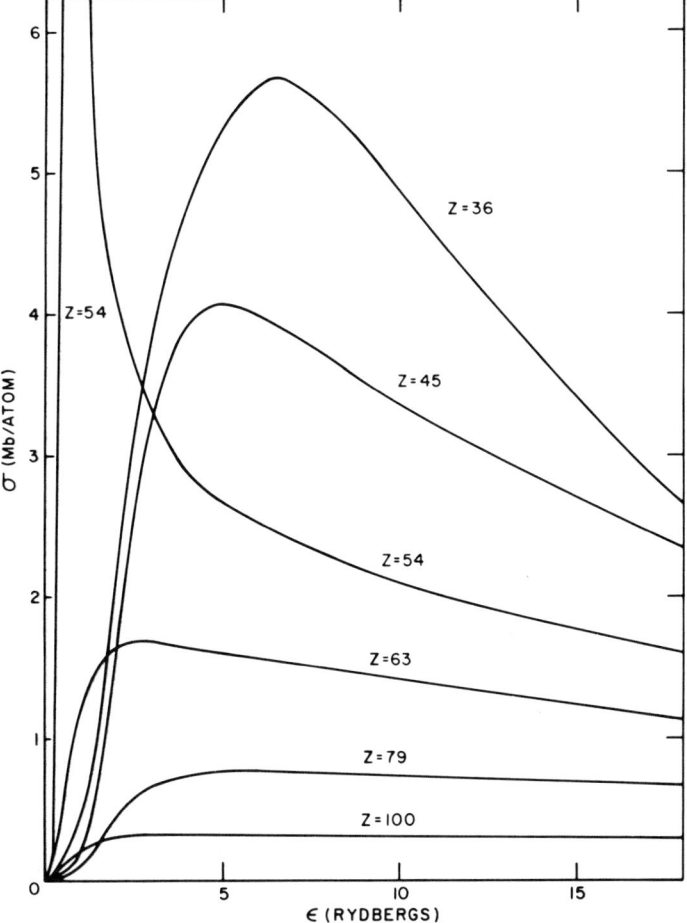

FIG. 4. Theoretical HS $3d \to \varepsilon f$ photoionization cross sections for $Z = 36, 45, 54, 63, 79$, and 100 (Manson and Cooper, 1968). The maximum for $Z = 54$ is 13.6 Mb at $\varepsilon = 0.6$ Ry.

culated using HS wave functions (Manson and Cooper, 1968), is shown over a wide range of Z. Note that each cross section is very small at threshold and rises to a "delayed maximum" well above threshold. This is in contrast to the monotone decrease of the cross section from threshold that is characteristic of hydrogenic results (Hall, 1936). Physically this occurs because the effective potential, as "seen" by an f wave electron, contains a large centrifugal (angular momentum) repulsion, which keeps the εf wave function from penetrating into the atom (where the 3d wave function has appreciable amplitude) at threshold. At threshold therefore, the 3d → εf dipole matrix element is quite small and thus the photoionization cross section is small. At higher energy, the εf continuum function can penetrate the atomic core more effectively, and the matrix element and cross section thus increases. This is illustrated in Fig. 5 for the 3d → εf photoionization in krypton. Here it is

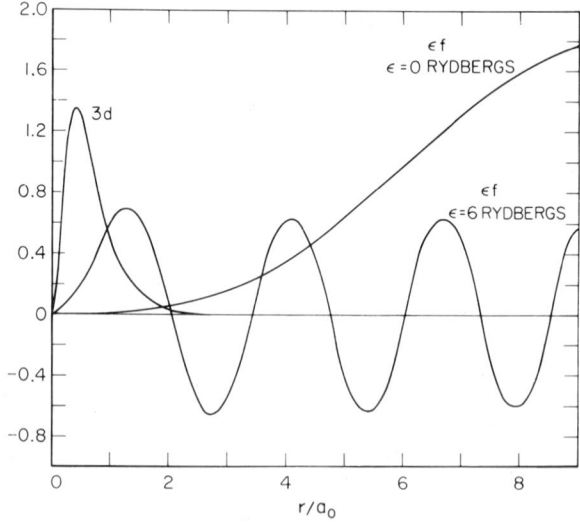

FIG. 5. The normalized 3d and εf wave functions in Kr ($Z = 36$).

seen that the overlap increases markedly in going from $\varepsilon = 0$ to $\varepsilon = 6$ Ry. For higher energies, the εf wave function continues to move in toward the nucleus and cancellation starts to occur owing to its oscillatory nature. This cancellation leads to a decrease in the cross section, with increasing energy, as shown in Fig. 4.

Thus, the delayed maximum phenomenon is related to the final angular momentum of the photoelectron and will be seen even more strongly for higher angular momenta, e.g., f → g transitions, but somewhat weaker for p → d photoionization channels, and almost nonexistent for s → p. The effective potentials over the periodic system have been studied by Rau and Fano

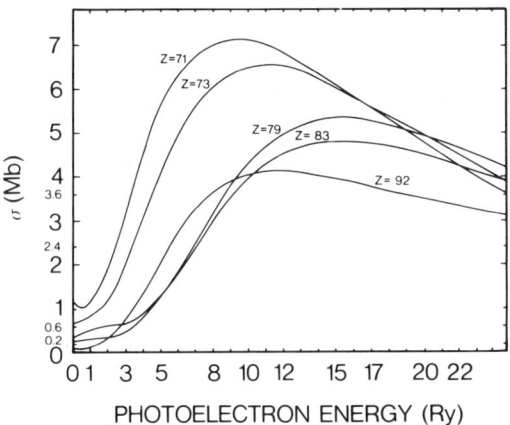

FIG. 6. Theoretical HS 4f photoionization cross sections for $Z = 71$, 73, 79, 83, and 92 (Combet Farnoux, 1971).

(1968), where the details of the angular momentum barriers are discussed.

The photoionization cross sections for a number of 4f subshells, including both $f \to g$ *and* $f \to d$ contributions, calculated using HS wave functions (Combet Farnoux, 1971), are shown in Fig. 6. The cross section is not vanishingly small at threshold due to the $f \to d$ channel, but in each case a very strong delayed maximum is in evidence 100–200 eV above threshold. The maximum is so far above threshold owing to the strength of the centrifugal barrier for g waves.

For the case of gold ($Z = 79$), the region of the photoionization cross section dominated by the 4f has been measured (Jaeglé and Missoni, 1966; Haensel *et al.*, 1968, 1969; Jaeglé *et al.*, 1969) and the results are compared with the HS central field calculations (Combet Farnoux and Heno, 1967; Manson and Cooper, 1968) in Fig. 7. From this comparison it is seen that theory agrees quite well with experiment *qualitatively*, despite the fact that the experiment was performed on the solid. The $4f \to \varepsilon g$ delayed maximum is too high and sharp, due to the only approximate inclusion of exchange in the HS calculation (Manson and Kennedy, 1970).

To include exchange more correctly, within the single-particle approximation, requires the use of HF theory, as discussed in I. Although no HF results have been published for the 4f subshell of gold, HF results for mercury ($Z = 80$) have been obtained (Shyu and Manson, 1975) and are shown in Fig. 8. From this result it is seen that the correct inclusion of exchange interactions (particularly the $4f \to \varepsilon g$ terms) causes the delayed maximum to be lower and broader. This is precisely what was needed to bring theory and experiment together in the case of gold, as discussed above.

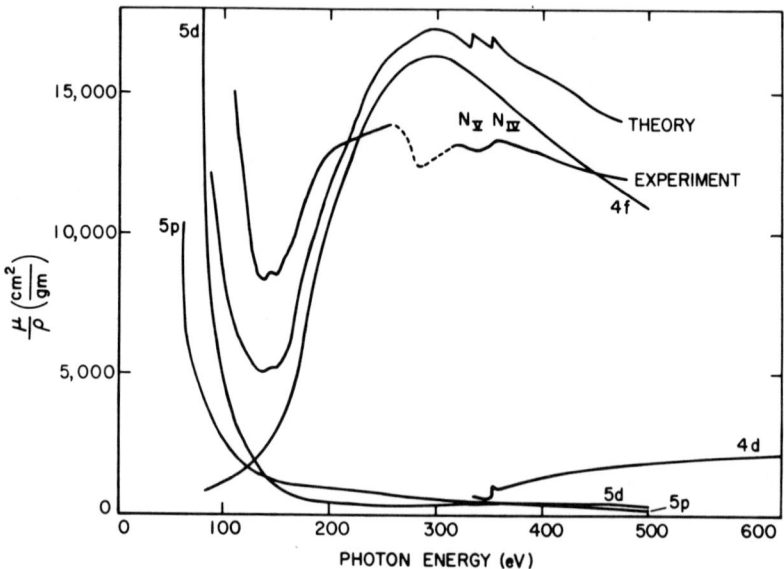

FIG. 7. Mass absorption coefficient of gold showing the theoretical HS subshell and total results (Manson and Cooper, 1968) and the experimental results (Jaeglé and Missoni, 1966). Note that the mass absorption coefficient is proportional to the photoionization cross section.

It turns out that the inclusion of exchange correctly, which can be done via HF or the intrachannel coupling formalism (Fano, 1961) as shown in I, generally has the effect of lowering and broadening the delayed maxima predicted by central field approximations (Starace, 1970; Combet Farnoux, 1970, 1972; Kennedy and Manson, 1972). This is illustrated in Fig. 9 for photoionization of 5d electrons in a number of cases (Combet Farnoux, 1972) where the lowering and broadening of the 5d → εf delayed maximum in HF as compared to HS is seen for $Z = 79, 83$, and 86. For $Z = 88$ a rather different effect, almost the opposite, is seen. Note that for both HS and HF results in Fig. 9, the maximum gets narrower and higher and closer to threshold, with increasing Z from 79 to 86. This behavior continues to $Z = 88$ for the HF calculation, but for the HS the maximum has moved below threshold (into the discrete range) at $Z = 88$. Thus the effect of including exchange correctly is to make the effective potential for f waves somewhat less attractive near the outer edge of the atom; in the HS calculation the field is strong enough to depress the delayed maximum into the discrete, while for the HF calculation, the field is weaker and the d → f maximum remains above threshold.

From the above examples, it is seen that HF (*intra*channel coupling) theory gives significant corrections to HS results in the energy region near

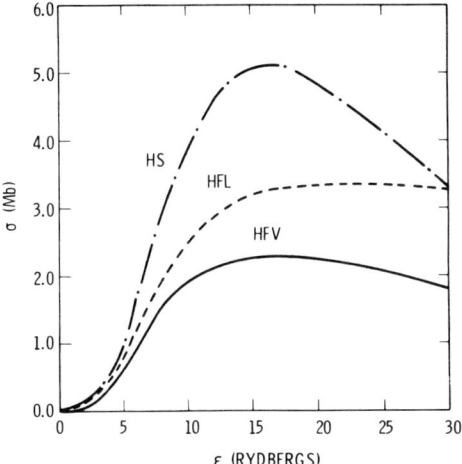

FIG. 8. Theoretical 4f → εg photoionization cross sections in mercury in HS and HF approximations (Shyu and Manson, 1975).

delayed maxima and Cooper minima. The differences between length and velocity results in HF theory, as seen for mercury 4f → εg in Fig. 8, indicate that electron–electron correlation effects (*inter*channel coupling) are non-negligible. In Fig. 10, the experimental photoionization cross section of xenon (Lukirskii *et al.*, 1964; Ederer and Tomboulian, 1964; Samson, 1966) is compared with HF length (HF-L) and velocity (HF-V) results (Kennedy and Manson, 1972). Here it is seen that fairly large discrepancies exist between HF-L and HF-V, and some quantitative disagreement with experiment near threshold (5p photoionization) and near $h\nu = 100$ eV (mainly 4d photoionization) is evident. This gives further credence to the notion that

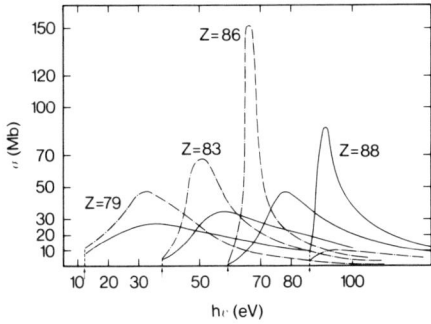

FIG. 9. Theoretical HS and HF 5d photoionization cross sections for $Z = 79, 83, 86$, and 88 (Combet Farnoux, 1972). The dot–dash curves are the HS results and the solid curves are the HF results in the length formulation.

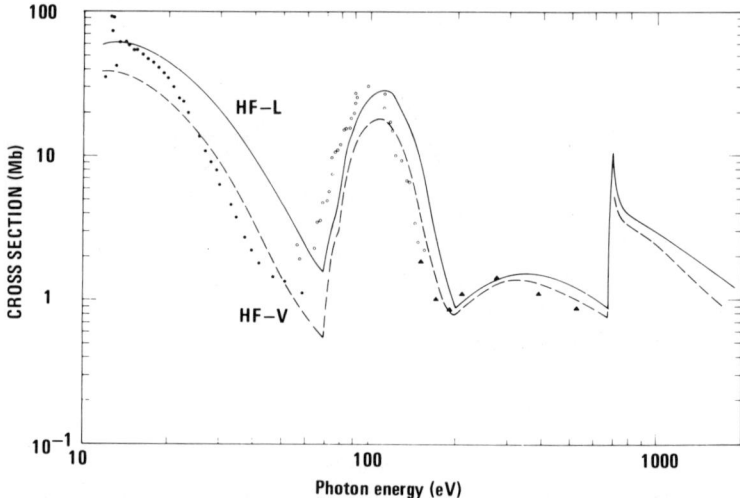

FIG. 10. Total photoionization cross section of xenon. The theoretical HF results are from Kennedy and Manson (1972) and the experimental points are from Samson (1966) (●), Ederer and Tomboulian (1964) (○), and Lukirskii et al. (1964) (△).

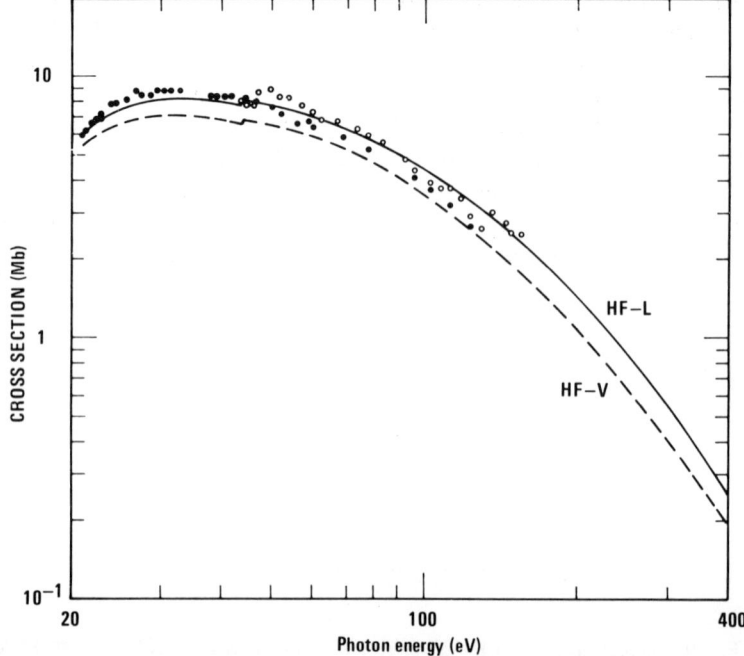

FIG. 11. Total photoionization cross section of neon. The theoretical HF results are from Kennedy and Manson (1972) and the experimental points are from Samson (1966) (●) and Lukirskii et al. (1964) (○).

correlation effects can be important, particularly in the neighborhood of delayed maxima and Cooper minima. Where these features are unimportant, the HF calculation is generally excellent. This is seen in Fig. 11, which shows the experimental neon photoionization cross section (Lukirskii et al., 1964; Samson, 1966) along with the theoretical HF results (Kennedy and Manson, 1972). The HF-L results differ from HF-V by less than 20% and agreement with experiment is to within about 10%. The central field HS calculation is generally fairly good in these regions as well (Manson and Cooper, 1968; McGuire, 1968).

Another example of a case where correlation (interchannel) effects are important is the photoionization of argon near threshold shown in Fig. 12.

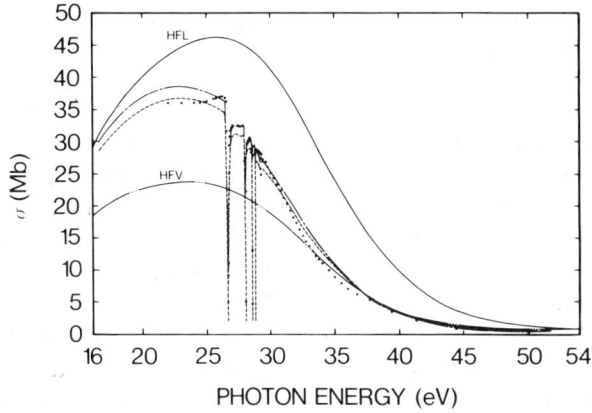

FIG. 12. Photoionization cross section of argon. The theoretical results of Kelly and Simons (1973) are HF calculations and MBPT calculations (shown in length and velocity formulations as dot–dashed and dashed curves, respectively) and the experimental results are from Samson (1966) and Madden et al. (1969).

The experimental points are from Samson (1966) and Madden et al. (1969) and the theoretical HF and MBPT results are due to Kelly and Simons (1973). Note that the MBPT length and velocity results are quite close to each other and agree extremely well with experiment. The HF results do not show good agreement between length and velocity or with experiment, similar to the xenon case discussed above. Actually these HF results do not include core relaxation effects; if they are included, agreement is somewhat better (Kennedy and Manson, 1972) but still not adequate.

The predominant effect of correlation on the argon photoionization cross section in the threshold region is the configuration interaction of the $3p^6$ initial state with states of the type $3p^4nd^2$. As a matter of fact, a calculation by Swanson and Armstrong (1977) that used a two-configuration ground state and an HF final state gives excellent agreement with experiment, while a close-coupling calculation (Lipsky and Cooper, 1967), which considered only final state interchannel coupling, did not improve significantly on the HF result. This ground state correlation effect is included in *R*-matrix theory (Burke and Taylor, 1975) and in RPA treatments (Amusia *et al.*, 1971; Chang, 1977); consequently, these calculations also give very good agreement with experiment.

The window resonances seen in the argon photoabsorption cross section (Fig. 12) between 25 and 30 eV are due to the autoionizing states leading up to the opening of the 3s photoionization channel. Autoionizing resonances are not included in HF theory; from a theoretical standpoint they are due to correlation effects in the form of an interchannel interaction in the *final* state between the open $3p \to \varepsilon d$ and $3p \to \varepsilon s$ channels on the one hand and the closed $3s \to \varepsilon p$ channel on the other. MBPT includes this correlation effect, and it is seen that agreement with experiment is excellent, both as to position and shape of the resonances.

A particularly interesting case is the 3s subshell of argon. Its cross section is small compared to the 3p so it does not show up very well in a photoabsorption measurement, as seen in Fig. 12. The argon 3s cross section has been measured by Samson and Gardner (1974) and Houlgate *et al.* (1976) via photoelectron spectroscopy and the results are given in Fig. 13. Note that the cross section shows a Cooper minimum, just like in the case of sodium shown in Fig. 1. The HF calculation (Kennedy and Manson, 1972), however, fails to give good agreement and shows no minimum in the continuum; the HF minimum occurs in the discrete range below threshold. Thus correlation effects are clearly of crucial importance. In particular, since the $3s \to \varepsilon p$ dipole matrix element is so small compared to the $3p \to \varepsilon d$ matrix

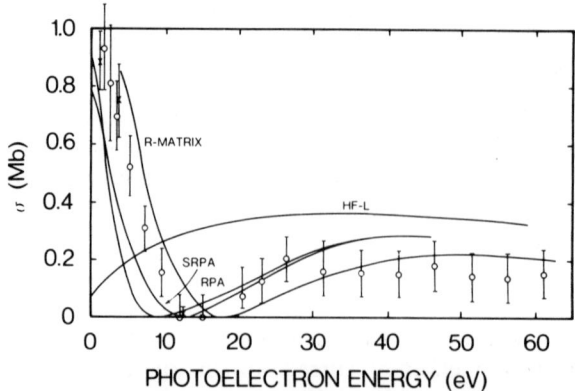

FIG. 13. Argon 3s subshell photoionization cross section. The theoretical results are *R*-matrix (Burke and Taylor, 1975), RPA (Amusia *et al.*, 1972a), SPRA (Lin, 1974), and HF-L (Kennedy and Manson, 1972); the experimental points are from Samson and Gardner (1974) (×) and Houlgate *et al.* (1976) (○).

element, a small amount of final state interchannel coupling between these channels causes a significant alteration of the 3s photoionization. The *R*-matrix theory calculation (Burke and Taylor, 1975), the RPA treatment (Amusia *et al.*, 1972a), and a simplified RPA theory (Lin, 1974) all include this interchannel coupling and thus show fairly good agreement with experiment, as seen in Fig. 13.

This type of final state interaction, which couples the $3s3p^6\varepsilon p\,^1P$ state to the $3s^23p^5\varepsilon d\,^1P$ state via the e^2/r_{12} electron–electron interaction for argon, is important whenever there is a subshell with a small cross section at a given $h\nu$ and a subshell having a much larger cross section at that $h\nu$. Thus, the outer ns subshells of krypton and xenon will also be strongly affected by final state correlation, as will the subshells lying near the outer d states in these elements, 3p and 3s in krypton, and 4p and 4s in xenon, since the cross sections for krypton 3d and xenon 4d are so large. In addition, this effect will be very widespread over the entire periodic table, but detailed information on systems other than noble gases is not yet available. Of course, the predominant effect will be on the subshell with the small cross section, and this very fact makes the effect unobservable in photoabsorption data, as in Fig. 12 for argon. This shows the power of photoelectron spectroscopy, as opposed to photoabsorption, in probing interactions within the atom.

For neon, the 2s photoionization cross section is not so much smaller than the 2p as the corresponding 3s and 3p results in argon: one order of magnitude difference in neon as compared to two orders of magnitude in argon. Thus the final state correlation effect on the 2s cross in neon will be less pronounced. The experimental neon 2s cross section obtained from photoelectron spectroscopy (Samson and Gardner, 1974; Wuilleumier and Krause, 1974; Codling *et al.*, 1976) is shown in Fig. 14 along with the theoretical HF-L results (Kennedy and Manson, 1972). The HF-V results are substantially the same as HF-L and are not shown. It is seen that the HF cross section is somewhat, but not dramatically, off as was the case for argon 3s. Calculations using RPA (Amusia *et al.*, 1972a) and *R*-matrix theory (Burke and Taylor, 1975) bring the theory into good agreement with experiment. These calculations indicate the importance of interchannel coupling in neon.

Exactly how some of these correlation effects affect complex open-shell atoms is not yet known. One very interesting feature has been uncovered, however, for photoionization of chlorine (Starace and Armstrong, 1976). RPA calculations, which included the ground state correlations that were so important in argon, were compared with HF results. The comparison is shown in Fig. 15 for the $3p^5(^2P) \rightarrow 3p^4(^3P)$ photoionization in chlorine. The RPA calculations included no interchannel effects and the HF and RPA results are in excellent agreement. Thus ground state correlations seem to be

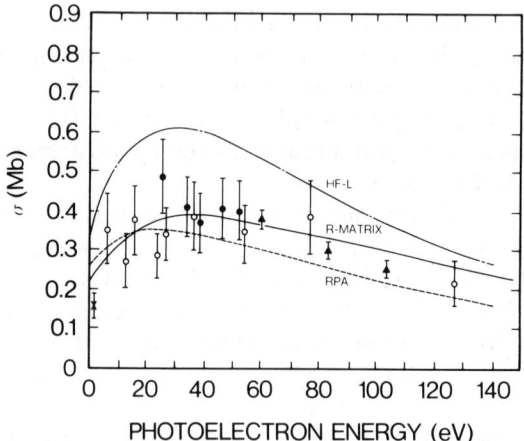

Fig. 14. Neon 2s subshell photoionization cross section. The theoretical results are R-matrix (Burke and Taylor, 1975), RPA (Amusia et al., 1972a), HF-L (Kennedy and Manson, 1972), and the experimental results are from Samson and Gardner (1974) (×), Wuilleumier and Krause (1974) (△), and Codling et al. (1976) (○,●).

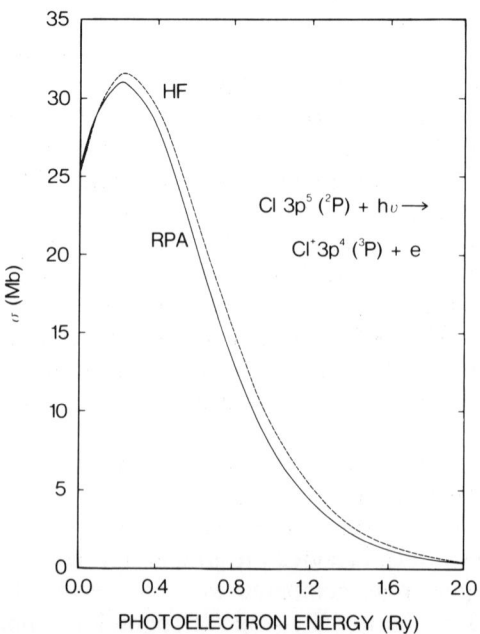

Fig. 15. Theoretical photoionization cross section of the $^2P \to {}^3P$ transition in chlorine showing HF and RPA results (Starace and Armstrong, 1976).

relatively unimportant in chlorine. No experimental photoionization cross sections have been reported for atomic chlorine, but preliminary indications of R-matrix theory results are that other correlation effects are quite important (Combet Farnoux, 1976). It thus seems that experience in the noble gases is not entirely indicative for open-shell atoms and much remains to be explored, both experimentally and theoretically.

II. Photoelectron Angular Distributions

Theoretical considerations for photoelectron angular distributions are given in I. In that treatment, it is shown that the angular distribution asymmetry parameter β is a sensitive function of the phase shifts of the various possible continuum waves in a given photoionization process, as well as of the sign of the various dipole matrix elements. In this section, we shall give examples that point out the types of phenomena observed, discuss the physics behind them, and compare experimental results with the predictions of some of the various types of calculations discussed in Section II of I.

For an electric dipole photoionization process, the parameter β contains all of the photoelectron angular distribution information, as discussed in I. Thus we shall consider only β and its variation with energy. As a first example, Fig. 16 shows β as a function of photoelectron energy for the 2p subshell of neon. The experimental results (Wuilleumier and Krause, 1974; Dehmer et al., 1975; Codling et al., 1976) are seen to agree very well with the calculations (Kennedy and Manson, 1972; Amusia et al., 1972b). Since neon is a closed shell, low-Z (where spin–orbit effects are negligible) atom, the Cooper–Zare (CZ) model (Cooper and Zare, 1968, 1969) given in Eq. (72) of

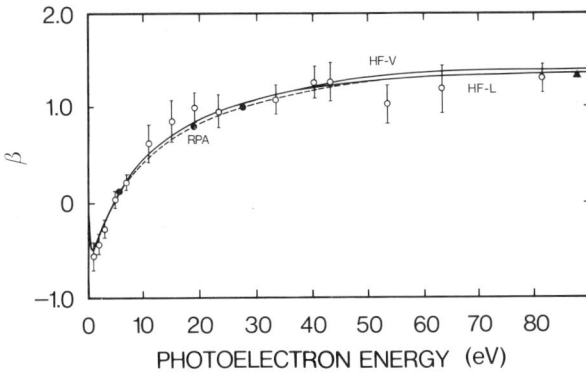

Fig. 16. Asymmetry parameter β for neon 2p. The theoretical HF-L and HF-V results are from Kennedy and Manson (1972), RPA from Amusia et al. (1972b), and the experimental points are from Wuilleumier and Krause (1974) (△), Dehmer et al. (1975) (●), and Codling et al. (1976) (○).

I is applicable. Both HF and RPA calculations, employing the CZ formulation, are seen to be in excellent agreement with experiment. Thus correlation effects are not of great importance in this case, confirming the conclusion based upon the cross section in the previous section. In addition, the HS results (Manson and Cooper, 1970; Kennedy and Manson, 1972), which are not shown, lie essentially on top of the HF prediction.

An interesting feature of the β vs ε curve of neon 2p shown in Fig. 16 is the rapid variation of β in the first few electron volts above threshold. This arises from the interference term in the CZ expression for β [Eq. (72) of **I**], which has a $\cos(\xi_{l+1} - \xi_{l-1})$ factor in it. Recall that the $\xi_{l\pm 1}$ are the sums of the Coulomb phase shifts, $\sigma_{l\pm 1}$ [= Arg $\Gamma(l \pm 1 + 1 - i\varepsilon^{-1/2})$], and the non-Coulomb phase shifts $\delta_{l\pm 1}$. The $\delta_{l\pm 1}$ are slowly varying over such a small energy range (Manson, 1969) that they cannot be the cause of the rapid variation. The $\sigma_{l\pm 1}$ are, however, very rapidly varying near threshold, $\varepsilon = 0$. Since we are interested only in the phase shift difference, this can be written explicitly (Manson, 1973) as

$$\sigma_{l+1} - \sigma_{l-1} = \tan^{-1}\left[\frac{-(2l+1)}{\varepsilon^{1/2}l(l+1) - \varepsilon^{-1/2}}\right] \quad (95)$$

For the neon 2p case, $l = 1$, this difference is π for $\varepsilon = 0$, and $\pi/2$ for $\varepsilon = 0.5$ Ry (6.8 eV) for a change of $\pi/2$, with two-thirds of the change coming in the first 0.25 Ry above threshold. This rapid variation in $\sigma_{l+1} - \sigma_{l-1}$ is seen to be independent of Z, so that the rapid variation of β near threshold will be a general feature. For d states, the $\pi/2$ change occurs by $\varepsilon = 1/6$ Ry above threshold and for f states by $\varepsilon = 1/12$ Ry above threshold as can be seen by application of Eq. (95). This phenomenon will thus be more pronounced for d and f states.

The asymmetry parameter β for argon 3p is given in Fig. 17. The experimental results (Dehmer et al., 1975; Houlgate et al., 1976) are in very good agreement with the RPA calculation (Amusia et al., 1972b). In addition, the HF results (Kennedy and Manson, 1972) are in good agreement with experiment from threshold to 30 eV; above 30 eV the agreement is only qualitative until about 70 eV, where all of the results shown, experimental and theoretical, are converging. The HS result (Manson and Cooper, 1970; Kennedy and Manson, 1972) is not shown, but it is only in qualitative agreement with experiment.

The energy variation of β for argon 3p is rather different than for neon 2p discussed above. The rapid variation of β near threshold is in evidence as was expected. A broad minimum in β for argon 3p is seen, which is quite different from the neon 2p results. This minimum owes its existence to the fact that the $np \to \varepsilon d$ channel has a Cooper minimum for argon 3p but none for neon 2p. At the Cooper minimum, the $3p \to \varepsilon d$ dipole matrix element

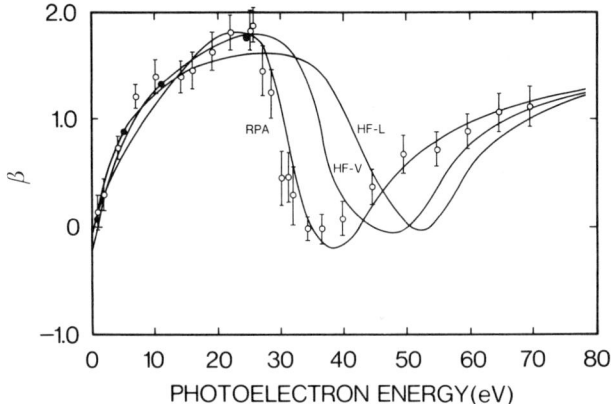

FIG. 17. Asymmetry parameter β for argon 3p. The theoretical HF-L and HF-V results are from Kennedy and Manson (1972), RPA (Amusia et al., 1972b), and the experimental points are from Dehmer et al. (1975) (●) and Houlgate et al. (1976) (○).

vanishes and the CZ expression for β, Eq. (72) of **I**, reduces to $\beta = 0$. Thus, to the extent that relativistic spin–orbit effects are small, the β vs ε curve allows us to accurately pinpoint the location of the Cooper minimum.

The major difficulty with the HF results, in predicting β quantitatively correctly, is that HF-L and HF-V have their Cooper minima at too large a photoelectron energy. It is interesting, however, that in the low-energy region, the HF-L and HF-V cross sections for argon 3p differ by about a factor of 2 as shown in Fig. 12, but the β's they predict are in excellent agreement with each other, RPA results, and experiment. This is because the expression for β is essentially a ratio, and thus β is relatively insensitive to the cross sections in any region far from a Cooper minimum (Manson and Kennedy, 1970); it is sensitive to the phase shifts primarily.

The principal effect of the ground state correlation, which is included in RPA, on β for argon 3p, is to move the Cooper minimum to the correct location. Therefore, the RPA results (Amusia et al., 1972b) agree well with the measured values.

The corresponding case for higher Z, that of xenon 5p, is shown in Fig. 18, where it is seen that the β's predicted by the HS calculation (Manson and Cooper, 1970), HF-L and HF-V calculations (Kennedy and Manson, 1972), and RPA calculation (Amusia et al., 1972b) are all qualitatively similar to each other and to the experimental results (Dehmer et al., 1975; Houlgate et al., 1976). Quantitative agreement is fairly good in the first 50 eV above threshold, with the RPA results being closer to experiment and predicting the minimum more accurately than the HF calculations; the fact

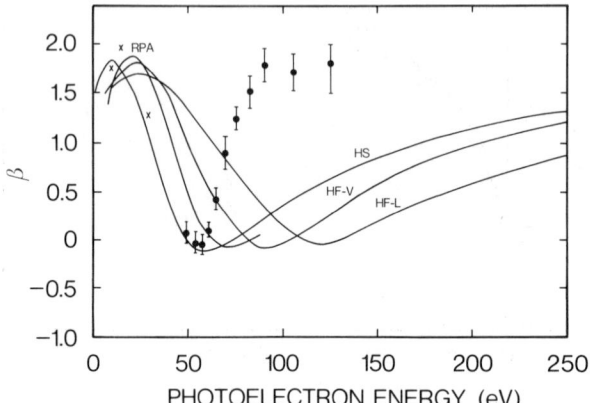

Fig. 18. Asymmetry parameter β for xenon 5p. The theoretical HS results are from Manson and Cooper (1970), HF from Kennedy and Manson (1972), RPA from Amusia et al. (1972b), and the experimental points are from Dehmer et al. (1975) ($\times$) and Houlgate et al. (1976) ($\bigcirc$).

that HS does even better than RPA must be considered accidental agreement. Above the Cooper minimum, the quantitative agreement between experiment and all of the calculations shown is quite poor. The reason for this lack of agreement is that none of the calculations reported include final state interchannel correlation effects. About 50 eV above the 5p ionization threshold in xenon is the 4d threshold. From Fig. 11, it is seen that the 4d photoionizing transition is very strong. Thus, in the region of 50 to 150 eV above the 5p threshold, the 5p photoionization matrix elements have become quite small, while the 4d matrix elements are quite large, particularly the 4d → εf. Then as was discussed in connection with the argon 3s photoionization cross section in the previous section, a small interchannel coupling can make a substantial difference in the small cross section. In this case, it is the xenon 5p photoionization that is strongly modified in both magnitude of the dipole matrix elements *and* phase shifts of the continuum waves, and this results in the asymmetry parameter behaving as shown in Fig. 18. Although no calculations have been reported that include this interchannel coupling, it is strongly felt that such would be in much better agreement with experiment than are the calculations shown.

Another possible source of disagreement between theory and experiment for the asymmetry parameter for xenon 5p is relativistic effects in the form of the spin–orbit interaction. This has several effects. First, the $5p^5$ residual ion core of Xe^+ is split into two levels with $J = 1/2$ and $3/2$, respectively. Thus, for photoionization of the xenon 5p subshell by photons of a given energy, two groups of photoelectrons are observed, separated in energy by the

1/2–3/2 splitting (∼1.3 eV). The β's derived, when plotted vs photoelectron energy, lie essentially on the same curve. Another effect of the spin–orbit interaction is on the photoelectron itself. The d wave can be either $\varepsilon d_{3/2}$ or $\varepsilon d_{5/2}$, each with *slightly* different wave function, phase shift, and dipole matrix element. The Cooper minimum for transitions to the two different continuum d states will be at slightly different energies. This effect was first discussed by Fano (1969) and, under certain conditions, can lead to spin-polarized photoelectrons (Heinzmann *et al.*, 1975a,b; Kessler, 1975). In addition, the two different continuum d waves can interfere with one another, leading to extra terms in the expression for β, i.e., the CZ expression is no longer valid and the general theory of β, discussed in I, must be used (Dill and Fano, 1972; Fano and Dill, 1972; Dill, 1973, 1975).

The spin–orbit effect on the continuum electron will have a dramatic effect on the angular distribution of photoelectrons from s states of closed-shell atoms. If the spin–orbit interaction is ignored, photoionization of the xenon 5s subshell can result in a single εp continuum state. Thus, no interference can occur and the asymmetry parameter β must be energy independent; its value must be $\beta = 2$. With the spin–orbit interaction, however, $\varepsilon p_{1/2}$ *and* $\varepsilon p_{3/2}$ occur and differ somewhat. The results of a relativistic HS central field calculation of β for xenon 5s (Walker and Weber, 1974) are shown in Fig. 19. These results show the dramatic deviations of β from 2 in the low-energy region and that as the photoelectron energy increases, $\beta \to 2$. This result, which treats exchange inexactly and omits entirely the very important effect of interchannel coupling with the 5p channel, is probably only qualitatively correct. As a matter of fact, a recent measurement (Dehmer and Dill, 1976a)

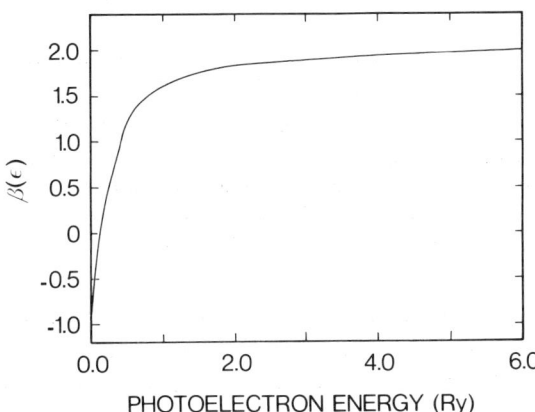

FIG. 19. Theoretical relativistic HS results of Walker and Weber (1974) for the asymmetry parameter β for the 5s subshell of xenon.

at $h\nu = 40.8$ eV, for the xenon 5s β, found a value of $\beta = 1.4 \pm 0.1$ as compared to $\beta = 1.7$ predicted in Fig. 19.

So far it has been shown that β will vary rapidly with energy near threshold and near Cooper minima. If the non-Coulomb phase shift varies rapidly, this too can cause oscillations in β. Such a case is the xenon 4d subshell whose asymmetry parameter is shown in Fig. 20. The value of β at threshold (not shown) is ~ 0.5 for all three theoretical results (Manson and Kennedy, 1970; Kennedy and Manson, 1972) and β increases to a maximum just above threshold. Thus the β vs ε curve has four turning points. The rapid variation near threshold is due to the Coulomb phase shifts, the last minimum in β is due to the Cooper minimum, and the extra oscillation between is the result of a shape resonance in the 4d $\rightarrow \varepsilon$f channel, i.e., the phase shift of the εf continuum function increases by π in a relatively small energy range in both the HS (Manson, 1969) and HF (Kennedy and Manson, 1972) calculations. The small differences between the HF-L and HF-V results are due primarily to the differences in the energies at which their Cooper minima appear (for d states, the CZ expression gives $\beta = 0.2$ at a Cooper minimum),

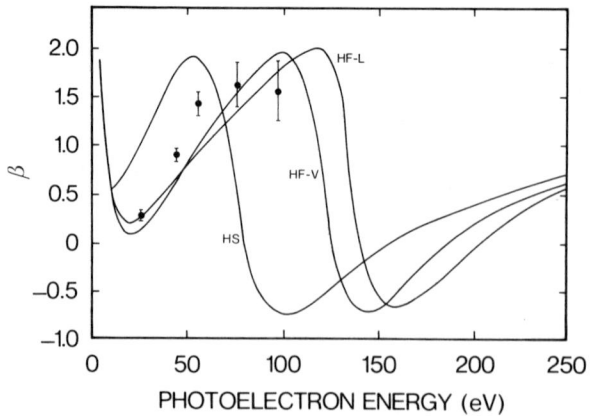

FIG. 20. Asymmetry parameter β for xenon 4d. The theoretical HS and HF results are from Kennedy and Manson (1972) and the experimental points are from Torop et al. (1976). Note that the threshold portion of all the theoretical results (not shown) are virtually identical and rise from ~ 0.5 at threshold to a maximum of almost 2.0 just above threshold.

and the differences between HF and HS are due to the Cooper minima being at differing energies as well as the rapidity with which the f wave phase shift increases. Note, however, that the qualitative shape of all three theoretical curves is the same. Note further that the agreement between all three calculations is excellent near threshold despite the fact that in this region the HS cross section is much too large, by an order of magnitude at certain energies

(Manson and Kennedy, 1970; Kennedy and Manson, 1972). This shows very dramatically how weakly β depends upon the cross section. Finally, it is seen in Fig. 20 that the experimental results (Torop et al., 1976) agree fairly well with the theoretical HF results. It would be useful to have experimental results at a wider range of energies, however, for a more exhaustive comparison.

The predictions of the CZ expression using HS central field wave functions have been studied extensively (Manson, 1973, 1977). Figure 21 shows the β's for photoionization of 4p electrons from several different elements. The general shape of β remains the same, as a function of Z, and is characteristic of a p subshell having a Cooper minimum such as was shown in Fig. 17 for argon 3p. The minimum is on the high-energy side of the peak in β and its position can be accurately deduced by noting where $\beta = 0$. It is seen from Fig. 21 that the Cooper minimum moves toward threshold as Z increases.

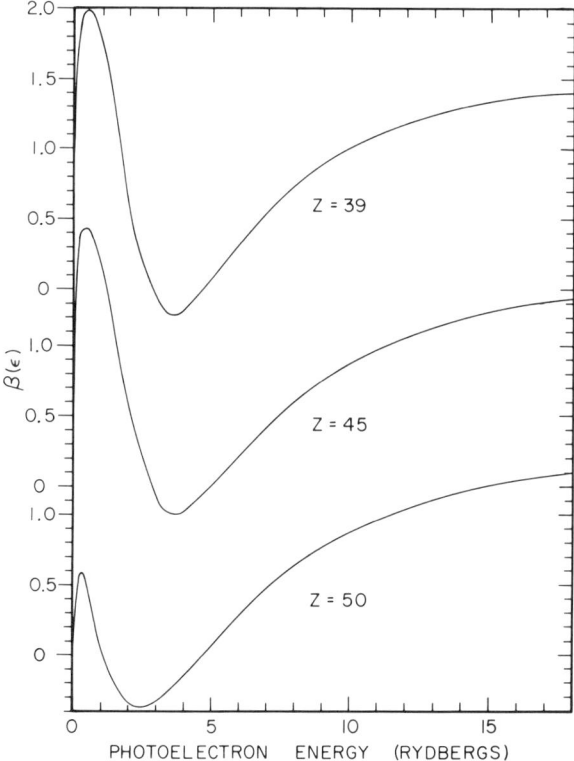

FIG. 21. Theoretical HS asymmetry parameter β for the 4p subshell of $Z = 39, 45$, and 50 (Manson, 1973).

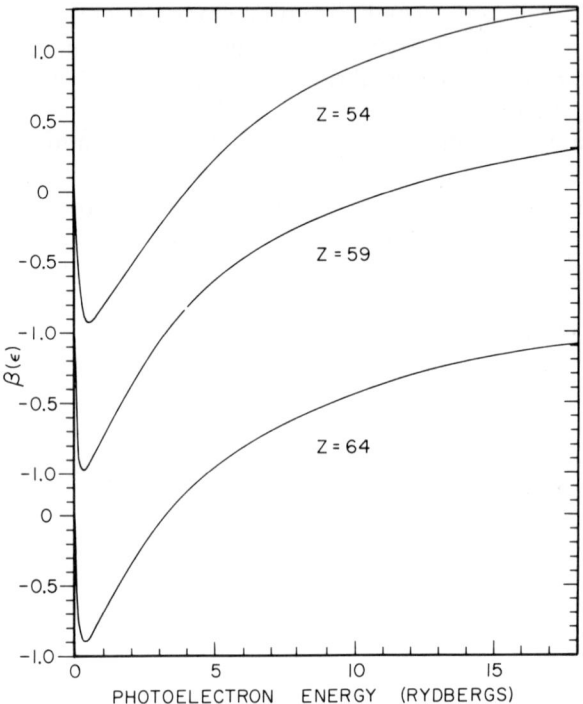

FIG. 22. Theoretical HS asymmetry parameter β for the 4p subshell of $Z = 54$, 59, and 64 (Manson, 1973).

Moreover, the rapid variation of β near threshold is also evident. The general shape of β for a number of cases that have no Cooper minima is shown in Fig. 22 for various 4p subshells at higher Z. These curves show only the rapid variation near threshold and are otherwise slowly varying. The curves show the general shape of β for a p subshell having no Cooper minimum as was seen in Fig. 16 for neon 2p.

It is thus seen that the asymmetry parameter for p states has one of two general shapes, depending upon whether or not the subshell has a Cooper minimum in the continuum. A further example of this is shown for 5p subshells in Fig. 23. Here it is seen that for $Z = 83$, a Cooper minimum is evident, but for $Z = 92$ and $Z = 100$, the Cooper minimum has moved into the discrete.

For d states, the form of β when the photoionization exhibits both a Cooper minimum and a shape resonance was given in Fig. 20 for xenon 4d. For higher Z elements, the asymmetry parameter for 4d subshells is given in Fig. 24 (Manson, 1977). These results, based upon the CZ expression and

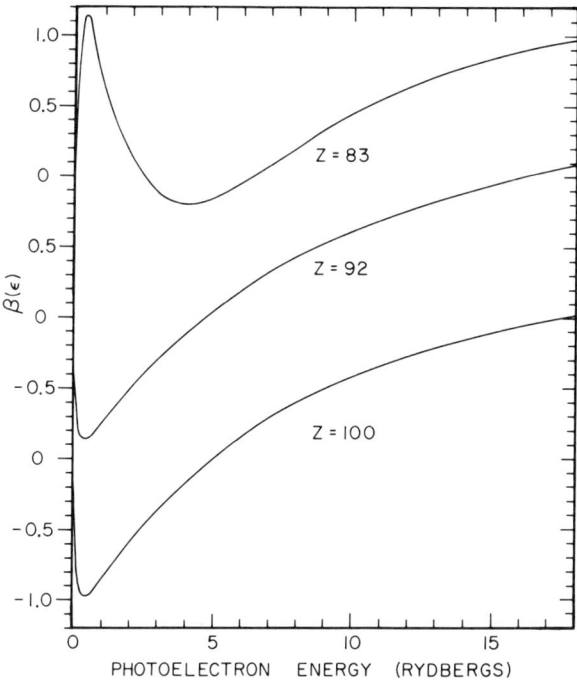

FIG. 23. Theoretical HS asymmetry parameter β for the 5p subshell of $Z = 83, 92$, and 100 (Manson, 1973).

HS wave functions, show that the first maximum in β, which was above threshold for xenon ($Z = 54$) is at threshold for $Z = 72$. This indicates that the shape resonance is less important here. At higher Z, the maximum has disappeared completely and is increasing from threshold for $Z = 79$. Interestingly, for $Z = 86$, a bit of the maximum still remains above threshold. This result shows that, although there are general trends with increasing Z since the strength of the electrostatic fields increase, the movement of the various features of β toward threshold is not necessarily monotonic since atomic fields do not vary monotonically but rather periodically (whence comes the term "periodic table"). For $Z = 92$, β shows a single maximum and is characteristic of a d subshell lacking a Cooper minimum, i.e., it is of the same general shape as the asymmetry parameter for 3d subshells since they are nodeless and have no Cooper minima.

The situation for 4f subshells is shown in Fig. 25 (Manson, 1977); again these results are calculated CZ HS β's. Over the range of Z shown, the general shape of β for 4f subshells remains quite similar. One interesting point is the lack of rapid variation of β near threshold for these cases. The

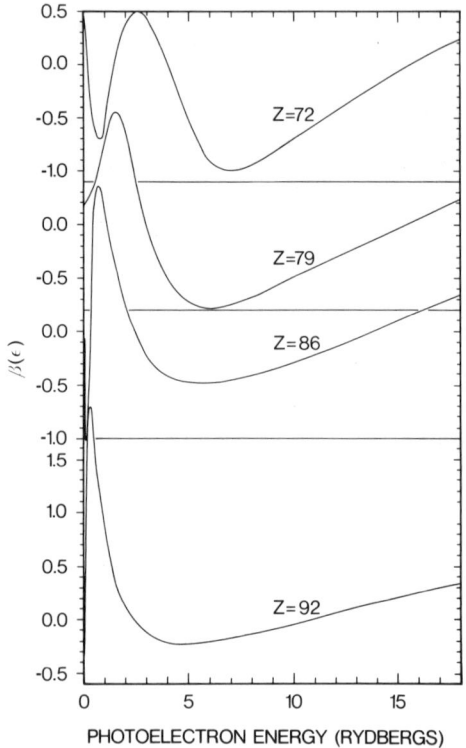

FIG. 24. Theoretical HS asymmetry parameter β for the 4d subshell of $Z = 72$, 79, 86, and 92 (Manson, 1977).

reason is that, near threshold, the 4f → εg dipole matrix element is extremely small, due to the huge angular momentum barrier for g waves, as discussed in the previous section. Thus, near threshold, the cross section is almost entirely from the 4f → εd transition, whose dipole matrix element is about two orders of magnitude larger than the 4f → εg. Since one of the matrix elements is almost vanishingly small compared to the other at threshold, β reduces to 2/7 (=0.29) as can be seen from Eq. (72) of **I**. Note that all of the β's for 4f subshells shown in Fig. 25 do have a value very close to 0.29 at threshold, indicating the validity of this analysis and that the interference term is vanishingly small. With increased energy, the 4f → εg matrix element increases extremely rapidly, but reaches an appreciable fraction of the 4f → εd only after the Coulomb phase shifts have completed most of their rapid variation, i.e., the interference term varies rapidly in an energy region where its coefficient renders it negligible, so that β is slowly varying. At higher energies, β reaches a maximum, in each case, showing that the phase

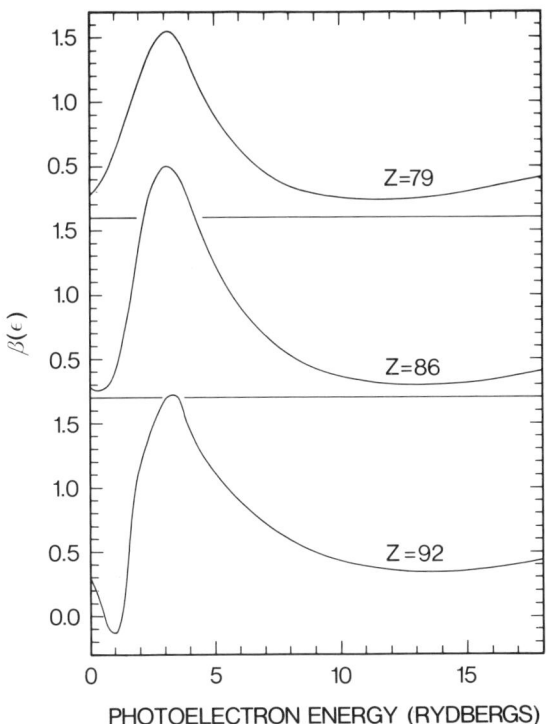

FIG. 25. Theoretical HS asymmetry parameter β for the 4f subshell of $Z = 79$, 86, and 92 (Manson, 1977).

shift difference is such as to make $\cos(\xi_{l+1} - \xi_{l-1})$ *negative* in this region. This, in turn, is related to the systematics of the Coulomb phase shift difference, Eq. (95), and the non-Coulomb phase shifts (Manson, 1969). The details of this relation are discussed elsewhere (Manson, 1977).

Unfortunately, no experimental photoelectron work has been yet reported on atomic f subshells. Thus it is of interest to compare some of the above HS results to the predictions of more sophisticated calculations such as HF. This comparison for the asymmetry parameter of the mercury 4f subshell (Shyu and Manson, 1975) is shown in Fig 26. Here excellent *quantitative* agreement between HF and HS is seen for the first 10 Ry (136 eV) above threshold. Above this energy region the β's start diverging slightly, but agreement is still fairly good; this despite the rather large (factor of two or more) discrepancies in the cross sections predicted by HF as compared to HS, as was seen in Fig. 8. Thus we have further evidence on the insensitivity of β to the cross section. The asymmetry parameter is very sensitive to the phase shifts, particularly δ_g in this case. These g wave phase shifts were

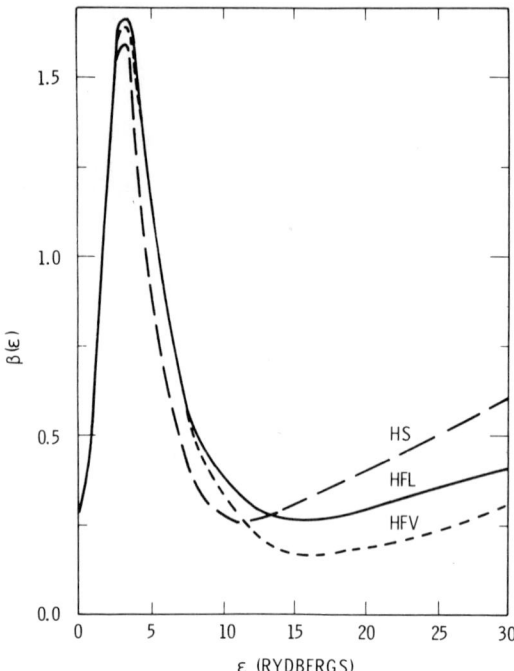

FIG. 26. Theoretical asymmetry parameter β for the 4f subshell of mercury in HS and HF approximations (Shyu and Manson, 1975).

shown to differ by less than 0.2 rad in the region from threshold to 10 Ry (Shyu and Manson, 1975), thus accounting for the excellent agreement of HF and HS β's in this region. At higher energies, the g wave phase shifts were found to diverge with δ_g^{HS}, being about 1 rad greater than δ_g^{HF} at 30 Ry. Thus, despite the fact that the cross sections for HS and HF-L agreed exactly at 30 Ry (as seen in Fig. 8), the β's they predict differ substantially there, as seen in Fig. 26, owing to the difference in g wave phase shifts.

The discussion and comparisons of this section so far have been restricted to fairly low photoelectron energies. This is primarily because most of the interesting phenomena occur in the low-energy region. At somewhat higher energies, however, the theoretical analysis is still valid. In fact, exchange and correlation effects will become fairly unimportant at the higher energies and the simple central field HS calculation will be adequate. An example, for the asymmetry parameter of the krypton 3p subshell, is shown in Fig. 27. The theoretical HS calculation (Cooper and Manson, 1969) is seen to agree quite well with experiment (Krause, 1969), the one discrepancy being at the lowest photoelectron measured, where exchange and correlation

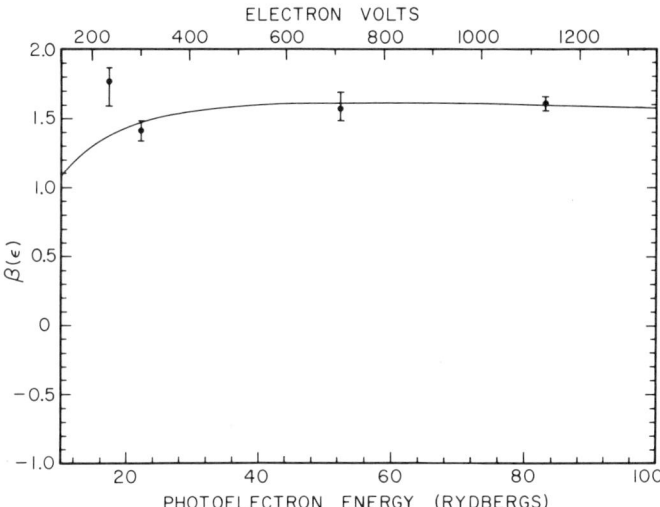

FIG. 27. Asymmetry parameter β for the 3p subshell of krypton at high energies. The theoretical results are from Cooper and Manson (1969) and the experimental points are from Krause (1969).

effects are not yet entirely negligible. Agreement between theoretical and experimental β's for other cases at higher energies is substantially the same as the krypton 3p case (Krause, 1969; Wuilleumier and Krause, 1974).

Open shell atoms present some particularly interesting features owing to the possibilities of exchange of angular momentum between the photoelectron and the ion (Fano and Dill, 1972; Dill and Fano, 1972; Dill, 1973; Dill et al., 1974, 1975) as discussed in detail in I. These angular momentum exchanges can only occur through a noncentral (anisotropic) atomic field, since a torque must be exerted to change angular momenta and central forces can, of course, exert no torque. In I, the theoretical situation for photoionization of a 3p electron from atomic sulfur in the ground $3p^4$ ^3P state was discussed and it was shown that the final ionic state of $S^+ 3p^3$ could be ^4S, ^2P, or ^2D. Owing to the differences in the anisotropic interaction of the photoelectron with each of the ionic terms, it was shown in I that β for each ionic term could differ as well. The result of HF calculations of the asymmetry parameter for sulfur 3p (Dill et al., 1974, 1975) is shown in Fig. 28. Note that, although the β curve for each ionic term has the same general shape, the value of β at a given photoelectron energy can be rather different for the different terms, e.g., just below 2 Ry, the ^4S β is ~ 1.5 larger than the ^2P β, which is a very great variation in a parameter that has a total range of 3, from -1 to 2. The detailed explanation for the results is given else-

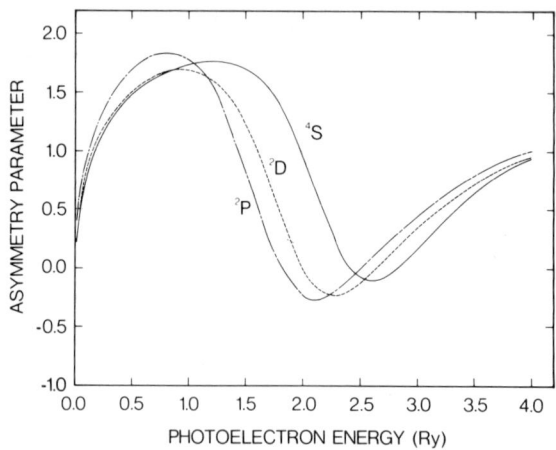

FIG. 28. Theoretical asymmetry parameter β for the $(3p^4)^3P \rightarrow (3p^3)$ ^4S, ^2D, and ^2P photoionizing transitions in atomic sulfur.

where (Dill et al., 1974, 1975) in terms of the phase shifts of the various possible continuum waves, discussed in I, and it is shown explicitly that for open shell atoms too, the phase shifts are the crucial factor in the determination of β. In fact, the situation for atomic oxygen (Manson et al., 1974; Starace et al., 1974), which lies in the same column of the periodic table as sulfur, is that the phase shifts are almost independent of final ionic term, indicating that anisotropic interactions are quite small and that the asymmetry parameters for the different ionic terms will hardly differ. Unfortunately, no measurements for angular distributions of photoelectrons from open shell atoms have been reported, but this area of photoelectron spectroscopy is seen as one that will be extremely fruitful in understanding the noncentral interactions in many-electron atoms.

The angular distribution of photoelectrons in the neighborhood of autoionizing resonances is a particularly interesting subject. This is because in going across a resonance, the phase shift changes by π, and, since the phase shifts are known to be the essential determinants of the asymmetry parameter, β will vary rapidly across a resonance as well. The asymmetry parameter for xenon in the neighborhood of the autoionizing resonances lying between the $^2P_{3/2}$ and $^2P_{1/2}$ threshold is shown in Fig. 29. The experimental results (Samson and Gardner, 1973) are seen to be in quite good agreement with theory (Dill, 1973) and both are extremely rapidly varying over the region. As mentioned, this rapid variation of β over a resonance will be a general feature over the entire periodic system, and further studies, both experimental and theoretical, should be extremely profitable.

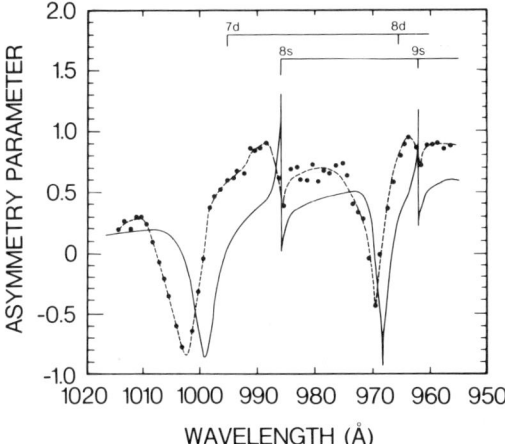

FIG. 29. Asymmetry parameter β for the 5p subshell of xenon in the neighborhood of the autoionizing resonances between the $^2P_{3/2}$ and $^2P_{1/2}$ thresholds. The solid curve is the theoretical work of Dill (1973) and the dashed curve and points are the experimental results of Samson and Gardner (1973).

III. Summary and Concluding Remarks

This review has presented a comprehensive overview of atomic photoelectron spectroscopy. Part I (Manson, 1976) dealt primarily with the theory of photoionization and photoelectron angular distributions and calculations at various levels of approximation. The focus was on the physical implications of each approximation to allow an approximate *a priori* assessment of how well a given calculation would do in a particular situation. In Part II (herein) the attempt was made to show the variety of phenomena uncovered, their physical explanation, and the predictive powers of the theory by comparison with experiment. In addition, we have tried to spotlight the areas, e.g., excited state or open shell atoms, for which more work is indicated. It is also hoped that the references, while not exhaustive, are extensive enough for use by those just entering or on the periphery on the field of atomic photoelectron spectroscopy.

From an experimental point of view, making measurements on gaseous atoms or molecules is quite similar, and many more measurements of molecules than atoms have been made since there are so many more gaseous molecules (Shirley, 1972; Caudano and Verbist, 1974). Therefore, we shall conclude this review with a brief discussion of some very successful recent calculations of molecular photoionization cross sections and angular distributions.

From a theoretical viewpoint, molecules differ from atoms in the very important respect of the multicenter nature of molecular fields, which affects both initial discrete *and* final continuum states. This is a crucial aspect of the photoionization of molecules and must be included even for *qualitative* accuracy as shown by recent calculations (Dill and Dehmer, 1974; Dehmer and Dill, 1975, 1976b; Davenport, 1976). The calculations are performed using the multiple scattering model of Johnson (1973) to take the multicenter aspects of the molecular field into account. The details of these calculations are beyond the scope of this review and the details can be found in the above references. Basically, the method involves partitioning the molecular field into closely packed spheres centered on each atomic site and one around the molecule. For simplicity the exchange effects are included via the Slater (1951) central field approximation to exchange, so that this model might be considered the molecular analog to HS for atoms. This calculation does include the correct boundary conditions at the nuclei and infinity and at the boundaries of each spherical region. Accurately describing the singularities at the atomic nuclei distributed throughout the molecular field, i.e., the multicenter nature of the field, is crucial to a correct description of the photoelectron–molecular ion interaction. This, in turn, ensures realistic continuum phase shifts, which are the essential ingredient in photoelectron angular distributions. The recent results of these calculations (Dehmer and Dill, 1975, 1976b; Davenport, 1976) are extremely encouraging.

Errata for Part I

1. The ψ on line 9 of p. 83 should be replaced by Ψ.
2. In Eq. (35) on p. 84, Y' and H' should be replaced by Y^1.
3. The reference Dill *et al.* (1974) should be to *Phys. Rev. Lett.*, not *Phys. Lett.*

References

Amusia, M. Y., Cherepkov, N. A., and Chernysheva, L. V. (1971). *Sov. Phys.—JETP* **33**, 90.
Amusia, M. Y., Ivanov, V. K., Cherepkov, N. A., and Chernysheva, L. V. (1972a). *Phys. Lett. A* **40**, 361.
Amusia, M. Y., Cherepkov, N. A., and Chernysheva, L. (1972b). *Phys. Lett. A* **40**, 15.
Burke, P. G., and Taylor, K. T. (1975). *J. Phys. B* **8**, 2620.
Caudano, R., and Verbist, J., eds. (1974). "Electron Spectroscopy." North-Holland Publ., Amsterdam.
Chang, T. N. (1975). *J. Phys. B* **8**, 743.
Chang, T. N. (1977). *Phys. Rev. A* **15**, xxx.
Codling, K., Houlgate, R. G., West, J. B., and Woodruff, P. R. (1976). *J. Phys. B* **9**, L83.
Combet Farnoux, F. (1969). *J. Phys. (Paris)* **30**, 521.
Combet Farnoux, F. (1970). *J. Phys. (Paris)* **31** (C4) 203.

Combet Farnoux, F. (1971). *J. Phys. (Paris)* **32** (C4) 7.
Combet Farnoux, F. (1972). *In* "Inner Shell Ionization Phenomena" (R. W. Fink, S. T. Manson, J. W. Palms, and P. V. Rao, eds.), CONF-720404, pp. 1130–1141. USAEC, Oak Ridge, Tennessee.
Combet Farnoux, F. (1976). Personal communication.
Combet Farnoux, F., and Heno, Y. (1967). *C.R. Acad. Sci., Ser. B* **264**, 138.
Cooper, J. W. (1962). *Phys. Rev.* **128**, 681.
Cooper, J. W. (1964). *Phys. Rev. Lett.* **13**, 762.
Cooper, J. W., and Manson, S. T. (1969). *Phys. Rev.* **177**, 159.
Cooper, J., and Zare, R. N. (1968). *J. Chem. Phys.* **48**, 942.
Cooper, J., and Zare, R. N. (1969). *In* "Lectures in Theoretical Physics" (S. Geltman, K. Mahanthappa, and W. Brittin, eds.), Vol. IIC, p. 317. Gordon & Breach, New York.
Davenport, J. W. (1976). *Phys. Rev. Lett.* **36**, 945.
Dehmer, J. L., and Dill, D. (1975). *Phys. Rev. Lett.* **35**, 213.
Dehmer, J. L., and Dill, D. (1976a). *Phys. Rev. Lett.* **37**, 1049.
Dehmer, J. L., and Dill, D. (1976b). *Proc. Int. Conf. Inner-Shell Ioniz. Phenom.*, 2nd, Invited Papers, pp. 221–238.
Dehmer, J. L., Chupka, W. A., Berkowitz, J., and Jivery, W. T. (1975). *Phys. Rev. A* **12**, 1966.
Dill, D. (1973). *Phys. Rev. A* **7**, 1976.
Dill, D. (1975). *In* "Photoionization and Other Probes of Many-Electron Interactions" (F. Wuilleumier, ed.), pp. 387–394. Plenum, New York.
Dill, D., and Dehmer, J. L. (1974). *J. Chem. Phys.* **61**, 692.
Dill, D., and Fano, U. (1972). *Phys. Rev. Lett.* **29**, 1203.
Dill, D., Manson, S. T., and Starace, A. F. (1974). *Phys. Rev. Lett.* **32**, 971.
Dill, D., Starace, A. F., and Manson, S. T. (1975). *Phys. Rev. A* **11**, 1596.
Ederer, D. L., and Tomboùlian, D. H. (1964). *Phys. Rev. A* **133**, 1525.
Fano, U. (1961). *Phys. Rev.* **124**, 1866.
Fano, U. (1969). *Phys. Rev.* **178**, 131.
Fano, U., and Cooper, J. W. (1968). *Rev. Mod. Phys.* **40**, 441.
Fano, U., and Dill, D. (1972). *Phys. Rev. A* **6**, 185.
Haensel, R., Kunz, C., Sasaki, T., and Sonntag, B. (1968). *Appl. Opt.* **7**, 301.
Haensel, R., Radler, K., Sonntag, B., and Kunz, C. (1969). *Solid State Commun.* **7**, 1495.
Hall, H. (1936). *Rev. Mod. Phys.* **8**, 358.
Heinzmann, U., Heuer, H., and Kessler, J. (1975a). *Phys. Rev. Lett.* **34**, 441.
Heinzmann, U., Heuer, H., and Kessler, J. (1975b). *Phys. Rev. Lett.* **34**, 710.
Heinzmann, U., Heuer, H., and Kessler, J. (1976). *Phys. Rev. Lett.* **36**, 1444.
Houlgate, R. G., West, J. B., Codling, K., and Marr, G. V. (1976). *J. Electron Spectrosc. Relat. Phenom.* **9**, 205.
Hudson, R. D., and Carter, V. L. (1967a). *J. Opt. Soc. Am.* **57**, 651.
Hudson, R. D., and Carter, V. L. (1967b). *J. Opt. Soc. Am.* **57**, 1471.
Jaeglé, P., and Missoni, G. (1966). *C.R. Acad. Sci., Ser. B* **262**, 71.
Jaeglé, P., Combet Farnoux, F., Dhez, P., Cremonese, M., and Onori, G. (1969). *Phys. Rev.* **188**, 30.
Johnson, K. H. (1973). *Adv. Quantum Chem.* **7**, 143.
Kelly, H. P., and Simons, R. L. (1973). *Phys. Rev. Lett.* **30**, 529.
Kennedy, D. J., and Manson, S. T. (1968).
Kennedy, D. J., and Manson, S. T. (1972). *Phys. Rev. A* **5**, 227.
Kessler, J. (1975). *In* "The Physics of Electronic and Atomic Collisions: Invited Lectures, Review Papers, and Progress Reports of the IX International Conference on the Physics of Electronic and Atomic Collisisons" (J. S. Risley and R. Geballe, eds.), pp. 112–125. Univ. of Washington Press, Seattle, and references therein.

Krause, M. O. (1969). *Phys. Rev.* **177**, 151.
Lin, C. D. (1974). *Phys. Rev. A* **9**, 171.
Lipsky, L., and Cooper, J. W. (1967). Personal communication.
Lukirskii, A. P., Brytov, I. A., and Zimkina, T. M. (1964). *Opt. Spectrosc.* (*USSR*) **17**, 234.
McGuire, E. J. (1968). *Phys. Rev.* **175**, 20.
Madden, R. P., Ederer, D. L., and Codling, K. (1969). *Phys. Rev.* **177**, 136.
Manson, S. T. (1969). *Phys. Rev.* **182**, 97.
Manson, S. T. (1973). *J. Electron Spectrosc. Relat. Phenom.* **1**, 413.
Manson, S. T. (1976). *Adv. Electron. Electron Phys.* **41**, 73.
Manson, S. T. (1977). *J. Electron Spectrosc. Relat. Phenom.* (to be published).
Manson, S. T., and Cooper, J. W. (1968). *Phys. Rev.* **165**, 126.
Manson, S. T., and Cooper, J. W. (1970). *Phys. Rev. A* **2**, 2170.
Manson, S. T., and Kennedy, D. J. (1970). *Chem. Phys. Lett.* **7**, 387.
Manson, S. T., Kennedy, D. J., Starace, A. F., and Dill, D. (1974). *Planet. Space Sci.* **22**, 1535.
Msezane, A., and Manson, S. T. (1975). *Phys. Rev. Lett.* **35**, 364.
Msezane, A., and Manson, S. T. (1977). To be published.
Rau, A. R. P., and Fano, U. (1968). *Phys. Rev.* **167**, 7.
Samson, J. A. R. (1966). *Adv. At. Mol. Phys.* **2**, 177.
Samson, J. A. R., and Gardner, J. L. (1973). *Phys. Rev. Lett.* **31**, 1327.
Samson, J. A. R., and Gardner, J. L. (1974). *Phys. Rev. Lett.* **33**, 671.
Seaton, M. J. (1951). *Proc. Roy. Soc., Ser. A* **208**, 418.
Shirley, D. A., ed. (1972). "Electron Spectroscopy." North-Holland Publ., Amsterdam.
Shyu, J. S., and Manson, S. T. (1975). *Phys. Rev. A* **11**, 166.
Slater, J. C. (1951). *Phys. Rev.* **81**, 385.
Starace, A. F. (1970). *Phys. Rev. A* **2**, 118.
Starace, A. F., and Armstrong, L. A., Jr. (1976). *Phys. Rev. A* **13**, 1850.
Starace, A. F., Manson, S. T., and Kennedy, D. J. (1974). *Phys. Rev. A* **9**, 2453.
Swanson, J. R., and Armstrong, L. A., Jr. (1977). *Phys. Rev. A* **15**, 661.
Torop, L., Morton, J., and West, J. B. (1976). *J. Phys. B* **9**, 2035.
Walker, T. E. H., and Weber, J. T. (1974). *J. Phys. B* **9**, 674.
Wuilleumier, F., and Krause, M. O. (1974). *Phys. Rev. A* **10**, 242.

Timekeeping and Its Applications

GERNOT M. R. WINKLER

U.S. Naval Observatory
Washington, D.C.

> *Horam non possum certam dicere:*
> *facilius inter philosophos*
> *quam inter horologia convenit.*
> SENECA
> (Apotheosis II, 2–3)

I. Introduction	34
A. Principles	34
B. History	39
C. Reviews, Literature, and Conferences	42
D. Terminology	43
II. Clocks	43
A. General Principles	43
B. Clock Performance	45
C. Clock Hardware	61
III. Time Determination	73
A. Classical (Optical) Methods	73
B. New Methods	75
IV. Remote Time Measurements—Synchronization	77
A. General Principles	77
B. Satellite Time Measurements	82
C. Time and Relativity	83
V. Applications	88
A. Celestial Navigation, Geodesy, and Space Tracking	88
B. Electronic Navigation and Positioning Systems in the Rho–Rho Mode	88
C. Pseudorandom Noise Systems	88
D. Communications Systems	89
E. Metrology, Scientific Applications	89
References	90

I. Introduction

A. Principles

The flow of time is a basic experience of every observer of nature. If one postulates that one can always say which of two local events is earlier, then one has arranged the events at a given location into an ordered sequence. One can use the real numbers as labels for the events and such a system of labeling is called a time scale. Evidently any mathematical transformation that leaves the order of events undisturbed is equally justified, i.e., the system of labeling is quite arbitrary. Time is therefore a purely conceptual quantity. It does not exist objectively by itself.

Any absolute objective reality of time, as opposed to the claimed subjective ideality, would have to imply that time by itself could produce effects. However, we do not allow time as such to play a causal role in science. We always invoke specific physical causes to explain changes in macroscopic nature. An example would be the law of inertia, which postulates that in the absence of forces (the causes of change) the momentum of a body remains constant. It is only in the realm of the "indeterminate" when we cannot identify physical causes that time appears to assume a pseudocausal role, since for spontaneous events, e.g., nuclear disintegration, no other notion is left as an explanation for change. Time is clearly connected with (idealized) causality. The cause must precede its effects, but causality is not merely succession or vice versa. On the other hand there is also a nontemporal, immutable element in causality. A basic postulate of science is that identical causes will evoke the same effects regardless of time; they always have done so and they always will do so in the future.* It is the cause, not time, that brings forth the effect. Time enters merely as an element of description of the effect in its relation to other events. We use time as an abstract concept to interconnect changes that we observe around us. It is a subjective ordering parameter, a cerebral product in its essential nature, which must be rigorously defined in its derivation and structure. The realization of this ideality of time that is of higher order than our usual abstract notions is necessary to protect us from unwarranted metaphysical beliefs, e.g., that

---

* The nontemporal part of causality means that the fundamental facts of nature must be assumed eternal. The electric force, e.g., as such is metaphysical: it is everywhere and whenever causes, the charges, call for its action. This nontemporal aspect of nature is doubly evident in the photon, since in its own frame of reference it does not experience time at all, i.e., the proper time between emission and absorption is always zero even if for us it may have been on its way for billions of years. Distance has vanished also and this is clearly evidence that time and space can be only means for the description of our experience but are not entities that exist by themselves.

absolute uniform time pervades the universe, i.e., that when it is 8 o'clock here, it must be 8 o'clock everywhere.*

Newton (1686) made this assumption axiomatically and it was sufficiently powerful to serve science for over 200 years. The refined measurements during the last century have, however, shown that this assumption is not tenable. Newton emphasizes that his absolute uniform time is a mathematical fiction, and he distinguishes it from the time of measurement (*hora, dies, annum*). A corresponding distinction is made by Russell (1936) between the physical events, which have a finite duration albeit arbitrarily small, and the instants of time, which are the real numbers on the time scale. In Russell's system events could overlap, but instants can not. Following general practice, we will not insist on this distinction in the following, except to point out that quantum-physical considerations, which discuss the possibility of a smallest time interval (of the order of 10^{-24} sec), evidently refer to the world of events and not to the instants. These and related topics are, however, beyond the scope of this article.† Similarly, we cannot consider whether all real numbers can be meaningfully assigned to represent physical events. This question must be left to cosmological and philosophical discussions.

Although our system of labeling events is in principle quite arbitrary, it is clear that a system of timekeeping in order to have widest applicability in the sciences and technology must be based on the following definition, which is

---

* The situation corresponds to our perception of space. There are three levels. The first is the necessary (a priori) form or category of experience. The second we learn from experience, which we idealize as Euclidean geometry and uniform time. The third is a purely logical construct based on postulates that can be quite arbitrary. Time is inherent to all experience; the space category applies only to visual, aural, vestibular, and tactile sense perception and to our respective memories. Kant did not distinguish between these levels and neither do most of his critics who point to the possibility of non-Euclidean geometry as proof that the categories cannot be a priori. This, however, is justified only in respect to the second and third levels, which deal with the idealized and with the abstract construct. Kant is quite correct in regard to the first, the intuitive background. Therefore we must specify the details of scientific time measurement by means of definitions and postulates. The a priori background can have only a local meaning. This corresponds to Riemannian geometry, where one accepts the idealization of Euclidean geometry as a postulate for the infinitesimal domain.

† As everywhere in science, an abyss opens when we look more closely at our simple postulate of the events being arrangeable into an ordered sequence. Wiener (1914) has considered the conditions that are necessary for the instants to form a series. Russell (1936) showed that the existence of instants can only be proved on the basis of assumptions that, he feels, we are not logically or empirically compelled to hold to be true. Such assumptions are that events can be well ordered; that every event has a first instant; that series of continually smaller events (Dedekind sections) can approach a point as their limit; that every event has a last instant, etc. If these assumptions should indeed have no basis in reality, then the existence of instants cannot be logically proved. They are then merely a logical ideal to which it is possible to approximate arbitrarily well: a situation that would still be good enough for our purposes. (The CIPM recommends "s" as the abbreviation for second.)

basic for our purposes: two time intervals are equal if it can be shown that equal processes took place during these two intervals. A time scale based on this definition and which uses equal intervals for its successive scale intervals will be called a uniform time scale.*

These definitions must be understood as idealizations that in practice can only be approximated. However, they are not the only basis of scientific time measurement. In celestial mechanics, time is understood as the independent argument in the equations of motion. This has led to the concept of ephemeris time, which is held by definition to be uniform. Mulholland (1972) equates the concept of uniform time with dynamical (ephemeris) time. We hold that to be too narrow a position, and not in the interest of greatest generality. The fact that our basic definitions refer to ideal limits is a great advantage compared to a concept of uniformity by definition since it leads to the possibility of determining the degree of uniformity by the intercomparison of many simultaneous equal processes in their realization as clocks. This view is of great historical significance since Newcomb (1874) established the nonuniformity of the rotation of the earth by comparing it with the orbital motion of several bodies.

On the other hand, one must also realize that our basic postulate is differential in its essential nature since it defines time intervals. This entails a number of disadvantages compared to what we may call the "functional" approach to time measurement as it is used traditionally in astronomy. Kovalevsky (1965) bases his criterion for uniformity on the satisfaction of the laws of mechanics. Time scales as used in astronomy are based, in general, on a measurable parameter $p(t)$, of a permanent and stable system. The evolution of such a parameter must be known theoretically. Only cosmic phenomena, e.g., the motions in the solar system, can provide a measure of time that covers the long past before high-performance artificial clocks became available. Kovalevsky shows convincingly that science

---

* Scale intervals are the defining units of the scale. They are the intervals between those instants designated by the integer numbers of the scale. In sidereal time the scale interval is the sidereal day that elapses between two meridian transits of the vernal equinox. The hour, minute, and second are the conventional (nondecimal) subunits. Similarly, the mean solar day is the time between two transits of the mean sun. Its subunits are the hour, minute, and second of our conventional time reckoning. The real scale unit of ephemeris time is the tropical year 1900, although for practical reasons the definition adopted (1956) refers to the ephemeris second as a fraction (1/31 556, 925.974 7) of the tropical year at the epoch 1900 January 0, noon ephemeris time (the tropical year is not strictly constant). In local atomic time the scale unit is the (S.I.) second, one of the base units of the international system of units. Minutes, hours, and days (atomic) are multiples that we use instead of a decimal count in the interest of easy comparability with the time units in public use. A general use of decimal time reckoning is not practical since a day has 86,400 sec, which is not a simple power of ten. The second is too firmly entrenched to be dropped.

cannot abandon such time scales, which can be extended backwards to the beginnings of history. Observations of celestial positions represent part of an on-going experiment. Such observations are of real value since about 1750, but all celestial observations going back 3000 years represent a wealth of information that must remain accessible to research by means of time scales that can be extended over long intervals.

A further distinction between the two basic approaches to time measurement appears in regard to the effect of the unavoidable errors, the noise in the measurement process. While errors can never be completely avoided, they nevertheless enter in the approach based on short time intervals in a fundamental way. Such intervals can be rigidly defined but can be realized only with some uncertainty, however small. Even in the absence of systematic changes in the duration of the basic intervals, i.e., in the ideal case of purely random errors, the summation process that is inherent in any clock operation will add this random noise to produce a random walk in the indicated clock time. This is the same effect as can be seen in the outcome (total gain or loss after n trials) of random coin-tossing experiments (see, e.g., Feller, 1957). In addition, however, actual clocks are never completely free from some internal correlation of their disturbances in rate and from systematic changes over long time intervals, which will increase the statistical dispersion of their readings beyond the dispersion of a random walk. In the light of this intrinsic involvement of clock timekeeping with statistics it is understandable that modeling of clock behavior and the study of algorithms for the production of highly uniform clock time scales has produced great interest during the last 15 years.* As a result of these studies and supported by highly sophisticated clock technology we now have clock time scales in existence with uniformities of 1 nsec over a day and about 1 μsec over a year.

The fact that time can be measured so precisely, far better than any other physical parameter, represents a technological asset of great importance. Correspondingly, time measurements have found many applications. Apart

* Campbell (1926) goes one step further in suggesting that time is only a macroscopic measure similar to temperature and that it would make no sense in principle to describe processes of a single atom in any temporal way. Time and its measurement is therefore fundamentally connected with statistics: and chance and time are inseparable. Derivations by Schroedinger (1931) that he bases on quantum-theoretical ideas also show that uncertainty is intrinsic to time measurement, i.e., that we must expect noise for fundamental reasons and not only because of particle counting or thermal noise processes. Processes in a system with mass m can at best be observed with a precision in time of $\tau_0 = h/4\pi mc^2$. Moreover, the uncertainty of the system's energy is necessary for the evolution of the system, i.e., that a clock changes its readings in a "lawful" manner. In a system with sharp energy all probabilities become time independent, i.e., it is similar to a system in thermodynamical equilibrium; we cannot observe any changes.

from the continuing requirements for time in support of celestial navigation, geodesy, and space technology, precise time or synchronization is needed by many electronic navigation and communications systems. The most demanding applications of timekeeping, however, still exist in astronomy. Very long baseline radio interferometry (VLBI) requires independent clocks at intercontinental distances to stay within parts in 10^{14} over several hours. Pulsar research needs a reference time scale to be uniform to within fractions of a microsecond over several months. Time can also be used as a substitute quantity to be measured instead of some other systems parameter. An important example is the possibility of expressing distance as light time, i.e., the possibility of having a unified standard for time and length with the velocity of light as a defined quantity as discussed by Halford et al. (1972).

Only the times of local events can be related to a clock without assumptions concerning the delay of the signals that relate the events to the clock. This raises the problem of simultaneity of distant events. Kant (1781) has pointed out that two events can only be called simultaneous if they are in immediate mutual interaction, i.e., if they are in spatial proximity. Mutual interaction means that either event could be said to be the cause of the other and it is clear that the concept of simultaneity, if based on causal interaction as the criterion for order in time, cannot be extended to distant events without ambiguity. If we assume the velocity of light to be the maximum speed at which a causal agent can travel through space, then we can distinguish two cases: two events are so separated in space–time that a light flash emanating from one event arrives at the location of the second either before or after the second event. In the first case, the second event could have been caused by the first, and we have fixed their temporal order unambiguously. In the second case, both events occur before light from the other event can arrive. In that case in which neither event could have caused the other, we can call the events quasi-simultaneous (they are separated by a spacelike interval). Their order in time depends entirely on the coordinate system of the observer.

The involvement of precision time measurement with relativity theory comes from different sides, depending on which principle we view as fundamental. Our basic postulate can in practice only be applied to local phenomena; clocks that keep proper time are combined to produce uniform proper time scales. Relativity corrections must be applied for the construction of spatially extended coordinate time scales. These are composite time scales, coordinate scales for the physicist but with sufficient precision still to be considered a proper scale for the astronomer, who makes his measurements with his observatory clock. The internationally coordinated scales UTC and TAI serve as long-term bench marks; in the moment of observation the clock is indicating local time and reductions must be made to the coordinate system that is used for the direct physical description of the phenomena observed. All these clock time scales, since they are originally local, have arbitrary epochs,* and only the desire for coordination suggests a common reference.

In distinction, the functional approach to time definition refers to an evolving parameter of an extended system. Any description of an extended process refers to general coordinates, i.e., a time scale so derived is of necessity a coordinate time. In addition, the epoch of the scale is not at all arbitrary but depends entirely on the theory used. Any changes in the constants of this theory bring about a change in the time scale as well.

Our basic definition mentions "equal processes" without specifying the processes in detail. This is not yet sufficient for the establishment of a unique standard. As has been pointed out by Synge (1960), an additional principle is needed to avoid ambiguities in time measurements. He uses a principle of "consistency" that requires that the ratio of the frequencies of two different atomic resonances will always be the same everywhere. A less stringent postulate is "monochronism" as discussed by Winkler and van Flandern (1977). It is the sole function of the time parameter in the sciences to relate the observed processes to each other. If a particular process requires a time scale different from the other phenomena in nature, then the theory of this process must be modified. In other words, it is necessary to adopt but one process as a time standard because it is conceivable that clocks based on different processes could generate mutually nonuniform time scales, even if each time scale by itself is in conformance with our basic definition of uniformity. This could be called a cosmological effect, but it would also be an indication that our theoretical understanding is insufficient for the processes involved. However, these considerations are important only as conceptual clarifications. In practice we must admit the usefulness of special time scales such as UT1, UT2, even though they are not strictly uniform by any definition. In addition, uniformity itself is a useful concept only if considered within one and the same reference system.

B. History

The standard of time has evolved historically from mean solar time, a measure of time that is linked to the rotation of the earth with respect to the average direction toward the sun. Several phenomena and principles are

---

* We adhere to the conventional meaning of "epoch" in astronomy. Epoch is the instant that is used as a reference or origin. It is given in a specific time scale. An example is the epoch of atomic time (1958 January 1, 0^h AT is approximately 1958 January 1, 0^h UT2) or of ET (1900 Jan. 0, 12^h ET).

here intermixed from the beginning and necessitate that any practical standard of timekeeping be a compromise. Until the end of the nineteenth century the rotation of the earth in space (in an inertial reference) was assumed to be perfectly uniform.* Any problems that existed had only to do with the motion of the reference point selected for the rotational period, i.e., the vernal equinox for sidereal time and the mean sun for mean solar time. The theories involved are those of the motion of the axis of the earth in space (precession and nutation) and of the earth around the sun (which includes the ellipticity of the orbit and the known perturbations). Since the mean sun in its motion has a small secular acceleration, Newcomb (1895) introduced a fictitious mean sun as an operational reference for mean solar time. This fictitious mean sun moves at a strictly constant speed in right ascension. Therefore, universal time, which is mean solar time plus 12 hours at the international meridian (IM, near the Greenwich meridian) is or should be uniform time, whereas sidereal time is not a uniform measure of time due to a small secular term in the precessional motion of the vernal equinox. This is true even of the so-called mean sidereal time, which refers to the mean equinox of date vs the true equinox, which is additionally affected by nutation.

Due to the existence of a mathematical expression for the right ascension of the fictitious mean sun, the clock time of transit of any celestial object with known position will yield UT0 after correction for longitude, aberration, parallax, nutation, and precession. However, this measure of time is still not strictly uniform because of three effects. First, the coordinates of the observatory are subject to small changes due to movements of the earth's axis with respect to the earth. After corrections are applied for this polar variation (PV), we arrive at a measure of UT that is applicable to the whole earth (UT1). The rotation of the earth is itself not strictly uniform but is slightly variable. There is a regular seasonal variation (SV) and irregular and unpredictable changes in the rotational period, which are not yet clearly understood. The Bureau International de l'Heure (BIH), which has coordinated international timekeeping since 1919, has adopted a formula for the SV that is used by all time services to reduce UT1 to UT2. UT2 still contains unpredictable variations (parts in 10^8) and a small (10^{-10} per annum) secular retardation that is somewhat uncertain and that probably has not been constant during geological times as summarized by Stoyko (1970) and Rochester (1973).

Since UT is so closely related to the orientation of the earth in space, it is

* Kepler seems to be the first who voiced doubts about the constancy of the rotation of the earth. Kant stated the nonuniformity of the rotation of the earth as a necessary consequence of tidal friction (Felber, 1974). These and some later contributions, however, remained mere speculation until Newcomb (1874) reported observational evidence.

indispensable for astronomy, navigation, geodesy, and space research. As an observed quantity it is known only after the fact. Therefore, clocks are necessary to extrapolate UT and to disseminate it as a time signal. This clock time UTC, which is UT as coordinated by the BIH, was originally kept quite close to UT1, later to UT2 (since 1960). This requires occasional adjustments of the time signals, which have been implemented since 1972 by means of the "leap second."

Newcomb (1895) derived the constants for the theory of the sun (the motion of the earth around the sun) on the basis of observations going back about 200 years. Since these observations were made in terms of mean solar time, the constants derived represent a connection between the average speed of rotation of the earth and a time parameter in which the dynamical laws of the revolution of the earth around the sun are valid. This time parameter is called ephemeris time (ET). It is accessible by comparison of the observed longitude of the mean sun with Newcomb's tables of the sun, which use ET as an argument (this is the present convention; Newcomb did not yet distinguish ET and UT). This is the best current example for the "functional" approach to timekeeping mentioned above. In actual practice, because of the available better precision, a lunar ephemeris together with the observed motion of the moon is used for the operational determination of ephemeris time. In either case, correction terms have to be added to improve the conventional tables and to correct for the presently used system of fundamental astronomical constants.

The first atomic clock time scale, A.1, was established at the U.S. Naval Observatory (USNO) in 1959 (Markowitz, 1962) with a rate that was set as closely as possible to agree with ET. Its epoch (origin) was nominally set to 1958 January 1, 0^h UT2. The scale (with extensions back to mid-1955) was used rather extensively as reference in the American Ephemeris and Nautical Almanac and for other scientific purposes. The measurement of the cesium frequency was performed in terms of ET with an accuracy of about 2×10^{-9} (Markowitz et al., 1958), limited by the resolution in ET. On the other hand, the standard deviation of the average rates of independently manufactured and adjusted cesium clocks is about 2×10^{-12}. Therefore, when the BIH started with its improved scale AT(BIH) on 1 January 1968, the scale was set to the average rate of the clocks of three establishments as an initial set (the French group "F," the Physikalisch Technische Bundesanstalt PTB, and the USNO). The scale was formally adopted as International Atomic Time (TAI) by the General Conference for Weights and Measures (CGPM) in 1970, and it is now computed from the individual contributions of more than seventy clocks in sixteen establishments.

TAI and its stepped version UTC, which is the basis for the time signal transmissions, have shown a rate stability of about 1×10^{-13} from year to year, and it is accessible through the published corrections to the reference clocks of its contributing time services. Its rate differs from the rate that would correspond to the "true" frequency of a local cesium beam frequency standard for two reasons. First, relativistic corrections have to be applied, and second, the initial rate, which has been maintained in the primary interest of uniformity, has now been found to be high in frequency by about 1×10^{-12}. After much deliberation and following a specific recommendation of the IAU (1976) a step of exactly -10×10^{-13} has been applied to the frequency of TAI on 1 January 1977 (its rate will be retarded by 86.4 nsec per day effective at 0^h TAI).

C. Reviews, Literature, and Conferences

Timekeeping and distribution have seen an enormous advance during the last 50 years. The work has been reviewed periodically and information is abundantly available. However, as usual, there is a striking difference between the primary literature, produced by the actual worker in the field, and the secondary production, which summarizes and compares. Some bias, lack of judgment, and lack of direct experience are unavoidable. Therefore, it is important to go back to the sources wherever possible. Direct experience is indispensable before critical decisions can be made. A small specific laboratory program will quickly familiarize the decision-makers and their consultants with the problems and capabilities at hand.

The Consultative Committee for the International Telecommunications Union (ITU), known as CCIR, has a study group exclusively responsible for questions of concern to the Standard Frequency and Time Signal (SFTS) Service. This study group VII, which brings its reports regularly up to date, therefore provides an important source of information in its "green book." A history of the quartz crystal clock is given by Marrison (1948). Bagley (1966) has given an excellent exposition of frequency and time measurements, which is still up to date except for details of radio transmissions and obvious instrumental improvements of more recent date. The Hewlett-Packard Company (1975) issued an "Application Note" with many useful details of frequency and clock measurements.

A major contributor to the field is the U.S. National Bureau of Standards (NBS), a leader in measurement technology. Many papers, reports, technical notes (TN), etc., are published regularly. Blair and Morgan (1972) and Blair (1974a) have edited highly valuable reference volumes.

There are several regular conferences dedicated exclusively to frequency and time measurement. The annual Symposium on Frequency Control (ASFC) is sponsored by the U.S. Army Electronics Command, Fort Monmouth, New Jersey. The Meeting is held in May in Atlantic City, New Jersey. The Planning Meeting for Precise Time and Time Interval (PTTI) is sponsored by NASA and DOD and takes place in December in Washington,

D.C. The International Congress of Chronometry (CIC) is held every five years (the latest meeting was in 1974) and is sponsored by the French, German, and Swiss Societies of Chronometry.

D. Terminology

The terms used by astronomers and geodesists in timekeeping and applications of time have been supplemented (and often confused) in recent years by terms coming from physics and electronics. This review is necessarily somewhat influenced by what the authors of the original papers used. However, there is clearly a need to be conservative if we want to avoid chaos. The CCIR Study Group VII has made a serious attempt to suggest a standard. However, Report 366-2 is at this time only a beginning. The efforts are extraordinarily difficult because of the wide range of the affected interests. Beehler et al. (1965) have given a list of definitions of terms such as stability and precision that are preferred by this author. We must also not forget that terms become established by usage in a selection process the same way as the other elements of language. In the past, any attempts to standardize terminology by a committee that ignores this fact have been doomed.

II. CLOCKS

A. General Principles

Chronographs can record the time of an event, i.e., of an external pulse, whereas a clock merely produces time markers together with identification of these markers.* The clock reading is reduced to an established time scale by adding the clock correction to the reading. The use of the clock correction (and of its complement, the clock error) is supplemented by the algebraic method of giving clock differences. The difference between two clocks A and B is given by the difference of their readings for the same event. In practice this will be obtained by measuring the time interval between two

---

* There is some recent confusion in the use of long-established terms. A clock is a device that indicates time by means of time markers and their identification. A time code generator provides this identification by means of an electric time code. In principle, a clock is an interpolation device and as such its reading can only approximate a time scale. This is analogous to the case of other physical scales, such as a temperature scale, which is approximated by a thermometer. The scale, however, is not defined by the thermometer but is defined by postulates. It is standard scientific practice to separate the concepts from the means of their realization. The term time standard is best avoided altogether. It could have various meanings that are unambiguously covered by well-established terms. Examples would be that salesmen use it with the implied vague meaning of a very good clock, such as an atomic clock. Other meanings include a clock that indicates Standard Time (e.g., EST), a reference clock, a master clock, or a time interval standard, i.e., a frequency standard with a 1 Hz output. In addition, we still have the term chronometer for applications that can use more bombastic language.

corresponding markers of the two clocks with clock A connected to the start input and clock B to the stop input of a time interval meter (counter). Let A and B be the corresponding readings of the two clocks; then their difference will be $A - B$. If UTC is the true time of the corresponding event, then the clock correction for clock A will be $UTC - A$, and the clock error $A - UTC$, both with respect to the UTC scale.

The rate of a clock is the increase of its correction per day. The quality of a clock is generally not affected by its rate. It is the variation of rate that is important as a measure of quality of the clock. The rate variations are indicative of a failure to generate strictly identical processes in the clock. In part we may not have succeeded in shielding the clock from the rest of the universe, in part we may see the effects of spontaneous relaxation effects inside the clock. The various types of precision clocks in use today differ in regard to the relative importance of these disturbances. It must be realized that all clocks show some disturbances in rate and only differ in regard to the magnitude of the disturbances. In that respect there is no principle difference between atomic and nonatomic clocks. Experience has shown that macroscopic phenomena and devices are in general more subject to unaccountable disturbances than atomic processes, which can be shielded very effectively. This is true at least for intermediate measurement times. For intervals of less than about 10 sec, nonatomic devices, e.g., quartz crystal clocks and superconductive cavity oscillators, show the greatest stability. On the long-term extreme the question becomes one of reliability and accessibility since atomic timekeeping is only some 20 years old (since mid-1955).

Atomic clocks, particularly the cesium beam devices, are superior in respect to rate accuracy. By this term we understand the ability of a clock to operate at the correct rate without reference or calibration to other devices. The best cesium clocks, laboratory instruments, generate seconds that are accurate (in agreement with the definition) to a few parts in 10^{13}. However, this capability is of very limited practical importance since all applications use clocks as parts of systems where initial calibration is available and desirable for other reasons. Precision, the ability to reproduce the externally given rate, is what is needed. The only exception to this occurs in the construction of clock time scales over very long intervals, particularly TAI, where accuracy information must be part of any algorithm input to prevent very long term drifts of the scale, i.e., to place a limit on the random walk in rate, which otherwise could become excessively large (Guinot, 1974).*

---

* The opinion seems widespread that this can only be done with large laboratory frequency standards. Hellwig et al. (1973) have shown, however, that any cesium beam clock can be evaluated with respect to its systematic frequency errors to about 5×10^{-13}. Averages over many such evaluations of the contributing clocks to the BIH would probably give better and more reliable results than what can be expected from a sole reliance on only three or four large devices.

The rate variations of a clock are generally a function of the measurement interval. The different types of clocks have various merits in this respect, so that the choice of the type of clock depends very much on the requirements of the application.

Since all clocks are subject to random noise rate variations in addition to systematic effects as mentioned above, it is clear that a perfect clock is an ideal case that can only be approached but not reached. Schroedinger (1931) points out that a clock, in order to give time exactly, would have to have an infinitely large energy uncertainty, i.e., also an infinite energy. The same conclusion is reached formally if we envision the phase of the perfect clock's output signal as modulated with noise. Improvements in clock performance require improvements in the signal to noise ratio of the carrier power to phase sideband power. A consequence of this is that requirements of low power consumption and extreme clock performance are contradictory, at least in principle. With the present state of the art, this has not yet produced an unsurmountable limitation in clock design.

However, as pointed out by Winkler (1972), the utility of the best clocks is often limited by the phase noise contributed by the transmission medium that separates distant clocks. Insufficient attention to the more practical aspects of clock use in large systems often leads to unnecessarily ambitious clock specifications that are not only costly but generally lead to less than optimum engineering compromises, with later consequences and disappointments in regard to reliability, complexity, maintenence, and operator training.

B. Clock Performance

1. General

We first consider a clock as generator of a nearly sinusoidal signal voltage

$$V(t) = [V_0 + \varepsilon(t)] \sin[\Phi(t)] \qquad (1)$$

For our purposes, $\varepsilon(t)$ will be very small or vary slowly and we can ignore it for most timing applications. t is the ideal time according to our basic postulate or time of a reference clock. We can write for the instantaneous phase

$$\Phi(t) = \Omega t + \varphi(t) \qquad (2)$$

where

$$F = \Omega/2\pi \qquad (3)$$

is the exact or nominal (reference) frequency, e.g., 5 MHz, Ω the circular frequency in rad/sec, and $\varphi(t)$ the phase error of our clock. The time error (clock error) is then

$$x(t) = \varphi(t)/\Omega \qquad (4)$$

and the instantaneous normalized (relative, fractional) frequency error (departure) is

$$y(t) = dx/dt = \Omega^{-1} \, d\varphi(t)/dt \tag{5}$$

The average normalized frequency error between t_1 and t_2 is given by

$$\bar{y} = \frac{x(t_2) - x(t_1)}{t_2 - t_1} = \frac{1}{t_2 - t_1} \int_{t_1}^{t_2} y(t) \, dt \tag{6}$$

All measures of clock performance will be related to $\bar{y}(t)$ or $x(t)$ in one way or another with the influence of age, environmental conditions, etc., as qualifying parameters. However, for various reasons the $\bar{y}(t)$ aspects were recently emphasized in their importance over the $x(t)$. Phenomenologically it is the $x(t)$ that form the primary time series since they are obtained directly from clock intercomparisons, which we can list in tabular form as shown in Table I.

TABLE I

MEASUREMENTS OF CLOCK ERROR IN FIXED INTERVALS

Time of measurement, $t_0 + k\tau$	Index, k	Clock error $x(k)$	Average relative frequency error, $\bar{y}(k) = [x(k) - x(k-1)]/\tau$	Change in $\bar{y}$ per unit of time, $\Delta(k) = [x(k) - 2x(k-1) + x(k-2)]/\tau^2$
t_0	0	$x(0)$		
$t_0 + 1\tau$	1	$x(1)$	$\bar{y}(1)$	
$t_0 + 2\tau$	2	$x(2)$	$\bar{y}(2)$	$\Delta(2)$
$t_0 + 3\tau$	3	$x(3)$	$\bar{y}(3)$	$\Delta(3)$
$t_0 + 4\tau$	4	$x(4)$	$\bar{y}(4)$	$\Delta(4)$
$t_0 + 5\tau$	5	$x(5)$	$\bar{y}(5)$	$\Delta(5)$
$t_0 + 6\tau$	6	$x(6)$	$\bar{y}(6)$	$\Delta(6)$
⋮	⋮	⋮	⋮	⋮

The basic problem in any measure of clock performance is the fact that the $x(t)$ as well as the $\bar{y}(t)$ may be unbounded as t goes to infinity. As a practical example we may consider the case of a quartz crystal clock. A logical procedure for the evaluation of clock performance would be to choose $\tau = 1$ day, i.e., to measure the clock error once per day at a fixed time with a time interval meter (counter). The start and stop inputs are connected to the clock and reference time marker pulses, respectively. The result of such measurements will be the time series $x(k)$, where k indicates the day of measurement ($\tau = 86{,}400$ sec). The relative frequency error averaged over a day will be given by $\bar{y}(k)$.

An estimate of the clock's error can be computed for a future date according to the extrapolation formula

$$x(t) = x(t_0) + [y(t_0) + \tfrac{1}{2} \Delta(t_0)(t - t_0)](t - t_0) \tag{7}$$

which is, of course, equivalent to the form that uses the clock correction $C(t)$ and that has been used extensively in the past,

$$C(t) = C(0) + R(0)t + at^2 \tag{8}$$

where $C(0)$ and $R(0)$ are the initial values for clock correction and rate, respectively, and a is the acceleration of the clock. The formula is often surprisingly good for a crystal clock because a can approximate the effect of aging. However, such formulas are deceptive. If measurements are dominated by noise then it becomes quickly obvious that neither this formula nor any other can ever be an exact representation of a clock's behavior, because we deal with a random process. Therefore the problem of characterization of a clock's performance takes on different aspects depending upon the applications. In general we can distinguish two main "philosophies":

1. The phenomenological method: We attempt to represent post facto the clock error of an individual clock with simple mathematical formulas. They may be of the type of formula (8) or they may consist of pieces of linear fits, etc. Since one does not know the causes for the variations of $\bar{y}$, the procedure for producing a mathematical representation of the systematic variations, i.e., the fitting of $x(t)$, is somewhat arbitrary and entirely phenomenological. As such it will tend to go all the way, i.e., until the residuals of $x(t)$ appear to be random. However, this can be approached only if the resolution is dominated by random phase noise. Otherwise one would attempt to represent a random walk in phase with a fitting procedure, which would eventually cause a random walk in rate. Barnes (1976) gives an excellent discussion on this question of the dangers of overmodeling deterministic or systematic effects. However, no assumptions are made whatever and the method has been used extensively.

2. The random process characterization: One considers all variations of $\bar{y}(k)$ as coming from a random process that can be characterized by its autocorrelation function, its spectrum, or other measures such as the various time domain variances. Measures so derived will be particularly useful for diagnostic purposes, for prediction of probable clock errors, and for specifications. Even in this case, however, one has to remove the most obvious deterministic features such as jumps in phase or frequency, and overall frequency drifts.

As a further development of the second method we can also list:

3. Clock modeling: This method uses well-known filtering procedures that have been developed for the detection of signals in the presence of noise, where the noise characteristics are known or can be measured. These methods as applied to the analysis of clock performance have been claimed

to be particularly useful for computer simulations, environmental correlations (diagnostics), and clock error prediction. The best known technique, the Box–Jenkins (1970) approach, is being used extensively in economic forecasting. The models are based on the assumption of an underlying purely random process of clock disturbances. A linear filter of the autoregressive moving average type (ARMA) is then to be found, which makes the model output statistically equal to the observed clock disturbances. A "prediction" of probable clock errors is based on the assumption that the clock (and its environment!) will continue to behave as the filter does. However, even if this stationarity assumption does not hold, nevertheless the first- or higher-order difference of the process will be free from trends, and the filter can still be useful. In this case, since integration must be added to the model, we speak of an integrated ARMA (ARIMA) model.

Some of the first serious attempts to characterize the performance of precision clocks have been reported by Scheibe and Adelsberger (1950), Smith (1953), Stoyko (1937), and Greaves and Symms (1943).

Smith (1953) describes in considerable detail the problems that were then encountered in evaluating high-quality quartz crystal clocks in terms of an inferior primary standard, i.e., the sidereal time determinations. Individual clocks were evaluated against a group of clocks with the average absolute second difference of their daily rates taken as a performance measure. In our notation this measure would be proportional to

$$V = \frac{1}{N-2} \sum_{k=3}^{N} |\Delta(k) - \Delta(k-1)| \qquad (9)$$

In today's language this would be described as a time domain measure of the random clock fluctuations.

An early example of the purely phenomenological approach can be found in Scheibe and Adelsberger's (1950) post facto fitting of parabolic arcs to their clock errors. The standard deviation of the residuals can be used in this case as the measure of a clock's "unusefulness." However, two questions will still have to be investigated: the distribution of the residuals (is it normal?) and their correlation in time (is it white noise?). We will therefore discuss briefly the general case of the random process characterization as it applies to clock errors and clock rates as time series.

A measure of dispersion of a random process that shows drifts in the means of its successive samples is not a unique problem. Von Neumann et al. (1941) investigated the use of the mean square successive difference as a measure of dispersion of successive values z_i:

$$\delta^2 = \frac{1}{N} \sum_{i=1}^{N} (z_i - z_{i-1})^2 \qquad (10)$$

where i gives the temporal order of a random variable z. δ^2 is not affected by slow drifts, whereas

$$s^2 = \frac{1}{N} \sum_1^N (z_i - \bar{z})^2 \qquad (11)$$

gives an estimate of variance σ^2 by using the observations independently of their order and therefore includes the effect of trends.

For the following we will make the assumptions of zero mean, of stationarity, and of ergodicity for our general time series $z(t)$, i.e.,

$$\mu \equiv E[z(t)] \equiv \int_{-\infty}^{\infty} zp(z)\, dz = \lim_{T \to \infty} \frac{1}{2T} \int_{-T}^{T} z(t)\, dt = 0 \qquad (12)$$

where the first integral represents the ensemble or stochastic average, which we designate as mathematical expectation $E[z]$, and the second integral is the time average variously designated with a bar or the symbol $\langle z(t) \rangle$. The probability density function (pdf) is $p(z)$. Stationarity in the strict sense would mean that the statistical properties of the series are independent of time. As a minimum that means that the pdf is independent of time, but for our purposes the lesser assumption of stationarity in the wide sense is sufficient, i.e., the autocovariance function (acvf) is only a function of the "lag" u, i.e.,

$$\gamma(u) = \lim_{T \to \infty} \frac{1}{2T} \int_{-T}^{T} z(t)z(t + u)\, dt \qquad (13)$$

We can follow the main expositions on general time series analysis (e.g., Papoulis, 1965; Jenkins and Watts, 1968), since they are most pertinent to our subject.* A time series is a function of time, ideally continuous but in practice given in discrete values, which has a random character. With a record of such a series it is not possible to predict future values. However, estimates can be made based on the statistical law as exhibited by the past values and on the assumption that this empirical law will not change. This is in other words the assumption of stationarity, which is crucial for a real understanding of the meaningfulness of certain measures of clock performance as well as for the use one can make of clock models. As has been pointed out by Barnes et al. (1971) and very emphatically by Barnes (1976), stationarity in any sense whatsoever must be viewed as a property of models.

* Sometimes we will have to distinguish between population parameters or ideal quantities such as μ, σ^2, γ, ρ, S, and the corresponding estimators or sample quantities $\bar{z}$, s^2, c, r, $\hat{S}$, for the mean, the variance, the autocovariance function, the autocorrelation function, and the spectral density. However, we can do this only inconsistently as part of our explanations, since most of the original contributions unfortunately do not make this distinction.

No actual time series can be assumed stationary as such if for no other reasons than that record lengths are limited and also devices have a finite lifetime. In judging to what degree a stationary model can represent actual observations, one can distinguish several cases:

    a. The available data can be assumed to be part of a stationary series.
    b. Pieces of the observed series can be treated as stationary.
    c. A filtered version of the series may be treated as stationary.
    d. There appears to be no way in which the data can be treated as stationary.

There are many forms of nonstationarity but usually one considers two main cases: (1) nonstationarity in μ, and (2) nonstationarity in μ and σ^2. One also distinguishes deterministic trends and stochastic trends. As an example of great importance to timekeeping we consider the case of an atomic clock that is only subject to purely random disturbances of rate. Let the normalized frequency errors $\bar{y}(t)$ be a purely random, normal process with a given mean $\mu_{\bar{y}}$ and variance $\sigma_{\bar{y}}^2$; then the clock error $x(t)$ will be a random walk with uncorrelated (orthogonal) increments:

$$x(t) = x(t - \tau) + \bar{y}(t)\tau \qquad (14)$$

which has as mean and variance

$$E[x(t)] = tu, \qquad \text{var}[x(t)] = t\sigma_{\bar{y}}^2 \qquad (15)$$

i.e., $x(t)$ is a nonstationary process. Such a process is shown in Fig. 1. Given the fact that some noise is intrinsic to the time measurement process, as discussed above, such a clock performance is the best one can hope to obtain. In all practical cases there will be additional disturbances that will increase the clock errors over and above the process considered in Fig. 1.

If the normalized frequency errors are uncorrelated this will be reflected in the data by the fact that the sample acvf is near zero for all lag values u except $u = 0$:

$$c_{\bar{y}}(u) = \frac{1}{N-1} \sum_{k=1}^{N-u} \bar{y}(k)\bar{y}(k+u) \approx \begin{cases} \sigma_{\bar{y}}^2 & \text{for } u = 0 \\ 0 & \text{for } u \neq 0 \end{cases} \qquad (16)$$

This relation is exact for the acvf:

$$\gamma_{\bar{y}}(u) = \begin{cases} \sigma_{\bar{y}}^2 & \text{for } u = 0 \\ 0 & \text{for } u \neq 0 \end{cases} \qquad (17)$$

From the above example it will be understandable that the $\bar{y}(t)$ can be expected to be generally more meaningful in the statistical sense than the $x(t)$, despite the fact noted above that the $x(t)$ represent our data more directly. Since the $x(t)$ represent at least the dispersion of a random walk (or

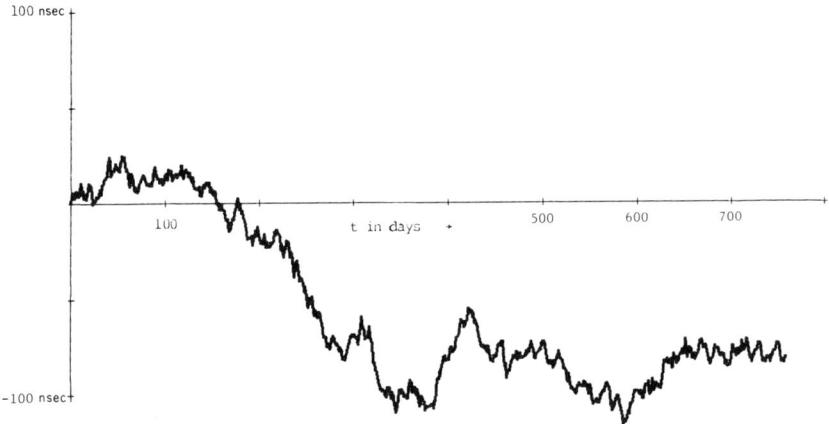

FIG. 1. Daily clock errors for white frequency noise. The data give an example for the performance of an ideal cesium clock with a $\sigma_y(3^h) = 1 \times 10^{-13}$ ("super" cesium clock).

Brownian motion) some authors have questioned the justification for concepts such as instantaneous relative frequency departure $y(t)$, since a mathematical Brownian motion does not generally have a derivative (Boileau and Picinbono, 1976). To this, Barnes (1976) has given a most useful, lucid, and complete reply. In all of our practical cases we deal with band-limited processes. There are no X-ray and gamma-ray components in the timing waveforms that we sample. We always have a practical upper frequency limit f_h of the oscillator circuitry. We also have a lower limit f_L, which is given by the data length T. We cannot hope to measure any frequency components with $f < f_L = 1/T$.

The acvf and its normalized version, the autocorrelation function (acf), characterize the degree of linear dependence between the values of the random disturbances at two points separated by the lag u. The spectral density $S(f)$, a function of the Fourier frequency f, contains the same information in the frequency domain. The two functions form a Fourier transform pair (S' is one-sided):

$$S'_{\bar{y}}(f) = 4 \int_0^\infty \gamma_{\bar{y}}(u) \cos(2\pi f u) \, du, \qquad \gamma_{\bar{y}}(u) = \int_0^\infty S'_{\bar{y}}(f) \cos(2\pi f u) \, df \quad (18)$$

A consequence of (18) is that

$$\gamma_{\bar{y}}(0) = \sigma_{\bar{y}}^2 = \int_0^\infty S'_{\bar{y}}(f) \, df \quad (19)$$

which shows that $S'(f)$ gives the information how the variance receives contributions from the different Fourier frequencies. The dimension of S is therefore variance/hertz.

For practical purposes of data analysis, the extensive literature must be consulted, especially Jenkins and Watts (1968) or Blackman and Tukey (1959), who give guidance on the problems of smoothing, aliasing, and statistical significance of the results. [See also Loeb (1972) for useful details.]

2. *Types of Clock Erratics (Clock Noise)*

The first recognition of different types of random clock disturbances was due to Greaves and Symms (1943), who distinguished erratics of type A, B, and C, which today we would call white phase noise (white PM), white rate noise (white FM), and random walk in rate. Formulas with second and third differences of the clock error were used to distinguish between the types of erratics. Barnes and Allan (1964) recommend measures of clock performance that do not depend critically on the data length or the "cut-off" frequencies f_L and f_h in the presence of drifts. They recommend two solutions to circumvent these problems. One could take an average of the variances of only two samples. This statistic is now generally accepted as a practical, simple, and reproducible measure in the time domain. It is designated as $\sigma_{\bar{y}}^2(\tau)$ and variously called "pair" variance or "Allan" variance. A second possibility would be the variance of second or higher differences of phase. They also recommend the power spectral density of phase and frequency variations as a measure of frequency stability. This has been criticized by some as unrealistic since phase or time errors can hardly ever be assumed stationary, as we have seen above in the case of a random walk in phase [Eq. (14)]. Therefore, it must be made completely clear that we must take off any trends and systematic effects or use some filtering before we attempt to characterize the random clock variations. Thus stationary models are a useful concept since we deal only with residuals in x or y and do not have to worry about what happens at $t = \infty$. Under these conditions standard statistical techniques should suffice, i.e., the determination of the mean and sample variance of the normalized frequency errors $\bar{y}$ in addition to the determination of the systematic parts.

However, fitting of clock data with mathematical functions is a somewhat arbitrary procedure, and Barnes and Allan (1964) are certainly right in their demand for measures that are not critically dependent upon either cut-off frequencies or residual trends in the data. The pair variance is defined as

$$\sigma_{\bar{y}}^2(\tau) = \left\langle \frac{1}{2n} \sum_{k=1}^{k=n} (\bar{y}_k - \bar{y}_{k-1})^2 \right\rangle = \frac{1}{2}\delta^2 \qquad (20)$$

Except for the factor $\frac{1}{2}$, it is identical with von Neumann's mean square successive difference. However, it was introduced as the limiting case with respect to N of a sample variance. We take only $N = 2$ samples per group but average over n groups.

Allan (1966) investigated also the dependence of the general sample variance of the $\bar{y}$ on the parameters T (dead time), τ (integration time), N (number of measurements in the sample), and f_h (measurement system bandwidth). Since the pair variance is just a special case of these generalized Allen sample variances,

$$\sigma_{\bar{y}}^2(\tau) = \sigma_{\bar{y}}^2(N, T, \tau) \quad \text{for} \quad N = 2 \quad \text{and} \quad T = \tau \tag{21}$$

a ratio of two variances, in particular the ratio

$$\chi(N, m) = \sigma_{\bar{y}}^2(N, \tau, \tau)/\sigma_{\bar{y}}^2(2, \tau, \tau) \tag{22}$$

will be a measure of the internal correlation of the $\bar{y}$ in time, i.e., of the noise type present. The parameter m is an indicator for the noise type since it can be shown that for a given N and a given noise type there exists a functional dependency of the sample variance upon the integration time of the form

$$\sigma_{\bar{y}}^2(N, \tau, \tau) \sim \tau^m \tag{23}$$

Allan (1966) has found that such a relation exists if the spectrum is of the form

$$S_y(f) = \begin{cases} h_\alpha f^\alpha & \text{for} \quad 0 < f \leq f_h \\ 0 & \text{for} \quad f > f_h \end{cases} \tag{24}$$

where α is a constant. Experience shows that generally clock performance can be represented very well by the expression

$$S_y(f) = \sum_{\alpha = -2}^{\alpha = 2} h_\alpha f^\alpha \tag{25}$$

Table II gives an overview of the five noise types that are of importance as contributors to clock noise. One type is predominant in a given region of the spectrum and also in a corresponding range of the sigma–tau plot. This allows an approximate conversion from one domain to the other. The details were worked out by Cutler and Searle (1966), Vessot et al. (1966), Barnes and Allan (1964), and Allan (1966), and are given concisely by Barnes et al. (1971). For a very useful and practically oriented review see Howe (1976). CCIR report 580 gives up to date references.

Baugh (1971) has described the use of the Hadamard variance for the analysis of frequency modulation. Rutman (1974b) suggests alternatives for the measurement and characterization of frequency stability supporting the general principles as outlined by Barnes et al. (1971). Several criticisms have,

TABLE II

Power Law Noise Models with Their Respective Stability Measures[a]

Noise type	Spectrum				Time domain, $\sigma_y^2(\tau)$		Slope	
	α	$S_y(f)$	Slope S_x	S_y		$\sigma_x(\tau)$	$\sigma_y(\tau)$	m
White phase	2	$h_2 f^2$	0	+2	$\dfrac{3 f_h h_2}{4\pi^2 \tau^2}$	0	-1	-2
Flicker phase	1	$h_1 f$	-1	+1	$\left[3 \ln(2\pi f_h \tau) - \ln 2 + \dfrac{9}{2} \right] \dfrac{h_1}{(2\pi\tau)^2}$			
White FM	0	h_0	-2	0	$\dfrac{h_0}{2\tau}$	$\dfrac{1}{2}$	$-\dfrac{1}{2}$	-1
Flicker FM	-1	h_{-1}/f	-3	-1	$2(\ln 2) h_{-1}$	1	0	0
Random walk FM	-2	h_{-2}/f^2	-4	-2	$\dfrac{4\pi^2}{6} h_{-2} \tau$	$\dfrac{3}{2}$	$\dfrac{1}{2}$	1

[a] Adapted from Howe (1976).

however, been voiced against the general approach. Boileau and Picinbono (1976) consider the use of f_L and f_h just as a mathematical trick to ensure convergence of (19), a trick that makes the actual measures $\sigma_{\bar{y}}(\tau)$ and $S_{\bar{y}}(f)$ unduly dependent upon the arbitrary choice of f_L and f_h. However, we must emphasize that the f_L and f_h constitute a physical reality and not a "trick." It is therefore necessary to account for this reality in (19), which must therefore include appropriate factors. Vessot (1976) gives a lucid discussion of the transformed time window due to the sampling time τ, which produces a factor

$$F_1 = \sin^2(\pi f \tau)/(\pi f \tau)^2 \tag{26}$$

There is a second factor F_2, which for finite N represents an effective high-pass filter, thereby eliminating worries about certain divergencies:

$$F_2 = 1 - [\sin^2(r\pi f N \tau)]/[N^2 \sin^2(r\pi f \tau)] \tag{27}$$

Cutler and Searle (1966) have given the complete formula for the conversion from frequency to time domain:

$$\sigma_{\bar{y}}^2(N, T, \tau) = \frac{N}{N-1} \int_0^\infty S_y(f) F_1 F_2 \, df \tag{28}$$

where r is the dead-time factor $r = T/\tau$.

Barnes (1968) has extended Allan's concept of the variance ratio $\chi(N, m)$ and introduced two "bias" functions B_1 and B_2, which allow the comparison of variances taken with any N and any dead time T. We note that our sampling scheme of Table I allows computation of the y without any dead time. Therefore most modern methods of frequency stability analysis start from phase or time error sampling. For very short integration times ($\tau < 1$ sec) it is more convenient to measure stability of an oscillator in the frequency domain directly (Allan, 1974; Rutman, 1974a; Howe, 1976). The practical application of the variance ratio method has remained limited.

Lindsey and Chie (1976) have recommended the use of Kolmogorov structure functions as a basis for frequency stability measures. The Cutler and Searle procedure bases the pertinent concepts of frequency stability upon the more commonly known aspects of a time series: the variance, the autocorrelation, and the power spectral density. The pair variance, the general Allen variance $\sigma_y^2(N, T, \tau)$, Kramer's curvature variance (Kramer, 1975), etc., are time domain variations of more or less the same kind. They are directly useful as objective measures and can be determined very easily. The pair variance is an excellent measure—practically a standard today for specifications. However, a fuller description of oscillator stability requires a separate treatment and removal of systematics before computation. Kramer's curvature variance is hardly more complicated but is not sensitive to linear drifts. The Hadamard variance, since it has a narrower transfer function than the pair variance determination, is more useful as a poor man's spectrum analyzer. It should be realized that the spectral density $S_y(f)$ would be the most meaningful information if it could only be determined more reliably and without the pitfalls of present-day digital spectral analysis.

Since the structure functions use exactly the same data, i.e., phase increments over the sampling time τ, the practical problems are largely the same. However, by virtue of the greater generality of the basic concept of a structure function vs the autocorrelation function (which is a special case of a structure function), the much greater and less transparent mathematical effort of this approach may pay off in some theoretical way. A question of concern ought to be that many derivations of Lindsey and Chie (1976) and some definitions depend on the validity of a particular model for the oscillator noise process.

Table III and Figs. 2–5 give a comparison and examples of the more frequently used stability measures. The cdf of Fig. 3 is plotted as a 95% confidence region. Figure 5 shows the normalized spectral density SD. We have to multiply SD with the variance in order to obtain $S_y(f)$. In other words, the SD has been computed from the autocorrelation function and not from the acvf. A Tukey lag window was used with a lag of 50 days (Jenkins and Watts, 1968, p. 244). The meaning of the pair variance in relation to the

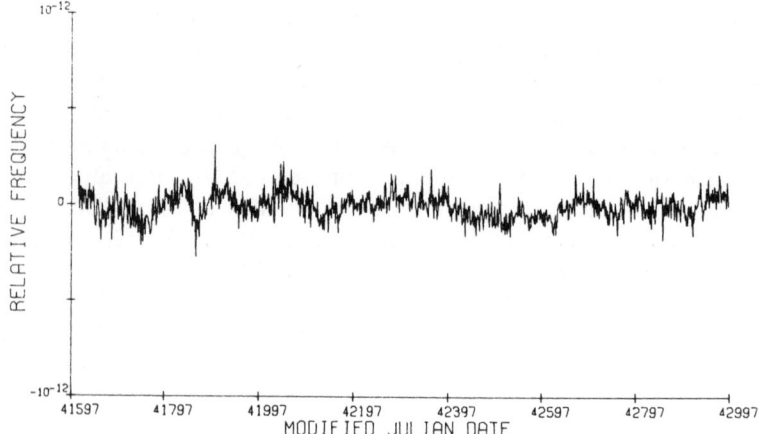

FIG. 2. Residual frequencies of cesium 571/1C-2 vs A.1 (USNO, mean), frequency drift removed. (Courtesy D. Percival, USNO.)

variance (which is given in Table III) can best be seen from the relation

$$\sigma_{\bar{y}}^2(\tau) \approx s_{\bar{y}}^2 - c_{\bar{y}}(u) \tag{29}$$

where $u = 1$ (which is here also τ). This can be obtained from the definition of the pair variance under the assumption of zero mean. From Eq. (29) and

$$\gamma(u) = \rho(u)\gamma(0), \qquad \gamma(0) = \sigma^2$$

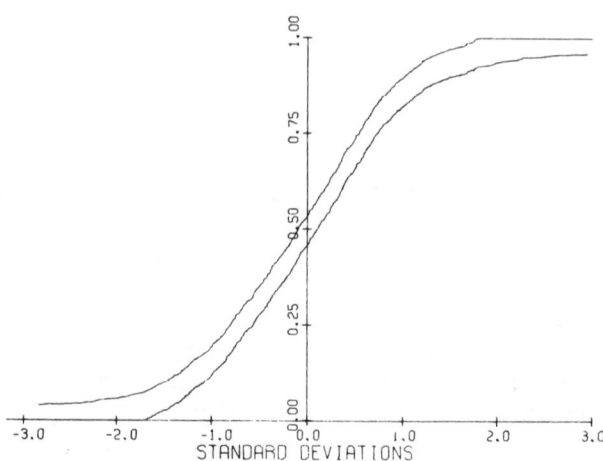

FIG. 3. Probability distribution function of the $\bar{y}(1^d)$ of cesium 571/1C-2 vs A.1 (USNO, mean).

TABLE III

Long-Term Performance of Ten Cesium Clocks at the USNO[a]

Oscillator[b]	Days	Linear frequency drift removed[c] (per year in 10^{-13})	Sample variance (after drift removed) (in 10^{-26})
Cs 346/1C	1200	+3.5	4.0
Cs 532/1C	1200	—	2.9
Cs 549/1	1200	—	2.9
Cs 571/1C-2	1185	+1.5	0.7
Cs 591/1	1138	+4.1	3.1
Cs 654/1C-2	972	−4.9	0.8
Cs 660/1C-2	952	—	6.4
Cs 783/1C-2	845	—	14.2[d]
Cs 834/1C-2	727	—	1.5
Cs 837/1C-2	726	−7.2	4.4

[a] The data refer to the $\bar{y}(1^d)$, which are measured in reference to A.1 (USNO, mean). (Data courtesy of D. Percival, USNO.)

[b] The units with a −2 designation are high-performance units (004).

[c] Dashes indicate that no significant drift was found in a straight-line fit.

[d] The large variance of Cs 783/1C-2 is caused by a large non-linear frequency drift which was not removed.

we can also obtain insight into the meaning of Allan's variance ratio (22):

$$\chi(N, m) \approx 1/[1 - r_{\bar{y}}(\tau)] \qquad (30)$$

It is clear that for any sample of size N, $\chi(N, m)$ gives information only about the very beginning of the autocorrelation function.

3. Clock Modeling

a. Trend modeling. We assume that at least a part of the $\Delta(k)$ is purely deterministic. Our distinction of deterministic from random variations can only be formal, i.e., it has nothing to do with causes, but is merely an expression of our ability to describe a process with a smooth mathematical function. Naturally the division cannot be sharp. By including more and more terms in a polynomial approximation we can reduce the variance of the residuals, but soon we reach a point of diminishing returns.

Another distinction is to be made between a step change due to a unique event in the history of the clock and tendencies that act over long intervals. If the steps are not accounted for separately, then they will impair the representation of the long-term tendencies. Therefore, sporadic disturbances

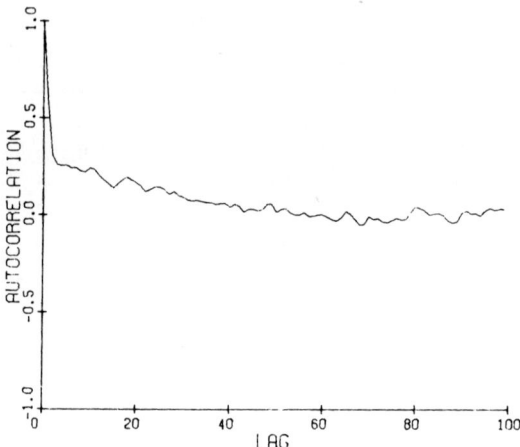

FIG. 4. Autocorrelation function of $\bar{y}(1^d)$ of cesium 571/1C-2 vs A.1 (USNO, mean).

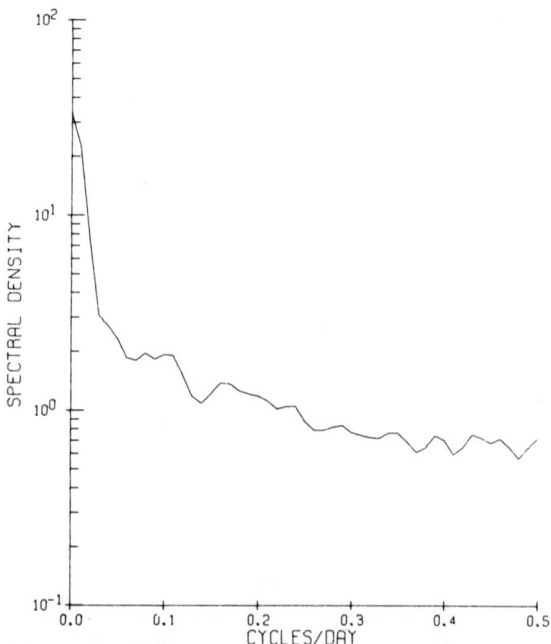

FIG. 5. Normalized spectral density function SD, of $\bar{y}(1^d)$ of cesium 571/1C-2 vs A.1 (USNO, mean).

must first be subtracted from clock measurements. In practice, the "undisturbed" parts of a clock record are best modeled separately. The separation of "disturbed" from "undisturbed" parts can be made by means of a clock group that can be represented solely with undisturbed or, in the case of weighting, less disturbed models of the contributing clocks, i.e., we are led to an iterative selection procedure (Winkler et al., 1970, "WHP").

In practice, cesium clocks have often been modeled with pieces of straight lines. Parabolas are used for rubidium clocks and cubic parabolas for quartz crystal clocks (which often exhibit a drift of drift). However, such procedures are only justified under certain conditions. The approximation of cesium clocks with pieces of linear fits has been criticized by Allan and Gray (1971) as less efficient for the determination of average normalized frequency errors as compared to a simpler endpoint procedure such as our Eq. (6). The argument is correct. However, WHP use the straight line fit in this case because of additional benefits of performance checks and error control. Barnes (1976) correctly points out that the use of linear fits as approximation for cesium clocks is dangerous since a white FM noise process produces a random walk in phase (Fig. 1). WHP use straight-line fits also for the computation of remote time scales where measurements are severely contaminated by white phase noise, and in this case the procedure is certainly justified. Barnes' (1976) warning that the variations in a Gaussian process are eventually self-correcting is correct in principle. One may not be able to wait that long in all cases but one must be as conservative as possible in the application of rate corrections. Percival (1976) gives objective criteria for the detection of significant frequency steps.

Another serious but quite popular pitfall is the attempt to ignore systematic disturbances altogether by incorporating them in the form of greater complexity of the model. What this means is easy to see in the case where, e.g., a crystal clock can be approximated quite well with a cubic parabola. Now assume that this clock has suffered a shock without our knowledge. An attempt to improve the fit with a fourth, fifth, or higher degree polynomial will inevitably lead to disastrous results if any extrapolation is attempted. In addition, as Barnes et al. (1971) have pointed out, the presence of just flicker noise will actually doom any simple mathematical model as more and more data become available. A recourse to higher differences (because these may be stationary!) is also of no practical help for prediction because each integration introduces additional walk in the integrated variable.

b. Optimum prediction. The principle of this method is shown in Fig. 6. Formally the problem of clock noise is reduced to the assumption of an underlying gaussian white noise process. If it is possible to construct a (digital) filter that can change this noise into the noise actually observed in

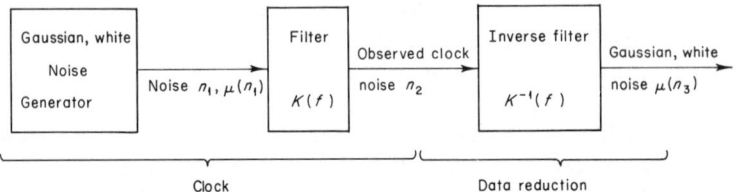

FIG. 6. Clock modeling, the filter approach. Procedure: Determine $K^{-1}(f)$ from data (whitening); determine $K(f)$ from $K^{-1}(f)$; determine initial conditions of $K(f)$ from final state of $K^{-1}(f)$; use mean $\mu(n_3)$ as input to $K(f)$ to give an "optimum" prediction (see Barnes, 1976).

the $\bar{y}(\tau)$, then one can use this filter with zero input (which is the average of the ideal process assumed) to produce an artificial output that will represent an optimum prediction of the actual clock performance. Unquestioningly this is a much more sophisticated way to go about the problem of prediction compared to simple extrapolation. Barnes (1976) discusses some of the details peculiar to the application in timekeeping and clocks.

Box and Jenkins (1970) and Anderson (1975) are most emphatic about the need to separate any and all visible trends, which really returns the problem to our discussion above. Here again, one must remind the neophyte that extrapolation is a very poor method of prediction, as is the assumption "if present trends continue." But these trends are now often incorporated in the filter functions $K(f)$ in Fig. 6.

There is, however, an area of unquestioned usefulness of this kind of modeling. Our filter $K(f)$ contains exactly the same information (but in the form of coefficients of the digital filter) as $S_{\bar{y}}(f)$ and therefore also as $\sigma_{\bar{y}}^2(\tau)$. In this direction, from filter to spectral density and to time domain measure, the conversion is relatively simple; in the reverse direction it is almost impossible. It appears then that $K(f)$ may be the most useful description of clock performance. Certainly this is true for computer simulations and possibly for some diagnostics. Environmental correlations, however, can simply be accomplished by cross-correlation techniques.

4. Time Scales

As we stated earlier, time scales for scientific and general use are based on our fundamental postulate, which must be implemented in a set of logical procedures that allow a determination of a measure of reliability and performance. Therefore, all major time scales are based on some averages of clocks or time determinations. There has been a considerable discussion of the merits of the various algorithms in use (Allan et al., 1975; Allan and Gray,

1971; BIH Annual Report, 1974; CCIR, 19 ; Angeli, 1974; Yoshimura, 1972; Markowitz, 1962; Mungall, 1971). It seems, however, that with the growing number of BIH contributing clocks, the question of the computation of TAI will become more and more academic. In other words, algorithms seem to be important only to those time scales that must be based on a small sample of clocks; but a small sample is really not a good basis for any statistical method.

Essentially we can distinguish three different approaches:

a. Emphasis on recognition and separation of sporadic and systematic disturbances before averaging with equal weights (Winkler *et al.*, 1970).

b. Emphasis on establishing individual statistical clock models for the assignment of weights to the contributing clocks (Allan *et al.*, 1975; Guinot, 1974; Huebner and Becker, 1974).

c. Use of Kalman filtering of the clock data (Angeli, 1974).

It appears in the light of Barnes' (1976) comments about the eventually self-correcting variations in a gaussian process that methods (b) and (c) will be more subject to the dangers of overmodeling than method (a).

The problem of long-term time scale computations from the readings of many clocks that individually have a random walk in clock error is interestingly related to the problem of least squares adjustment of leveling networks. The identification of single blunders and systematic scale distortions has direct parallels in composite clock time scales (Borre and Meissl, 1974).

In the past, the physical implementation of a composite time scale was sometimes attempted with phase averaging equipment. However, the trend is now toward complete independence of the contributing clocks. A computer may initiate and record the measurements of clock differences but the time scale is realized in the form of tabulated corrections for the contributing clocks (Putkovich, 1972). This approach is the best in regard to questions of reliability of time scales that have not yet met great interest with few exceptions where a start has been made (Percival and Winkler, 1975).

C. Clock Hardware

1. *Quartz Crystal Oscillators*

These devices play a fundamental role in timekeeping since they are almost indispensable sources of stable signals in all systems including atomic clocks. They are used either free running or they are locked to a reference signal with time constants ranging from milliseconds to days. Commensurate with such a wide range of usage is the range of performance characteristics, which may go from 10^{-4} to 10^{-13} in frequency stability per day.

Volume, power consumption, etc., also range over many orders of magnitude. Details of design, testing, sensitivity to shock, vibration, static acceleration, temperature change, radiation, etc., fill the volumes of the proceedings of the ASFC in Atlantic City, where most of the progress has been reported. Gerber and Sykes (1966) give a rather recent and complete review, which, however, does not yet cover the great progress over the last few years in the areas of improved spectral purity, temperature compensated crystals, total power consumption, and reduced sensitivity to static acceleration. A fundamental paper is the contribution by Hafner (1966) on the effects of noise in oscillators. Walls and Wainwright (1974) have shown a method to measure the quartz resonator in a bridge that makes the measurement insensitive to the noise coming from the signal source. They obtained conclusive evidence that noise in the electronics seriously degrades the stability of the conventional crystal oscillator. This is compatible with the old rule that a high drive level (1 mW) is necessary for a good short-term performance. On the other hand, a low drive level (< 1 μW) is required to minimize long-term drift. Good spectral purity is needed for the oscillator in atomic clocks because sideband noise power is converted in various ways into offsets from the "true" resonance frequencyof the atoms used in a passive device. Good long-term stability is of less importance in such applications of a quartz oscillator. However, for integration periods of a few seconds the beam-tube exhibits generally a higher noise level than the crystal. Therefore one would like to avoid such short time constants, i.e., the crystal must have very good performance up to integration times of tens of seconds. By use of special low-noise components in the circuitry, considerable progress has been achieved in short-term performance of the latest oscillators available on the market. This progress is very significant because there are now new requirements in some applications of timekeeping that need both very good short- and long-term stability. This is the case in certain wideband communications systems and in frequency measurements in the optical range. Such requirements can best be met by the use of a high spectral purity device locked to a good long-term reference.

2. *Cesium Beam Atomic Frequency Standards*

a. General comments. The idea of using atomic resonances in an "atomic" clock is due to Rabi, who suggested it in an APS lecture in January 1945. The details of the early developments are given by Ramsey (1972) and Beehler (1967) with different emphases. The first atomic clock in actual metrological use was built by NPL and was described by Essen and Parry (1957). This is the standard with which the cesium resonance frequency was measured in collaboration with the USNO (Markowitz *et al.*,

1958). Mockler (1961) has given an excellent review on atomic beam frequency standards with a good discussion of the physics involved. Beehler et al. (1965) report under a misleading title on all atomic frequency standards and the methods of their intercomparison. The paper is valuable because it gives a concise picture of the state of the art at that time, including an extensive list of references.

  b. *Large laboratory cesium beam frequency standards.* In contrast to a clock that is only required to produce uniform time intervals, the purpose of a frequency standard is the generation of a frequency that is accurately known, i.e., a frequency known with respect to the definition of the standard. The presently adopted standard is a certain hyperfine transition of the atom ^{133}Cs ($F = 3$, $m_F = 0$ to $F = 4$, $m_F = 0$), which is assumed to be 9,192,631,770 Hz in a magnetic field of $H = 0$.

  It is necessary for the operation of a frequency standard of highest accuracy to evaluate all perturbing influences that, under actual operating conditions, prevent the exact realization of the definition. For the accuracies that are being achieved today (1×10^{-13}) this is no small task.* For a resonance linewidth of the order of 90 Hz (10^{-8}) a desired accuracy of 10^{-13} requires the splitting of the line to 10^{-5}. This entails the use of extremely stable oscillators, a sufficiently long integration time, and the need for a good signal to noise (S/N) ratio in the beam output signal. It also requires a very high symmetry of the line. This is adversely affected by any phase difference between the two Ramsey cavities, i.e., the microwave injection points. Figure 7 shows the new standard of the National Research Council (NRC) in Ottawa as an example for the design of a very large laboratory cesium beam frequency standard.

  Ramsey (1956) has introduced the double cavity concept that bears his name. His book, which is the standard reference for atomic beam spectroscopy, gives the details of the various disturbances, particularly due to the presence of microwave power near the resonance frequency. This is the origin of the spectral purity requirement for the microwave power source. The effect of a phase difference between the Ramsey cavities is proportional to the beam velocity. This offers a possibility to determine the phase error by changing the velocity of the beam. The effect has been studied in detail in all operating standards, e.g., by Mungall for the Canadian standard NRC Cs III (Mungall, 1972).† The definition of the second refers to zero magnetic field, but the resonance cannot be observed near zero field because in this case the

---

  * The term accuracy when used with a numeric value has been criticized by some colleagues. Since numbers go down in magnitude with increased accuracy, some people prefer the term inaccuracy. Others specify an uncertainty in the above application.
  † See also Schroeder (1974) for a description of the phase adjustment.

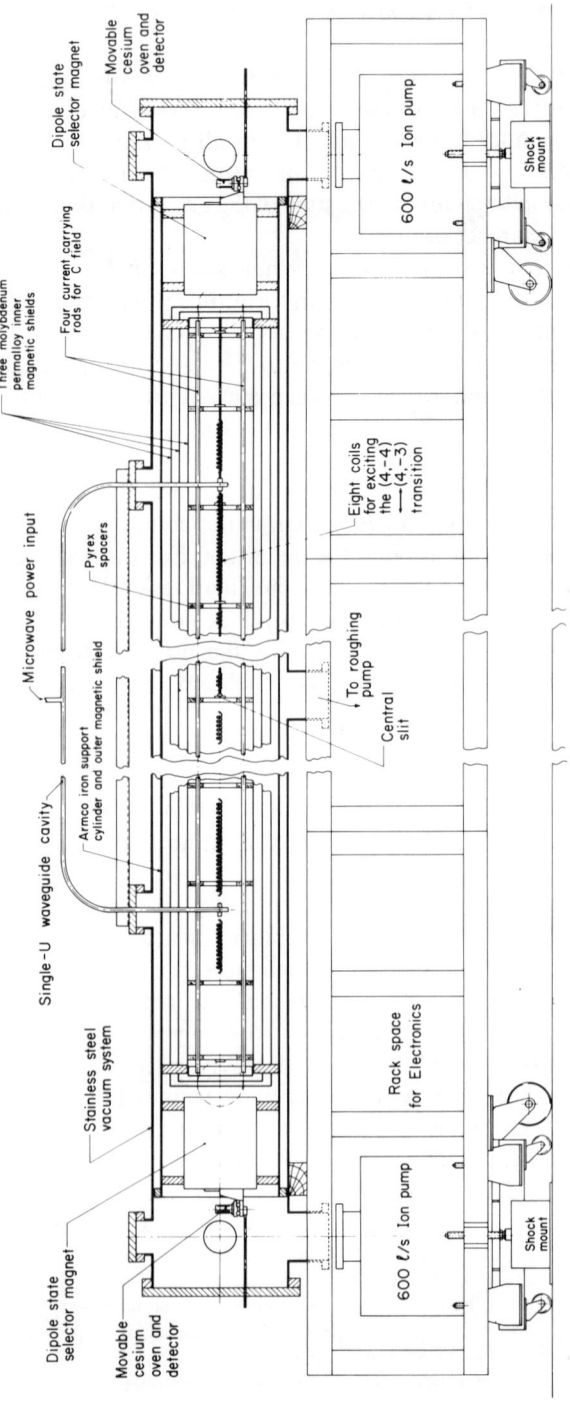

FIG. 7. NRC Cs V line drawing. From Mungall et al. (1975).

other atomic states with $m_F \neq 0$ would not be sufficiently separated energetically and would contribute to the signal in an unaccountable way. The resonance frequency of the standard transition does contain a quadratic (and negligible higher order) term in the Breit–Rabi formula (Mockler, 1961)

$$f_r = 9{,}192{,}631{,}770 + 427.18 H_C^2 \quad \text{Hz} \tag{31}$$

which describes the dependence upon the magnetic field in the transition region (H_C in oersteds). It is clear that the homogeneity and exact knowledge of the H_C field are of major importance.

The cavity phase shift and the uncertainty in H_C usually give the two largest contributions to the error of the standard. A comparison of the performance of the various standards that contribute presently to the accuracy (but not yet to the stability) of TAI is given in Table IV, which has been adapted from material kindly given by Prof. Becker, PTB.

TABLE IV

FREQUENCY UNCERTAINTY DUE TO PARAMETER UNCERTAINTIES[a]

Parameter	PTB Cs 1	NRC Cs V	NBS #6
Second-order Doppler	1	20	138
Phase difference	20		
Spectral impurities	4	20	4
Millman effect	—		—
C Field	13	40	5
Cavity pulling	1	2	1
Servo system	2	20	15
Adjacent transitions	1		20
Majorana transitions			3
Random uncertainty			31
Total uncertainty stated	24	53	90

[a] In 10^{-15}.

A systematic effect, listed as a source of uncertainty for the NRC standard, has been experimentally found by Mungall (1975) by reversing the H_C field. Based on a paper by Hahn (1975) this has been tentatively interpreted by Mungall as the Millman effect. This has not found general agreement among the specialists. On theoretical grounds it has been argued that a Millman effect of noticeable magnitude can only be expected for $\Delta m_F \neq 0$

transitions. However, there is after all experimental evidence for an effect of the right magnitude. It can be separated from cavity phase shift effects because those depend only on beam direction but are independent of the H_C direction. Mungall is therefore emphatic in his recommendation that standard evaluation procedures include C field reversals. The matter is unresolved at this time.

In order to avoid a slow random walk of the rate of TAI it is obviously important to obtain accurate measurements of its rate from time to time (Azoubib *et al.*, 1976). It is therefore very fortunate that we have available today four operating standards as contributors to the BIH, of widely different designs. NBS-4 and NBS-6 have undergone recent modifications and their present performance has been investigated by Wineland *et al.* (1976). The new NRC (Ottawa) standard Cs V is shown in Figs. 7 and 8. The PTB

FIG. 8. NRC Cs V. From Mungall *et al.* (1975.)

standard Cs 1 (shown in Fig. 9) differs from the others in several important aspects. Instead of dipole state selector magnets it uses hexapole magnets in two different pairs, which can be quickly interchanged. This changes the effective beam velocity from a slow group (130 m/sec, $\Delta v/v = 25\%$) to a faster group (250 m/sec, $\Delta v/v = 40\%$). This allows a measurement of the resonance cavity phase difference (Fischer, 1974). In addition, however, this important source of errors is also being determined by the usual beam reversal method. This requires an exchange of the cesium oven and the

Fig. 9. PTB Cs 1. (Courtesy Prof. Dr. G. Becker, PTB.)

detector under vacuum with the condition that the beam must go through the same trace in both directions, since the phase gradient in the two Ramsey cavities is not known accurately enough. The hexapole deflection magnets allow a relatively narrow velocity range selection, in contrast to the dipole magnets. This is claimed to be an advantage for the computation of the transversal (relativistic) Doppler correction. However, one also loses signal intensity in this way. Hellwig *et al.* (1973) have demonstrated a very elegant method for the determination of the beam velocity distribution on any cesium beam device. Becker (1974) also uses a C field that is parallel to the beam, which should produce a greater homogeneity of the field as compared to a perpendicular orientation. Becker (1976) has measured the frequency of TAI since 1969 with the PTB Cs 1 standard and finds a downward trend between the dates of approximately MJD 40500 and 42000 of 4.5×10^{-16} per day or a total of little less than 7×10^{-13}.

In conclusion, it appears that the requirements for clock operation and the operation as a frequency standard are somewhat conflicting. All standards mentioned above except PTB Cs 1 are now operated as clocks. But this is in principle incompatible with the disturbances, interruptions (however short), and interpolations with other devices (hydrogen masers) that the evaluation process requires.* Modifications on an operating clock are an even greater blemish in respect to the ideal concept of a clock, which must be as isolated from the environment as possible. In addition, a desire to obtain a very high short-term stability for the benefit of clock operation by means of an intense beam is undesirable because it magnifies the systematic uncertainties. These can best be avoided by making the resonance device as "passive" as possible.

Laboratory frequency standards have reached a high accuracy, the highest of any standard of measurement. However, the only way to show that this accuracy is not only internal but absolute is again a consensus of standards of different design in different locations. Mungall et al. (1975) are correct in pointing to the current frequency difference of 3.5×10^{-13} between AT(NBS) and PT(NRC) as a cause of concern, and their suspicion is shared by others that new and unsuspected sources for frequency shifts will still be discovered. But this should pose a dilemma. TAI and its associated clock time scales must be as uniform as possible. Strictly speaking, accuracy would then be, at these borders of our capability, a more relative notion, reflecting our momentary state of knowledge rather than anything truly absolute.

c. *Commercial cesium clocks.* These devices are currently the instruments of choice where highest long-term stability is required in systems applications. The major causes for the observed long-term frequency changes (see Table III) are changes in the magnetic C field, phase changes of

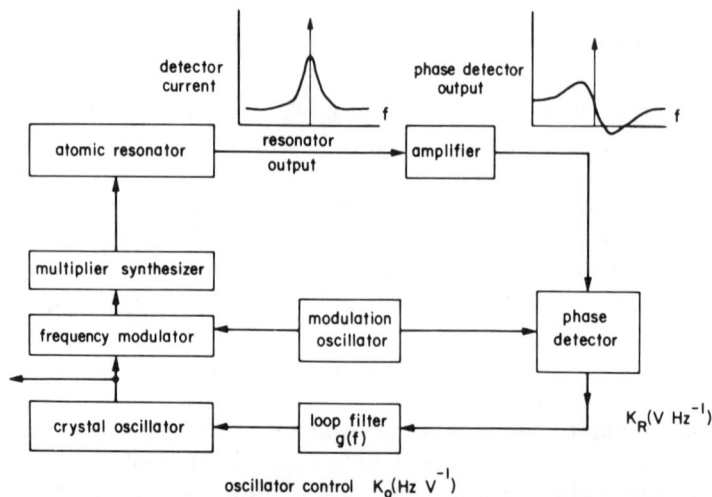

FIG. 10. Passive frequency control servo loop. From Vessot (1976.)

* Evaluations are also not really independent unless a complete break is made.

the Ramsey cavities, and velocity distribution changes due to cesium oven temperature fluctuations, beam optic changes (mechanical relaxations), and microwave power changes. Figure 10 gives a block diagram of a typical passive frequency control loop, which produces an output spectral density

$$S_y(f) = {}^cS_y \frac{1}{|1+G(f)|^2} + {}^aS_y(f) \left|\frac{G(f)}{1+G(f)}\right|^2 \qquad (32)$$

crystal   atomic resonator

where $G(f)$ is the servo-loop gain function, with K_R the lock-in detector sensitivity near the resonance in V/Hz; K_O is the oscillator response in Hz/V (referred to the resonance frequency):

$$G(f) = K_R K_O g(f) \qquad (33)$$

The filter is usually an operational amplifier–integrator with an overall response*

$$g(f) = f_c/if \qquad (34)$$

(f_c is a cut-off frequency). In the experiment of Alley et al. (1977), a second integrator was added by Cutler to the clocks used in order to eliminate a small but noticeable frequency variation due to the mechanical stresses of the crystal oscillator arising from environmental shocks. Under normal operating conditions this (type II servo) is not needed and the crystal's drift can be compensated manually as part of a regular maintenance routine every three to six months.

The ${}^aS_y(f)$ of the atomic resonator is essentially white FM, i.e., shot noise, which decreases with the square root of the beam intensity, while ${}^cS_y(f)$ of the crystal is white PM ($\alpha = 2$). By proper choice of f_c one can therefore adjust somewhat the overall noise performance of the clock. For frequencies $f > f_c$ one sees essentially only the crystal's signal. However, it is necessary to buffer all output circuits well enough to prevent a possible contamination of the microwave drive signal with external noise. For this reason it is better to clean up the signal with an external low noise phase-locked oscillator wherever extreme spectral purity is required (see Walls and Wainwright, 1974).

Commercial cesium tubes operate with an output current of about 10^{-8} A, with a signal sensitivity of 10^{-11} to 10^{-10} A/Hz and a magnetic C field of about 60 mOe. This field must be kept stable to better than about

---

* This is a simplification. There are really two break frequencies, one associated with the filter and the other with the finite gain of the amplifier. In a severe environment a large servo-bandwidth is necessary to avoid loss of lock. One pays for this with beam tube noise contributions in a τ area, where a good crystal oscillator is still superior to any cesium tube stability. It is clear that clock specifications for systems use must be weighted very carefully on this and other points.

0.05% for an allowed frequency change of 1×10^{-13}. This requires excellent shielding, which is itself sensitive to shocks and large external fields, and outstanding long-term stability of the current source. The low output current imposes very stringent long-term stability requirements upon the operational amplifier and phase detector (lock-in) circuits, i.e., offset voltages must be kept to fractions of 1 mV over years.

It is clear that instruments that contain such sophisticated electronics would present serious maintenance and training problems if it were not for the excellent reliabilities achieved as reported by Percival and Winkler (1975). A replacement type of maintenance–repair logistics has therefore been recommended.

McCoubrey (1966) has given an extensive and thorough discussion of the state of the art of all types of atomic clocks including the cesium devices with an assessment of their various merits. Hellwig (1974, 1975) has brought this information up to date. Percival (1973) has reported the typical stability of high-performance clocks as 3×10^{-14} for $\tau = 1^d$. CCIR Report 364-2 gives a committee's judgment of the comparative merits of all types of high-stability clocks currently available. In this author's experience, this information does not do justice to the commercial cesium atomic clocks.

3. Hydrogen Masers

The maser is the instrument of choice for applications where the utmost stability is needed for integration times from 10 sec to 10 days regardless of cost. In contrast to the cesium clock, the maser is an active device and therefore can be used as a reference in a phase-lock loop. Peters (1973) has given an excellent review of the device, written from first-hand experience. He discusses in detail the disturbing influences such as cavity tuning, wall-shift, and magnetic field. Hydrogen masers must be operated with fields of about 500 μOe in order to keep the magnetic disturbances coming from field variations at an acceptable level. The clock transition is the transition $F = 1$, $m_F = 0$ to $F = 0$, $m_F = 0$, with a transition frequency

$$f_H = 1{,}420{,}405{,}751.768 \text{ Hz} + 2750\, H_C^2 \tag{35}$$

(Reinhardt and Lavanceau, 1974). The coefficient of the quadratic term is quite large, relatively more than 41 times as large as in the case of cesium. In addition, the resonance cavity and the transition region is larger but with a better form factor as compared to the cesium beam apparatus. Shielding must therefore be done with multiple shields. Cavity tuning is a very critical parameter because the maser, being an active device, is pulled in frequency by a cavity detuning in the ratio of the Q of the cavity over the Q of the hydrogen linewidth. Typical values are $Q_1 = 5 \times 10^9$ and $Q_c = 5 \times 10^4$.

With this factor of 10^5 the cavity tuning must be stable to better than 10^{-9} for a maser stability of 10^{-14}. This can be done with CERVIT* cavities. The hydrogen storage bulb has an inside coating of Teflon† to provide a "soft" enclosure for the hydrogen atoms. During the average storage time of about 1 sec the atoms bounce off the wall about 10,000 times. The result of this is the wall shift, a small frequency shift of 2×10^{-11}. Unfortunately, the effect cannot be modeled theoretically but must be determined experimentally by changing the volume of the bulb. This has been cited as an argument against the use of the maser as a primary frequency standard. However, Peters (1975), Vessot and Levine (1970), and Vanier et al. (1975) have demonstrated several ways to overcome the problem. Hellwig and Bell (1972) have demonstrated a hydrogen beam system that eliminates the cavity problem for practical purposes. The major difficulty with the beam, however, is the detection of the atoms. Detection efficiencies actually obtained were only about 10^{-8}. A nonoscillating maser with a sufficiently low Q_c is therefore a more practical solution. The tuning can be made automatic, and experience with an automatically tuned maser does indeed show an improved long-term stability. Morris and Nakagiri (1976) describe the instrumentation at the NRC, where the masers are used in conjunction with the accuracy evaluations of the primary cesium standard. Petit et al. (1974) report a reproducibility of maser tuning of 2×10^{-14}. Kleppner et al. (1965) give a full discussion of the various physical effects governing the maser's behavior along with detailed technical considerations. In particular, various different tuning methods are intercompared. In practice these do not yield exactly the same results. Menoud et al. (1967) report on maser work in Switzerland.

Ramsey (1968) gives a beautiful lecture on the maser's principles and scientific background, which makes it immediately clear why the hydrogen maser has gained such high interest among physicists. In principle, it is a much simpler clock than any other. The transition used is the simplest possible: the transition between the parallel to the antiparallel spins of electron and proton in the atomic ground state of hydrogen. This emission frequency can be related theoretically to the fundamental quantum-mechanical constants of nature, which seems almost impossible for cesium. The maser performance as a clock is outstanding, and yet there are more than 20 times as many cesium clocks in operation as hydrogen masers.

Cost and operational reliability are the two main reasons for this discrepancy between scientific interest and practical applications. Hydrogen masers are about ten times as expensive as cesium clocks and their manufacture and repair are still essentially laboratory operations. Many of the causes

---

* CERVIT is a registered trademark of the Owens-Illinois Company.
† Teflon is a registered trademark of the E.I. du Pont de Nemours Company.

for failure can be traced to vacuum problems and problems associated with the electric gas discharge. The breakdowns of the electronics can be mostly attributed to the fact that hydrogen maser work is being done by more than 15 laboratories, which has spread out the effort too much to allow for a completely matured and debugged electronics to emerge. A new and major developmental effort is currently underway to overcome these problems (Easton, 1976).

4. Rubidium Vapor Cells

The rubidium clock is the best device where a relatively low-cost clock is needed that has better stability than a quartz crystal clock. The rubidium vapor cell is a very simple and compact instrument. (There is also a rubidium maser with excellent short-term stability, but it has not found any applications outside the research laboratory.) The rubidium clock can reach a stability of 1×10^{-13} under the best conditions, but it is intrinsically subject to temperature- and pressure-induced frequency variations. It also shows a moderate amount of aging ($< 10^{-12}$/day) in addition to magnetic sensitivity, which is common to all atomic clocks that use a magnetic transition (the ammonia maser, however, uses an electric transition and is more sensitive to electric fields). Alley *et al.* (1977) reported on the performance of a small commercial rubidium clock that has meanwhile found very wide usage as part of avionics systems. The availability of such a low-cost atomic clock will eventually remove the incentive for a reduction of long-term drift in quartz crystal oscillators. Ringer *et al.* (1975) describe the design and performance of a clock for the GPS satellites. Stabilities of 2×10^{-13} per day have been achieved with this latest rubidium clock.

Vessot (1976), compares the typical frequency stability of the various clocks in a double logarithmic plot of $\sigma_y(\tau)$, which shows the different noise types as pieces of straight lines (Fig. 11; see also Table II).

5. Others and Outlook

Hellwig (1975) and Audoin and Vanier (1976) mention in their reviews such potential designs as magnesium beams and ion storage devices. Hafner (1975) discusses in some detail the CO_2 resonance fluorescence cell where the resonance at 10.6 μm would be observed as a dip in the 4.3 μm fluorescence signal that is emitted when the CO_2 molecules relax to the ground state. The use of lasers for clock stabilization is also envisioned for the future, but at this time the difficulties of frequency multiplication into the optical region of the spectrum make such schemes unrealistic.

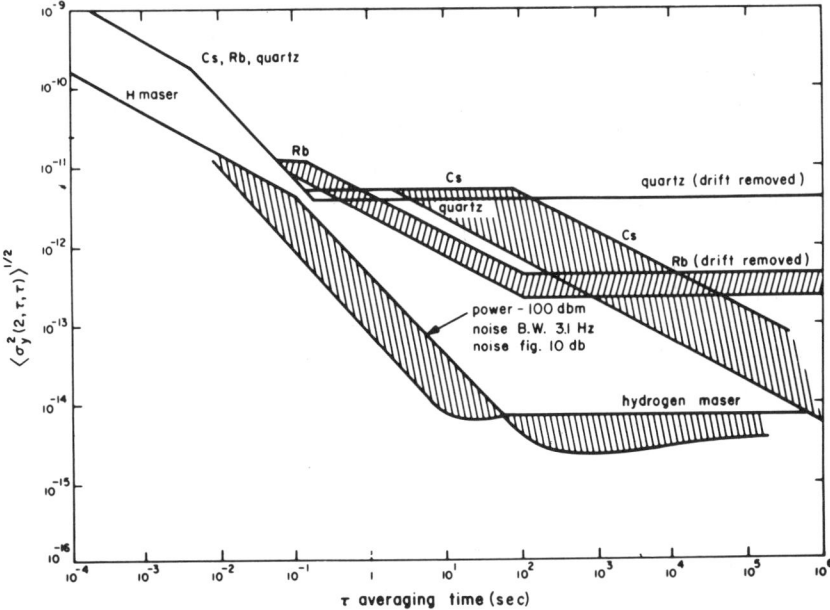

FIG. 11. Composite plot of clock stabilities. From Vessot (1976.)

Stein (1975) reports stabilities of parts in 10^{16} for superconductive cavity oscillators. These devices, which are nonatomic resonators, are therefore the most stable oscillators in existence. Their enormous environmental sensitivity through acoustic coupling precludes practical applications outside the laboratory at this time, but it is expected that this difficulty could eventually be overcome.

III. Time Determination

A. Classical (Optical) Methods

1. *Observations*

UT is determined with transit instruments, the Danjon astrolabe (Danjon, 1960), and the photographic zenith tube (PZT) (Markowitz, 1960). The methods used are those of classical spherical astronomy as presented in standard textbooks (Mueller, 1969; Woolard and Clemence, 1966; Danjon, 1959) or compilations of formulas and tables for the practitioner in the field (Albrecht, 1967). The determination of UT and concurrently of the instantaneous latitude (which is variable due to polar motion, PM) is made with reference to computed coordinates α and δ of the observed stars (given for a fixed date with annual changes in the fundamental star catalog FK4) and with reference to the local vertical as established by a mercury surface or a spirit level. Photographic and visual observations are carried out in a limited range of the spectrum and are affected by turbulence and refraction. Zenith observations minimize these disturbing effects but cannot completely eliminate them. The procedures for combination of the observations of a single night (up to some 40 star transits) and for the combination of several instruments are documented by McCarthy (1976). The results of these efforts are predictions accurate to about 10 msec for a couple of weeks in advance and after the fact knowledge of UT to about 1 msec as published by the BIH as final values in the Annual Report and with only slightly less accuracy as bulletins available every week.

Corrections for PM are necessary for the determination of UT because the local geographical coordinates are affected. This is related but different from nutation and precession, which refer to the motion of the earth's axis in space and which changes the coordinates α and δ of the stars.

The major present limitations of accuracy are fourfold:

1. errors in the star catalog ($1\sigma \sim 0''1$),
2. movements of the local vertical due to earth tides (Melchior, 1974) and tectonic disturbances, earthquakes, and microseisms,
3. instrumental errors,
4. refraction.

Considerable progress can still be expected in all of these areas if modern instrumental technology is applied to them. As an example one could list the planned simultaneous operation of instruments in the blue and red spectral ranges, which should allow the determination of better refraction corrections. The present state of the art must therefore be compared with caution with the expected capabilities of new methods. There is no question, however, that the classical methods are close to real limitations. On the other hand, they will have to continue to satisfy most current applications in fundamental astronomy, navigation, geodesy, and certain aspects of space research, all of which need UT, which gives the rotational orientation of the earth with respect to the system FK4.

2. The Rotation of the Earth

The first observational evidence of a repeating seasonal variation in the rotation of the earth was obtained with the excellent crystal clocks at the PTR (Scheibe and Adelsberger, 1936). A prior report about observed changes by Pavel and Uhink (1935) must, in retrospect, be judged as based on spurious evidence. These variations were later confirmed by Stoyko (1936, 1937), Finch (1950), and Smith and Tucker (1953). Markowitz (1959) used the Washington PZT in conjunction with NBS crystal clocks, an Essen ring-crystal oscillator, and the NPL cesium standard. Lunar tidal variations were found to be in reasonable agreement with theory. For the first time the nature of the so-called irregular variation of the rotation of the earth was demonstrated also with clocks. A more recent report is due to O'Hora (1975), Guinot (1970), and Markowitz (1976).

Up to this time the research in the long-term variation of the rotation of the earth was intimately tied up with the motion of the moon, which was the only clock available with sufficient precision (Spencer-Jones, 1956; Munk and McDonald, 1960). It is an area where the availability of atomic clocks has had a most stimulating effect and has brought together the most heterogeneous disciplines, such as celestial mechanics (Woolard, 1953; Kinoshita, 1975), ancient history (Newton, 1970; Muller and Stephenson, 1975), paleontology (Wells, 1963; Panella, 1975; Runcorn, 1975), earthquakes, the motion of the earth's magnetic field, and others (Rochester, 1973).

B. New Methods

1. General Comments

All of the new methods for the determination of UT and PM have in common that they are independent of three major problems of the classical determinations: atmospheric refraction, star catalog errors, and spurious deflections of the vertical. However, there are other problems, albeit smaller. (See Table V.)

TABLE V

CURRENT ESTIMATED CAPABILITIES OF VARIOUS METHODS FOR UT AND PM

Technique	PM (")	UT (msec)
PZT	0.04	4
Astrolabe	0.06	4
IPMS	0.01	
BIH	0.015	1
Doppler	0.012	
SBL	0.01	1
VLBI	0.01	1
Laser ranging	0.01	0.7

2. The Doppler Method

Anderle (1976) gives a state of the art report for this currently most advanced method, which is already an active contributor to the BIH. The Doppler satellite system uses as reference a network of stations distributed over the globe. Since the system does not have a fixed reference in space it entirely depends on the dynamics of the satellites. Its use will therefore likely remain that of a high-resolution interpolation between the noisier classical determinations. Changes in the constants used for station coordinates, the gravity field, etc., have prevented this method up to now from producing longer undisturbed records. The BIH with French support has recently initiated a supplementary test program of satellite tracking (project MEDOC). The same Doppler satellites are to be used but with different data reductions as a study for a possible PM service in the future. There is some suspicion that a small seasonal effect may also exist in this system due to changes in the ionosphere.

3. Lunar Laser Ranging

Laser pulse ranging with large telescopes ($\phi > 60$ cm) to several lunar retroreflectors can give UT, PM, corrections to the theory of the motion of the moon, its libration, data concerning the theory of general relativity, earth tides, continental drift, plate motion, etc. There is no question that this is one of the most interesting scientific experiments because all of the above effects can, in principle, be observed with outstanding precision (< 12 cm). However, at the same time the fact that all these effects are intermixed in the observational results has raised questions about how practical this method would be for the routine provision of PM and UT data on a service basis. A pilot test has therefore been organized with BIH cooperation to make routine determinations of PM and UT during the years 1977 and 1978. A capability of giving a precision of a fraction of 1 msec in UT has been demonstrated during favorable weather periods (Silverberg, 1974).

4. Very Long Baseline Radio Interferometry (VLBI)

Counselman (1976) gives an up to date review of this all-weather technique. The rotational position of the earth is measured with respect to a plane parallel wavefront, which comes from a cosmic radio source that is so far away that it can be safely assumed to be without proper motion. However, for practical applications the connection between the system FK4 and system of VLBI sources must be established and then monitored. VLBI can give not only UT and PM with excellent resolution, but at the same time of observation, a synchronization between the participating observatories is also obtained as a "fringe" benefit and with a precision of a fraction of a nanosecond. Meeks (1976) has several chapters on the technical details of the observations. One question being investigated is the sometimes observed change in brightness distribution of certain sources. This is of course equivalent to the proper motion problem in the classical determination of UT and PM. It is clear, however, in this author's opinion, that VLBI is by far the most promising method for future routine observations of UT and PM. Johnston (1974) emphasizes the advantages of conventional (SBL) radio-interferometry, and there seems to be agreement that in terms of actual present capabilities SBL is much simpler, less expensive, and, at least at this time, equally powerful for the astrometric observations of interest.

IV. Remote Time Measurements—Synchronization

A. General Principles

There is no single system available that could distribute time and fulfill all requirements of accuracy, coverage, economy, etc. The HF Standard Frequency and Time Signal (SFTS) stations constitute the only service exclusively designed as such. They provide time to 1 msec with worldwide coverage. Up to date lists of operating stations, their frequencies, etc., are available from BIH in the Annual Report, from the USNO (Time Service Announcement Series 1), and from CCIR Report VII/267-3. HF STFS stations satisfy the requirements of the public, the celestial navigator, the geodesist, in fact, the vast majority of all users. For the user it is by far the most economical service thinkable, since radio receivers are available everywhere for other purposes anyway. The service is indispensable for provision of "coarse" time, which is needed by users of precision methods to eliminate the ambiguities inherent in systems that were not specifically designed for time dissemination.

CCIR Report 363-3 gives a useful overview of the various systems that are available for remote synchronization. Blair (1974b) gives a very extensive discussion including many tables for comparison and a most useful list of references. The major aspects common to all methods could be summarized as follows:

1. Geometric Path Delay

Fixed users can compute their distance on the sphere or more precisely, on the ellipsoid following standard geodetic practice such as given by Kalliomäki and Kakkuri (1974). Such calculations will yield an accuracy of about 6 m in the best case over large distances. For a precision of 15 km a sphere with radius 6370 km will approximate the ellipsoid very well (Winkler, 1972). Distance determination is a problem for moving users. In such cases time must be obtained either in a simultaneous two-way link or from a synchronized electronic navigation system such as LORAN C, Transit, or the new Global Positioning System (GPS) as described by Parkinson (1974).

2. Delay and Dispersion in the Medium

This is a more or less serious problem for all systems except for simultaneous two-way transmissions at UHF (Goldberg, 1976). Within its limitations of bandwidth LORAN C is relatively less affected because only surface wave transmission is selected by time of arrival (TOA) discrimination against skywave components of the signal (Potts and Wieder, 1972).

3. Bandwidth (BW), Signal to Noise (S/N), and Averaging Time (τ)

The BW of the signal determines the precision with which a time of arrival (TOA) measurement can be performed. A narrow BW will not only unduly impair the rise time (RT) of the timing waveform (RT $\sim$ 1/BW) and thereby magnify trigger jitter but, almost as important, also increase the group delay variations in the circuitry employed. The problem requires, of course, that in any dispersive channel the pulse envelope must be optimized with respect to the available BW and the medium characteristics. However, with given modulation characteristics and the resulting spectrum occupancy (Prabhu, 1976) there is also an optimum receiver BW that minimizes the stretching of the detected modulation envelope (Wait, 1969). The carrier phase is delayed by a phase delay t_p of the channel, while the envelope is delayed by the group delay t_g. Several timed systems use both phase measurements and envelope timing, e.g., LORAN C. In such cases the envelope is used only to identify a particular cycle of the carrier and the actual timing is extracted from the carrier phase. This cycle identification is somewhat of a problem in LORAN C because it is influenced by any envelope distortion.

4. Pseudorandom Noise (PRN)

PRN modulation techniques offer great advantages for timing in various ways. They employ great signal BW (not necessarily also great information BW) thereby offering high timing precision and very low sensitivity particularly to coherent interference. In addition, every bit contributes to the timing and one can easily compensate for a very poor S/N ratio by increasing the averaging time τ. The use of orthogonal codes (codes that give a very small cross correlation) also offers the possibility of piggy-backing a timing signal of low amplitude upon a high-level communications channel without having to worry about special frequency assignments and without the pitfalls of narrow BW timing schemes (phase instability and cycle ambiguity). Such a piggy-backing has been used successfully over the Defense Satellite Communications System (DSCS) on circuits between California and the Central and Western Pacific. The "Modems" for the timing signal have been quite simple and relatively inexpensive. They can also be used on the 70 MHz IF of microwave duplex links (Murray, 1971). For the time "customer" the use of PRN on an electronic system is not only the vehicle through which one can obtain remote timing with very high precision, i.e., usually to much better than 10% of the PRN bit length, but also one pays for this capability with much greater complexity. However, since the PRN circuitry is inherently digital one must consider PRN systems as the timing systems of the future. A useful introduction to PRN systems is given by Dixon (1976).

The use of PRN on electronic systems has another consequence. A presynchronization of the receiver within a suitably designed "acquisition window" allows very fast acquisition of the signal and thereby constitutes a powerful argument for the use of better clocks as local time reference and for appropriate timekeeping procedures even if no external use of the timing capability is initially contemplated. This is a good example for the general rule (suggested by J. Barnes) that a requirement for time can also be turned into a capability to provide time. This practical rule is usually not fully recognized by the designers of the system. However, up to now every one of these time-ordered systems has come to realize in their implementation this dual role of customer as well as provider of time. In other words, coordination of system time with UTC offers two-way benefits. In fact, time is usually the only common interface between the most heterogeneous electronic systems.

5. *Coordination*

Within the timing community, systems and clocks are never synchronized but rather coordinated with UTC with a tolerance Δt_i and a time constant τ_i, which depend on the individual requirements of the *i*th system and the performance of the clocks available. Coordination implies a free-running operation, in contrast to a synchronized system such as TV. This is possible by virtue of the superior stability of the clocks available today as local time references.* While the system clocks are free running, their differences with each other and with external coordinated clocks are measured regularly by all means available. On the basis of all this information small adjustments are made from time to time (in some cases as rarely as once per year) to keep within the specified tolerance Δt_i. In the ideal case no immediate adjustments are ever necessary in response to a single measurement if the station clocks have been designed with sufficient margins of performance (within τ_i) and reliability (back-ups). On the other hand, if immediate time adjustments are at all envisioned or permitted, then the use of expensive clocks cannot be justified.

In other words, the coordinated mode of operation of a time-ordered system gives the station and/or the system a great degree of independence thanks to the "inertia" of the clocks, which brings benefits of less vulnerability, greater stability of operation by integrating the measurements or disturbances over τ_i, spectrum and/or power conservation since no special synchronizing channels have to be provided, and more flexibility because everybody is presynchronized with everybody else. As examples one could consider the individual LORAN C chains. Every station operates from a cesium clock through a phase/rate adjustment device. The monitor station can integrate over many measurements of the time difference master–slave before a small adjustment (50–100 nsec depending on the chain) is made at the slave. The chain as a whole needs to make a small rate adjustment only every couple of months in order to stay within a few microseconds of its external reference, UTC(USNO, MC). The benefit of using cesium clocks at the LORAN C stations is therefore mainly one of much better precision of operation because of the available long integration time of the order of a day. Fewer adjustments also means fewer possibilities of error.

For the user, existing coordination of the time services means that almost all SFTS stations are within 1 msec of UTC. Most of the major contributors to the BIH are within ± 10 μsec of UTC. The BIH publishes the exact differences of the contributing services in its Circular D. Any of these clocks provides therefore an access to UTC. Circular D is available on request; in the USA it is distributed by the USNO.

Table VI is adapted from Blair (1974b), who gives exhaustive documentation. VLF is used only in the "phase-lock" mode as discussed by Reder *et al.* (1972), Reder and Winkler, (1960), Becker (1973), and Becker *et al.* (1973). Doherty *et al.* (1960) give data on LORAN C timing capabilities. HF time

---

* Strictly speaking, the difference between a synchronized system and a coordinated system is not qualitative but merely consists in a vast difference of τ_i as used in the two cases (fractions of a second in TV vs days in a typical coordinated system). Because of the smallness of τ_{TV} the oscillators used in TV can be very cheap.

TABLE VI
Time Dissemination Capabilities

System	Precision potential	Coverage range (km)	Bandwidth used	Ambiguity	Estimated receiver cost (k$)	Main use	Disadvantages, problems
VLF/Omega	> 1 μsec rel.	Global, 10,000	50 Hz	1 cycle, 50 μsec	4	Low-cost frequency lock	Cycle ambiguity
LF-CW	> 1 μsec rel.	Regional, 2000	50 Hz	1 cycle, 12 μsec	4	Regional standard F (diss.)	Modal interference, interference, ambiguity
LF-LORAN C	0.1 μsec rel. 2 μsec abs.	Regional, 2000	±10 kHz	10 μsec	7	Medium-cost, high precision	Cycle identification
HF	0.1–2 msec	Global, 10,000	±5 kHz	24^h	0.1	Coarse time reference	Interference, low precision, ionosphere
TV	1 nsec local 1 μsec reg.	100 1000	Video	16.7 msec (in USA)	0.5	Local time comp.	Limited range, delay variations
Microwave	1–10 nsec	50	Video	—	20	Point to point	Cost, range, fixed
Cable	1–10 nsec	5	100 MHz	—	0	Local links	Limited range
Portable clock	1–300 nsec	Global	—	0	0	System calibration	Infrequent use
VLBI	< 1 nsec	15,000	Video	0	200	Intercontinental calibration	Point to point
Satellite	A 1 μsec B 10 nsec C 1 nsec	Continental Global Global	10 kHz 10 MHz 40 MHz	— — —	1 20 500	Broadcast (WWVS) Global timing Trunks to PTRS	Variable delay High cost Scheduling

signals are discussed by Stanley (1972) and in NBS TN 668, which gives an exemplary and complete instruction for the use of these indispensable services.

B. Satellite Time Measurements

Easton *et al.* (1976) have given an excellent review of the three principal methods of using satellites for synchronization.

Method A uses geostationary satellites in a broadcast mode with a view toward eventually replacing many of the present HF SFTS emissions. NBS has made extensive tests with ATS and other repeater-type satellites. An encoded time signal is already included in the NOAA GOES satellites and allows timing to about 25 μsec, the limitation being in the narrow bandwidth of the signal Cateora *et al.* (1976).

Method B uses clock-carrying satellites that are part of navigation services. Since the position of the satellites is known very accurately, the major problem of a satellite synchronization system, the path delay, is easily solvable.

Method C uses two-way repeaters on communications satellites. This is in principle the most precise of all synchronization methods, and several experiments have given indications of an available precision of 1 nsec (Chi and Byron, 1975; Saburi *et al.*, 1976; Yamamoto *et al.*, 1975). In operational use by USNO since the early 1970s, the accuracies are presently limited to about 0.1 μsec, since this is the resolution of the time interval meters at the ground stations currently in use. Other examples of this method are the French–Canadian and French–German experiments that began in late 1976 and use the communications satellite SYMPHONIE.

Looking into the future, it appears that the GPS satellites, which carry rubidium or cesium clocks on board, will allow timing globally with a precision sufficient for all but the most extreme requirements. The broadcast satellites of the WWVS type will eventually replace many HF time signals and will satisfy the large majority of users. However, it must be recognized that it will be quite a few years until the present widespread use of HF can be completely and economically replaced with a UHF satellite service. Communications satellite use will remain limited to those users who are close to a satellite ground station. In this case, however, timing of a large number of users can be accomplished very economically and with very high precision. We may call this mode of utilization "trunk-line" timing.

Table VII gives this author's summary views in contrast to those of Easton *et al.* (1976), which appear a little optimistic. CCIR Study Group VII report 518-1 deals with satellite timing specifically and gives many references.

TABLE VII

An Overview of Satellite Timing Systems

System type	Communication	R&D/weather	Navigation	Advanced R&D
Type of operation	Repeater, two way	Repeater, one way, BC	Clock carrier, one way, BC	Clock carrier, Laser reflector
Accuracy range	100–1 nsec	25–1 μsec	25 μsec–< 100 nsec, 10 nsec?	0.1 nsec
Expected use	Trunkline timing to PTRS	Replacement of HF time signals	Global timing	Research
Experiments	ECHO, RELAY II, ATS, SYMPHONIE	NBS–ATS, GOES	TRANSIT, GEOS, TIMATION, NTS, NDS, NAVSTAR-GPS[a]	French proposal
Operational	DSCS	Plan: NBS WWVS	TRANSIT, GPS[a]	

[a] An excellent overview of the Navstar GPS program is given by Parkinson (1974).

C. Time and Relativity

As pointed out by Shapiro (1976), many modern experimental and observational techniques require corrections to the observed measures in order to account for effects predicted by the special theory of relativity (SR) and the general theory of relativity (GR). The SR has been considered as a firmly established part of modern physics for more than 50 years and its predictions have been verified in many experiments. However, despite the fact that a great number of textbooks exist on the subject, a controversy on its logical foundations has never been laid to rest. The controversy has been revitalized by the recent availability of clocks with a performance sufficiently great to check the relativistic effects upon clock-time measurements directly as opposed to experiments with particles and electromagnetic radiation. Modern time measurements are therefore directly confronted with the controversy, which has even somewhat affected the adoption of standards and definitions.

The fact that a number of highly regarded scientists have actively contributed to the confusion has amply demonstrated that the SR is conceptually very difficult. In addition, it is strange that the major critics oppose each other, but agree that all make valuable contributions as long as they keep the controversy open and criticize the theory (Essen, 1971, Ch. 12).

The so-called clock paradox is not a paradox at all because there is no symmetry as claimed by the critics. As has been demonstrated many times, perhaps most clearly by Schild (1959), the notorious twin situation has only one of the twins at rest in an inertial system S. A traveling clock can be related to S by means of a Lorentz transformation but not vice versa, because travel and return necessitate some accelerations. This fact disqualifies the traveling clock as a legitimate reference within the SR. Therefore, when Dingle (1972) asks "What decides in SR which clock is slow?" the answer is that the clock that is at rest in an inertial system will always measure an interval Δt, between two events at the location of this clock, which is larger than any other measure with clocks that move to and from this location. It is disturbing to find such celebrities as Pauli (1958, Ch. 24, p. 73) contributing to the confusion, albeit by obvious oversight. In the famous encyclopedia article that was translated and published as our reference, we find from the very beginning the error that the inertial clock always measures the *smallest* interval. The basis for this most regrettable confusion, which has persisted for so long, seems to be his Fig. 3.

GR enters only if we deal with gravitational fields and/or if we want to insist on the above-mentioned reverse transformation (which is illegitimate in SR). As Fock (1959) has elaborated, such a reverse transformation needs to invoke the principle of equivalence (about which he has misgivings) in order to represent the acceleration by means of a virtual gravitational potential. Acceleration itself does not directly enter the transformation formulas because effects take place only as a function of eventual velocity differences. It is therefore misleading to say that accelerations are the cause of the observed clock differences. Clocks always measure proper time and the reason for their differences when they meet again must be seen in the space–time metric. A direct physical effect upon a clock could only be expected if structural changes take place as a result of acceleration. Such changes can produce only secondary effects via cavity detuning, etc., unless the clock loses lock. The atomic reference frequency is generally not affected since the resonating atoms remain in free fall in a beam machine. Enormous tidal forces that can change the structure within the atom would be required for a real change in the resonance frequency. Such tides could be expected near hypothetical "black holes."

There is no question that SR is appropriate to deal with general motions of a clock as long as we use an inertial reference system and we do not deal with gravitational potentials. In such a frame S, with coordinates t, x, y, z; $\Delta l = (x^2 + y^2 + z^2)^{1/2}$, and $v = dl/dt$, we compute as the proper time interval $\Delta\tau$ of a moving clock between two events at times t_1 and t_2

$$\Delta\tau = \int_{t_1}^{t_2} [1 - (v/c)^2]^{1/2} \, dt \tag{36}$$

since the interval

$$ds^2 = c^2 \, dt^2 - dl^2 \tag{37}$$

is invariant and $ds = c\, d\tau$. It is clear that we will always have

$$\Delta \tau \le (t_2 - t_1) \tag{38}$$

The basis for Eq. (37) is the experimental evidence that c is a constant, i.e., that a light flash produces a spherical wave front after a time t in any inertial system S' with coinciding origins at the flash. Thus the equation

$$c^2 t^2 = x^2 + y^2 + z^2 \tag{39}$$

must be invariant for all systems S'. It can easily be shown [Fock (1959) shows it in a rather elaborate way] that there is only one linear transformation that leaves

$$s^2 = c^2 t^2 - l^2 \tag{40}$$

invariant, i.e., the Lorentz transformation. A linear transformation is needed if we want to leave rectilinear motions rectilinear. One can recognize the transformation as a formal rotation of axes in a formal four-dimensional space with

$$R^2 = x^2 + y^2 + z^2 + t^{*2} \tag{41}$$

remaining constant (in the case of light, R is zero) where we substitute $t^* = ict$. The choice of coordinates in S is therefore governed by the requirement that s remain invariant, i.e., that the propagation of light can be described in the same way as in any other inertial system S'. The interval s^2 can be zero for arbitrarily distant events, positive (timelike), or negative (spacelike).

Noninertial systems cannot be used as equivalent references because in the presence of a gravitational field or of equivalent accelerations (37) cannot be kept invariant in Euclidean coordinates except in an infinitesimal space–time element. The presence of a clock in a gravitational potential U gives rise to a local measure of time, the proper time τ, which differs from the virtual measure t in an inertial system outside the gravitational potential. t is designated as a coordinate time and we have with sufficient approximation

$$d\tau = [1 + (2U/c^2)]^{1/2}\, dt \tag{42}$$

i.e., τ is a smaller measure than t or, in other words, the τ clock runs at a lower rate (since U is a negative quantity).

The "trick" of using a formal four-dimensional space with an imaginary fourth coordinate, due to Minkowski, has given rise to all kinds of unwarranted statements, such as that space and time form a union. The union is only formal in the mathematical sense since clocks will always measure time and meter sticks distance. More exactly, a timelike interval will always be timelike, and a spacelike interval, spacelike. It must also be admitted freely that the experimental evidence of the constancy of c in all inertial systems is not understandable in terms of our a priori notions of space and time. However, this does not make a formal mathematical representation of these facts, which is the SR, illogical as claimed by many critics who try to construe an internal inconsistency of SR from this clash of experimental evidence with in-borne feelings. Nature has no obligation to conform with the human brain, which developed in a very "local" (in a relativistic sense) selection process. Another point not duly emphasized is the essentially geometric character of space with its implied free mobility. Time, however, is only a formal coordinate and really a unidirectional parameter, the abstract measure of change. D'Alembert (1759) was the first to mention the possibility of considering time as another dimension on a purely mathematical basis. Finally, we should remember that Dingle (1972) is right with many of his comments. It is true that we do not know the basic structure of the universe and we should not claim to when we know only very little. It does appear that the experimental basis of relativity theory is often forgotten in speculations that go vastly beyond present physical evidence. Therefore, caution is recommended. That c is a limiting velocity seems a necessary conclusion only for electromagnetic energy and dynamics, but it is conceivable that fields exist for which this is not true. Similarly, GR is only experimentally verified as a description of our experience in weak gravitational fields where the inherently nonlinear equations can be approximated by adding terms.* As an example, the effects of the potential and velocity differences upon the time indicated by a traveling clock can be simply added:

$$\Delta t = \int_{\text{path}} \left[1 - \frac{g(\phi)h}{c^2} + \frac{1}{2}\left(\frac{v}{c}\right)^2 + \frac{(\boldsymbol{\omega} \times \mathbf{r}) \cdot \mathbf{v}}{c^2} \right] d\tau \qquad (43)$$

This expression for the coordinate time interval Δt as a function of the traveling clock's measure of proper time contains the gravitational term as a function of altitude h above sea level and the acceleration of gravity as a function of latitude. The third term is the same (with additional approximation) as our formula (36), and the fourth term is a cross term that accounts for the combined effects of the rotation of the earth and the clock's velocity ($\mathbf{r}$ is the vector from the center of the earth and $\boldsymbol{\omega}$ is the rotational velocity vector of the earth). All terms are functions of τ.

Hefele and Keating (1972), and with much greater resources Alley *et al.* (1977), have measured these effects with atomic clocks in airplanes. The predicted results are evidence that the abstract formalism of GR gives a better description of nature than the classical concepts do. Ashby (1975) has

---

* There has been a report at the recent Pavia (Italy) meeting on relativity (1976) that the existence of the cross terms must be considered established by the results of lunar laser ranging.

provided a very useful summary of the various terms of possible interest for practical time measurements on the earth. He also emphasizes the result of Cocke (1966) that all clocks on the terrestrial equipotential surface (on the geoid) run at the same rate to a very high degree of approximation. This has been questioned recently, and we believe wrongly, from several sides. Briatore and Leschiutta (1976) postulate a latitude effect due to the earth's rotation and they claim to have found some evidence in the rates of the atomic time scales of several institutes. In an earlier paper by Cannon and Jensen (1975) the mistake was made of finding "evidence" in the rates of the UTC clocks of these establishments, clocks that are, of course, deliberately adjusted to remain close to UTC(BIH). The "discovery" became therefore just one more item in the relativity controversy, which has in the past provided ample material for periodicals such as *Science* and *Nature*. Sexl (1976) also postulates a latitude effect but of a solar-seasonal nature. An experimental test is marginally possible but has not yet been performed.

The needs of high-precision timing applications have lately induced astronomers to redefine or actually replace the old ET with a "dynamical" time scale for use in precision ephemerides and tracking of deep-space probes. The new definition is consistent with relativity but is not explicit about it. The IAU General Assembly of 1976 (Grenoble) has adopted, as part of a new system of constants and definitions, the following recommendations:

a. At the instant 1977 Jan. $01^d00^h00^m00^s$ TAI the value of the new time scale for apparent geocentric ephemerides be 1977 Jan. 1^d.000 372 5 exactly.

b. The unit of this time scale be a day of 86400 SI sec at mean sea level.

c. The time scales for equations of motion referred to the barycenter of the solar system be such that there be only periodic variations between these time scales and that for the apparent geocentric ephemerides.

d. No time-step be introduced in TAI.

As explained by Winkler and van Flandern (1977), these recommendations imply that a different (virtual) value for the cesium standard frequency be assumed for astronomical coordinate times as compared to the one that defines the second. This is necessary in order to avoid a secular run-off of these coordinate time scales. According to Eq. (42) a higher rate for the t clock would result if the same definition of the second were assumed. The periodic variations contain the combined effects of the nonuniform motion of the terrestrial clocks in the varying gravitational potential of the sun and other celestial bodies as the earth moves in its orbit. The seasonal effect due to the ellipticity of the earth's orbit is the largest and reaches about 3.3×10^{-10}, TAI running slow in January (Clemence and Szebehely, 1967).

Time as a basic physical parameter has therefore uniqueness only as a local measure, the proper time τ. The coordinate time t has to be understood in GR as representative of a family of possible times t_i (Adler *et al.*, 1975). The choice i is dictated by the choice of the spacelike hypersurface considered, i.e., the coordinates chosen.

V. Applications

The applications of precise time and frequency measurements fall into several broad categories.

A. Celestial Navigation, Geodesy, and Space Tracking

Operations in this group require a knowledge of the rotational position of the earth, i.e., UT 1 and the polar coordinates x and y. The accuracies needed are modest for navigation (0.1 sec at the time of observation), intermediate for geodesy (10 msec after the fact) and high for space tracking (< 1 msec, soon after the fact). The last requirement as stated by Bender (1974) is also valid for advanced geodynamics/geodesy. Its satisfaction will depend heavily on the new methods for time determination. The navigation requirement, although modest, does include tens of thousands of navigators who need a reliable time signal at unpredictable times and most urgently when they are in a crisis. This is the major reason for the need to keep UTC, the basis for the time of the time signals, close to UT 1 (Chi and Fosque, 1972). Users such as astronomers who must set a telescope accurately or navigators who use more sophisticated instruments than a sextant have to apply a more accurate correction to the time signal. The DUT1 code (DUT1 $\approx$ UT1 $-$ UTC) gives only a precision of 0.1 sec. The time services provide extrapolations in advance, which are accurate to a few milliseconds.

B. Electronic Navigation and Positioning Systems in the Rho–Rho Mode

In this category of applications a synchronized clock is used for the measurement of times of arrival of radio signals that are transmitted from beacons. The timing accuracy needed depends upon the desired accuracy of fix via the velocity of light ($c = 299{,}792{,}458$ m/sec in vacuo). In this category we find the most exacting timing requirements (Bender, 1974) down to the nanosecond range.

C. Pseudorandom Noise Systems

As discussed in Section IV,A,4, these systems create rather moderate needs for timing (10 μsec to tens of milliseconds) but it can be expected that the use of this technique and the concomitant use of clocks will spread to many electronic systems in the future.

D. Communications Systems

Folts (1972) has given an overview of the applications of time and frequency in communications technology. The advantages of timing are mainly in the stabilization of time ordered networks, time division multiplex, and the greater economy that can be achieved in the use of buffers, etc.

Applications of these principles can be found in Stone et al. (1972), who describe the use of time in Naval FSK transmissions. Saltzberg and Zydney (1975) report on the current network synchronization (in precise frequency) in the Bell system. Stover (1974) discusses in detail a time reference distribution concept with emphasis on the advantages of controlling tolerances on time rather than frequency. This paper is fundamental in its general importance. In another paper, Stover (1975) emphasizes the importance of time for modern communications systems by considering time as the natural consequence of modern digital techniques in all its applications.

E. Metrology, Scientific Applications

The increased utilization of digital techniques has brought about an increased utilization of time and frequency as intermediary quantities in the measurement process. This applies not only to distance measurements as dealt with by Bender (1974), who gives extensive references. The speciality of pulsar research (see Huguenin, 1976) provides an extremely interesting application of timekeeping. It also poses the most stringent requirements for long-term stability and reduction to an extraterrestrial coordinate time. Finally, VLBI, as reviewed by Counselman (1976), is a user of timekeeping as well as a potential provider of data that will be important to other time users. The VLBI requirements for highest-quality clocks have given the largest initial support for the improvements in hydrogen masers.

ACKNOWLEDGMENTS

I have received much help and information from my colleagues at the USNO and elsewhere. Special thanks are due to Don Percival, USNO, for data and analyses; to Drs. J. Barnes, H. Hellwig, and D. Allen, NBS, for discussions; and to Professor Dr. G. Becker, PTB, and Drs. A. Mungall, NRC, L. Cutler, Hewlett-Packard, Inc., and R. Vessot, SAO, for data and material.

References

Adler, R., Bazin, M., and Schiffer, M. (1975). "Introduction to General Relativity." McGraw-Hill, New York. [Pages 122–128 give a lucid explanation of proper vs. coordinate time.]

Albrecht, T. (1967). "Formeln und Hilfstafeln für Geographische Ortsbestimmungen." Akademie-Verlag, Berlin.

Allan, D. W. (1966). Statistics of atomic frequency standards. *Proc. IEEE* **54**, 221–230.

Allan, D. W. (1974). The measurement of frequency and frequency stability of precision oscillators. *Proc. PTTI Conf., 6th*; also as *Nat. Bur. Stand. (U.S.), Tech. Note* No. 669.*

Allan, D. W., and Gray, J. E. (1971). *Metrologia* **7**(2), 79–82.

Allan, D. W., Hellwig, H., and Glaze, D. J. (1975). An accuracy algorithm for an atomic time scale. *Metrologia* **11**, 133–138.

Alley, O. C., Cutler, L. S., Reisse, R. A., Williams, R. E., Steggerda, C. A., Rayner, J., Mullendore, J., and Davis, S. (1977). Atomic clock measurements of the general relativity time differences produced by aircraft flights using both direct and laser pulse time comparison. *Phys. Rev. Lett.* (submitted).

American Ephemeris and Nautical Almanac ("Astronomical Ephemeris"). US Gov. Printing Office, Washington, D.C. (Annual publication.)

Anderle, R. J. (1976). "Polar Motion Determined by Doppler Satellite Observations," Rep. to Comm. 19. Int. Astron. Union, Grenoble. (*Bull. Geodes.*)

Anderson, O. D. (1975). "Time Series Analysis and Forecasting; The Box-Jenkins Approach." Butterworth, London.

Angeli, M. T. (1974). The use of Kalman filter techniques in constructing the UTC(IEN) time scale. *Proc. Cagliari Int. Meet. Time, 2nd* pp. 37–61.

Annuaire du Bureau des Longitudes. Gauthier-Villars, Paris. (Publication annuelle.)

Ashby, N. (1975). An earth-based coordinate clock network. *Nat. Bur. Stand. (U.S.), Tech. Note* No. 659.

Audoin, C., and Vanier, J. (1976). Atomic frequency standards and clocks. *J. Phys. E* **9**(9), 697–792.

Azoubib, J., Granveaud, M., and Guinot, B. (1976). Estimation of the scale unit duration of time scales. *Metrologia* (submitted).

Bagley, A. S. (1966). Frequency and time measurements. *In* " Handbuch der Physik " (S. Flügge, ed.), Vol. 23, pp. 289–371. Springer-Verlag, Berlin and New York.

Barber, R. E. (1971). Short-term frequency stability of precision oscillators and frequency generators. *BSTJ* **50**(3), 881–915.

Barnes, J. A. (1968). Tables of bias functions. *Nat. Bur. Stand. (U.S.), Tech. Note* No. 375.

Barnes, J. A. (1976). Models for the interpretation of frequency stability measurements. *Nat. Bur. Stand. (U.S.), Tech. Note* No. 683.

Barnes, J. A., and Allan, D. W. (1964). Effects of long-term stability on the definition and measurement of short-term stability. *Proc. IEEE-NASA Symp. Short-Term Freq. Stabil.*; *NASA Spec. Publ.* No. 80, pp. 119–123.

Barnes, J. A., and Mockler, R. C. (1960). The power spectrum and its importance in precise frequency measurements. *IRE Trans. Instrum.* **I-9**, 149–155.

Barnes, J. A., Chi, A. R., Cutler, L. S., Healey, D. J., Leeson, D. B., McGunigal, T. E., Mullen, J. A., Smith, W. L., Sydnor, R. L., Vessot, R. F. C., and Winkler, G. M. R. (1971). Characterization of frequency stability. *IEEE Trans. Instrum. Meas.* **IM-20**(2), 105–120.

* Information concerning availability of the Proceedings of Annual PTTI Internal Planning Meeting may be obtained from Technical Information Division, Code 250, Goddard Space Flight Center, Greenbelt, Maryland 20771 [Telephone: (301)982-4488].

Baugh, R. A. (1971). Frequency modulation analysis with the Hadamard variance. *Proc. ASFC, 25th* pp. 222–225.*
Becker, G. (1973). Frequenzvergleiche mit dem Primaeren Frequenznormal etc. *PTB—Mitt.* **83**(5), 319–326.
Becker, G. (1974). Das Primaere Zeitnormal Cs1. *CIC* **A1**, 1–13.†
Becker, G. (1976). Recent progress in primary Cs beam frequency standards at the PTB. *CPEM. IEEE Trans. Instrum. Meas.* **25**(4), 458–465.
Becker, G., Fischer, B., and Hetzel, P. (1973). Langzeituntersuchungen ueber die Unsicherheit von Zeit und Frequenzvergleichen mittels Laengstwellen. *PTB—Mitt.* **73**(4), 222–231. See also *Kleinheubacher Ber.* **16**, 5.
Beehler, R. E. (1967). A historical review of atomic frequency standards. *Proc. IEEE* **55**(6), 792–805. See also in Blair (1974a), pp. 85–109.
Beehler, R. E., Mockler, R. C., and Richardson, J. M. (1965). Cesium beam atomic time and frequency standards. *Metrologia* **1**(3), 114–131.
Bender, P. L. (1974). Applications of PTTI to new techniques for determining crustal movements, polar motion, and the rotation of the earth. *Proc. PTTI Conf., 6th* pp. 39–55.
BIH Annual Report (1974). Bureau International de l'Heure, 61, avenue de l'Observatoire, 75014 Paris.
Blackman, R. B., and Tukey, J. W. (1959). "The Measurement of Power Spectra." Dover, New York.
Blair, B. E., ed. (1974a). "Time and Frequency" (Red Book). *Nat. Bur. Stand. (U.S.), Monogr.* No. 140.
Blair, B. E. (1974b). In Blair (1974a), pp. 233–314.
Blair, B. E., and Morgan, A. H., eds. (1972). "Precision Measurement and Calibration" (Green Book) *Nat. Bur. Stand. (U.S.), Spec. Publ.* Vol. 5, No. 300.
Boileau, E., and Picinbono, B. (1976). Statistical study of phase fluctuations and oscillator stability. *IEEE Trans. Instrum. Meas.* **25**(1), 66–75.
Borre, K., and Meissl, P. (1974). "Strength Analysis of Leveling-Type Networks; An Application of Random Walk Theory," Meddelelse No. 50. Geodaetisk Institut, Copenhagen (ISBN 8774500171).
Box, G. E. P., and Jenkins, G. M. (1970). "Time Series Analysis, Forecasting and Control." Holden-Day, San Francisco, California.
Brenner, N. (1976). The fast Fourier transform. In Meeks (1976), pp. 284–295.
Briatore, L., and Leschiutta, S. (1976). Verifying the gravitational shift due to the earth's rotation (Letter). *Nuovo Cimento* **15**(6), 203.
Campbell, N. (1926). Time and chance. *Phil. Mag.* p. 1106.
Cannon, W. H., and Jensen, O. G. (1975). *Science* **188**(April 25), 318.
Cateora, J. V., Davis, D. D., and Hanson, D. W. (1976). A satellite-controlled digital clock. *Nat. Bur. Stand. (U.S.), Tech. Note* No. 681.
CCIR (1976). Study Group VII Documents. Int. Telecommun. Union, Geneva.
Chi, A. R., and Byron, E. (1975). Two way time transfer experiment using a synchronous satellite. *Proc. PTTI Conf., 7th* pp. 357–375.
Chi, A. R., and Fosque, H. S. (1972). A step in time. *IEEE Spectrum* **9**(1), 82–86.
Clemence, G. M., and Szebehely, V. (1967). Annual variation of an atomic clock. *Astron. J.* **72**(10), 1324–1326.

* Copies of the Proceedings of the Annual Symposium on Frequency Control are available from Electronic Industries Association, 2001 Eye Street, NW, Washington, D.C. 20006.
† International Congress of Chronometry Proceedings are available from Herausgeber, Deutsche Gesellschaft für Chronometrie e.V., Stuttgart, West Germany.

Cocke, W. J. (1966). *Phys. Rev. Lett.* **16**, 662, 779, 1233.
Counselman, C. C. (1976). Radio astrometry. *Annu. Rev. Astron. Astrophys.* **14**, 197–214.
Creer, K. M. (1975). On a tentative correlation between changes in the geomagnetic polarity bias and reversal frequency and the earth's rotation through phanerozoic time. In Rosenberg and Runcorn (1975), pp. 293–318.
Cutler, L. S., and Searle, C. L. (1966). Frequency fluctuations in frequency standards. *Proc. IEEE* **54**(2), 136.
D'Alembert, J. L. R. (1759). Dimension. *In* "Encyclopédie," Vol. IV, p. 1010.
Danjon, A. (1959). "Astronomie Generale." Sennac, Paris.
Danjon, A. (1960). The impersonal astrolabe. *In* "Telescopes" (G. Kuiper and B. Middlehurst, eds.), Stars and Stellar Systems, Vol. 1, pp. 115–137. Univ. of Chicago Press, Chicago, Illinois.
Dingle, H. (1972). "Science at the Crossroads," p. 86. Martin Brian & O'Keefe, London.
Dixon, R. C. (1976). "Spread Spectrum Systems." Wiley, New York.
Doherty, R., Hefley, G., and Linfield, R. (1960). Timing potential of Loran C. *Proc. ASFC, 14th* p. 276.
Easton, R. (1976). The Hydrogen Maser Program for Navstar- GPS. *Proc. PTTI Conf., 8th* (submitted).
Easton, R., Fisher, L. C., Hanson, D. W., Hellwig, H., and Rueger, L. (1976). Dissemination of time and frequency by satellites. *IEEE Proc.* **64**(10), 1482–1493.
Edson, W. A. (1960). Noise in oscillators. *Proc. IRE* **48**, 1454–1466. [Defines a "period of coherence" for 1Rd phasedeviation.]
Enslin, H., and Proverbio, E., eds. (1974). *Proc. Cagliari Int. Meet. Time*, Edizioni Anastatiche, Cagliari, Italy.
Essen, L. (1971). "The Special Theory of Relativity (A Critical Analysis)." Oxford Univ. Press, London and New York.
Essen, L., and Parry, J. V. L. (1957). The cesium resonator as a standard of frequency and time. *Phil. Trans. R. Soc. London, Ser. A* **250**, 45–69.
Felber, H.-J. (1974). Kants Beitrag zur Frage der Verzoegerung der Erdrotation. *Die Sterne* **50**(2), 82–90.
Feller, W. (1957). "An Introduction to Probability Theory and Its Applications," Vol. 1. Wiley, New York.
Finch, H. F. (1950). On a periodic fluctuation in the length of the day. *Mon. Notic. R. Astron. Soc.* **110**(1), 3–14.
Fischer, B. (1974). Die Geschwindigkeitsverteilung etc. *CIC* **A3**, 1–12.
Fock, V. (1959). "The Theory of Space, Time, and Gravitation." Pergamon, Oxford.
Folts, H. C. (1972). Time and frequency for digital telecommunications. *Proc. PTTI Conf., 4th* pp. 194–202.
Gerber, E. A., and Sykes, R. A. (1966). State of the art: Quartz crystal units and oscillators. *Proc. IEEE* **54**(2), 103–116. See also in Blair, 1974a, where the authors give a brief update of the more recent developments in a Part B.
Goldberg, B., ed. (1976). "Communication Channels; Characterization and Behavior," 762 pp. IEEE Press, New York. [A most valuable literature reference. See particularly Part 3 (VLF/LF), Part 4 (HF and Ionosphere), and Part 8 (Satellite Communications).]
Greaves, W. M. H., and Symms, L. S. T. (1943). The short period erratics of free pendulum and quartz clocks. *Mon. Notic. R. Astron. Soc.* **103**(4), 196–209. [Outstanding early work on the characterization of clock noise.]
Guinot, B. (1970). Short period terms in universal time. *Astron. Astrophys.* **8**, 26–28.
Guinot, B. (1974). The accuracy of the atomic time scales. *Proc. Cagliari Int. Meet. Time* pp. 15–35.

Hafner, E. (1966). The effects of noise in oscillators. *Proc. IEEE* **54**(2), 179-198.
Hafner, E. (1975). Outlook for precision frequency control in the 1980's. *Proc. PTTI Conf.*, *7th* pp. 101-118.
Hahn, S. L. (1975). *Bull. Acad. Pol. Sci.* **23**, 249.
Halford, D., Hellwig, H., and Wells, J. S. (1972). Progress and feasibility for a unified standard for frequency, time, and length. *Proc. IEEE* **60**, 623-625.
Hefele, J. C., and Keating, R. E. (1972). *Science* **177**, 166-168.
Hellwig, H. (1974). Atomic frequency standards; a survey. *Proc. ASFC, 28th* pp. 315-339. See also *Proc. IEEE* **63**(2), 212-229.
Hellwig, H. (1975). A review of precision oscillators. *Proc. PTTI Conf., 6th*; also as *Nat. Bur. Stand. (U.S.), Tech. Note* No. 662 [supplements Hellwig (1974)].
Hellwig, H., and Bell, H. E. (1972). Experimental results with atomic hydrogen storage beam systems. *Proc. ASFC 26th* pp. 242-247.
Hellwig, H., Jarvis, S., Jr., Halford, D., and Bell, H. E. (1973). Evaluation and operation of atomic beam tube frequency standards using time domain velocity selection modulation. *Metrologia* **9**, 107-112.
Hewlett-Packard Company (1975). "Time Keeping and Frequency Calibration," Application Note 52-2.
Howe, D. A. (1976). Frequency domain stability measurements. *Nat. Bur. Stand. (U.S.), Tech. Note* No. 679.
Huebner, U., and Becker, G. (1974). Eine Alternative etc. *CIC* **A4**, 1-7.
Huguenin, G. R. (1976). Pulsar observing techniques. In Meeks (1976), pp. 77-91.
IPMS, Reports of the Central Bureau of the International Polar Motion Service, Mizusawa, Japan.
Jenkins, G. M., and Watts, D. G. (1968). "Spectral Analysis and its Applications." Holden-Day, San Francisco, California.
Johnston, K. J. (1974). Radio astrometry. *Proc. PTTI Conf., 6th* pp. 373-379.
Kalliomäki, K., and Kakkuri, J. (1974). "Time Keeping Methods Applied in Finland," Report of the Finnish Geodetic Institute, No. 74(2) (ISBN 951-711-009-x).
Kant, I. (1781). Critik der reinen Vernunft. (See "Third Analogy, Principle of Simultaneity.")
Kinoshita, H. (1975). Theory of the rotation of the rigid earth. Celestial Mechanics (submitted). (SAO Preprints No. 443).
Kleppner, D., Berg, H. C., Crampton, S. B., Ramsey, N. R., Vessot, R. F. C., Peters, H. E., and Vanier, J. (1965). Hydrogen maser principles and techniques. *Phys. Rev. A* **188**(4), 972-983.
Kovalevsky, J. (1965). Astronomical time. *Metrologia* **1**(4), 169-180.
Kramer, G. (1975). Digitale Bestimmung von Schwankungsspektren. *PTB—Jahresber.* Pt. 2, p. 157.
Lindsey, W. C., and Chie, C. M. (1976). Theory of oscillator instability based upon structure functions. *Proc. IEEE* **64**(12), 1652-1666.
Loeb, H. W. (1972). Efficient data transformation in time domain spectrometry. *IEEE Trans. Instrum. Meas.* **21**(2), 166-168.
McCarthy, D. D. (1976). The determination of universal time at the U.S. Naval Observatory. *U.S. Nav. Observ. Circ.* No. 154.
McCoubrey, A. O. (1966). A survey of atomic frequency standards. *Proc. IEEE* **54**(2), 116-135.
Markowitz, W. (1959). Variations in rotation of the earth etc. *Astron. J.* **64**, 106-113.
Markowitz, W. (1960). The photographic zenith tube and the dual-rate moon-position camera. *In* "Telescopes" (G. Kuiper and B. Middlehurst, eds.), Stars and Stellar Systems, Vol. 1, pp. 88-114. Univ. of Chicago Press, Chicago, Illinois.
Markowitz, W. (1962). The atomic time scale. *IRE Trans. Instrum.* **I-11**, 239.

Markowitz, W. (1976). "Comparison of ILS, IPMS, BIH and Doppler Polar Motions with Theoretical," Rep. to Comm. 19 & 31. Int. Astron. Union General Assembly, Grenoble.

Markowitz, W., Hall, R. G., Essen, L., and Parry, J. V. L. (1958). Frequency of cesium in terms of ephemeris time. *Phys. Rev. Lett.* **1**(3), 204 (105–106).

Marrison, W. A. (1948). The evolution of the quartz crystal clock. *BSTJ* **27**, 510–588.

Meeks, M. L., ed. (1976). "Astrophysics, Part C: Radio Observations" (L. Marton, ed.-in-chief), Methods of Experimental Physics, Vol. 12. Academic Press, New York.

Melchior, P. (1974). Earth tides. *Geophys. Surv.* **1**, 275–303.

Menoud, C., Racine, J., and Kartaschoff, P. (1967). Description et performances actuelles des masers a hydrogene du L.S.R.H. *J. Suisse Horlog.* **6**(7), 265–280.

Mockler, R. C. (1961). Atomic beam frequency standards. *Adv. Electron. Electron Phys.* **15**, 1–71.

Moran, J. M. (1976). Very long baseline interferometer systems. In Meeks (1976), pp. 174–197.

Morris, D., and Nakagiri, K. (1976). The frequency stability of a pair of auto tuned masers. *Metrologia* **12**, 1–6.

Mueller, I. I. (1969). "Spherical and Practical Astronomy as Applied to Geodesy." Ungar, New York.

Mulholland, J. D. (1972). Measures of time in astronomy. *Publ. Astron. Soc. Pac.* **84**(499), 357–364.

Muller, P. M., and Stephenson, F. R. (1975). The accelerations of the earth and moon from early astronomical observations. In Rosenberg and Runcorn (1975), pp. 459–534.

Mungall, A. G. (1971). Atomic time scales. *Metrologia* **7**(4), 146–153.

Mungall, A. G. (1972). Cavity phase dependent frequency shifts in cesium beam frequency standards. *Metrologia* **8**(1), 28–32.

Mungall, A. G. (1975). The Millman effect. *Proc. PTTI Conf.*, 7th pp. 195–211.

Mungall, A. G., Daams, H., Morris, D., and Costain, C. C. (1975). Performance and operation of the NRC primary cesium clock CsV. *Proc. PTTI Conf.*, 7th pp. 165–189. See also *Metrologia* **12**, 129–139.

Munk, W. H., and McDonald, G. J. F. (1960). "The Rotation of the Earth." Cambridge Univ. Press, London and New York.

Murray, J. A., Jr. (1971). Mini-modem for PTTI dissimination. *Proc. PTTI Conf.*, 3rd pp. 16–18. See also *Proc. PTTI Conf.*, 4th pp. 182–193.

Newcomb, S. (1874). On the possible variability of the earth's axial rotation. *Am. J. Sci. Arts., Sec. 3* **8**(45).

Newcomb, S. (1895). *Astron. Pap. Am. Ephemeris* **6**, Pt. 1, 9.

Newton, I. (1686). "Philosophiae Naturalis Principia Mathematica." (See Definitiones, Scholium.)

Newton, R. R. (1970). "Ancient Astronomical Observations and the Accelerations of the Earth and the Moon." Johns Hopkins Press, Baltimore, Maryland.

O'Hora, N. P. J. (1975). The detection of recent changes in the earth's rotation. In Rosenberg and Runcorn (1975), pp. 427–443.

Panella, G. (1975). Paleontological clocks and the history of the earth's rotation. In Rosenberg and Runcorn (1975), pp. 253–284.

Papoulis, A. (1965). "Probability, Random Variables and Stochastic Processes." McGraw-Hill, New York.

Parkinson, B. W. (1974). Navstar: Global positioning system. *Proc. PTTI Conf.*, 6th pp. 469–495.

Pauli, W. (1958). "Theory of Relativity." Pergamon, Oxford.

Pavel, F., and Uhink, W. (1935). Die Quarzuhren des Geodaetischen Institutes in Potsdam. *Astron. Nachr.* **257**, 367–390.

Percival, D. B. (1973). Statistical properties of high performance cesium standards. *Proc. PTTI Conf.*, *5th* pp. 239–263.

Percival, D. B. (1976). A heuristic model of atomic clock behavior. *Proc. ASFC*, *30th* pp. 414–419. [Duscusses a frequency step detector based on objective criteria and on the use of sum charts.]

Percival, D. B., and Winkler, G. M. R. (1975). Time keeping and the reliability problem. *Proc. ASFC*, *29th* pp. 412–416.

Peters, E. H. (1973). Characteristics of advanced hydrogen maser frequency standards. *Proc. PTTI Conf.*, *5th* pp. 283–316.

Peters, H. E. (1975). The concertina hydrogen maser. *Proc. ASFC*, *29th* pp. 362–370.

Petit, P., Viennet, J., Barillet, R., Desaintfuscien, M., and Audoin, C. (1974). Development of hydrogen masers as frequency standards at the Laboratoire de l'Horloge Atomique. *Metrologia* **10**, 61–67.

Potts, C. E., and Wieder, B. (1972). Precise time and frequency dissemination via the Loran C system. *Proc. IEEE* **60**(5), 530–539.

Prabhu, V. K. (1976). Spectral occupancy of digital angle modulation signals. *BSTJ* **55**(4), 429–453.

Putkovich, K. (1972). Automated timekeeping. *IEEE Trans. Instrum. Meas.* **21**, 400–405.

Quesada, V. (1974). Accord entre les echelles coordonné et intégré. *CIC* **A7**(1), 1–13. [In Europe, LORAN C allows time coordination about fifty times more precisely than LF stations such as MSF or DCF77.]

Ramsey, N. F. (1956). "Molecular Beams." Oxford Univ. Press (Clarendon), London and New York.

Ramsey, N. F. (1968). The atomic hydrogen maser. *Am. Sci.* **56**(4), 420–438.

Ramsey, N. F. (1972). History of atomic and molecular standards of frequency and time. *IEEE Trans. Instrum. Meas.* **21**(2), 90–99.

Reder, F. H., and Winkler, G. M. R. (1960). World wide clock synchronization. *IRE Trans. Mil. Electron.* **4**(2/3), 366–376.

Reder, F. H., Hargrave, J., and Crouchley, J. (1972). Interpretation of VLF phase data. *Proc. PTTI Conf.*, *4th* pp. 324–344.

Reinhardt, V. S., and Lavanceau, J. (1974). A comparison of the cesium and hydrogen hyperfine frequencies by means of LORAN C and portable clocks. *Proc. ASFC*, *28th* pp. 379–383.

Ringer, D. E., Gandy, J., and Jechart, E. (1975). Spaceborne rubidium frequency standard for Navstar GPS." *Proc. PTTI Conf.*, *7th* pp. 671–696.

Rochester, M. G. (1973). The earth's rotation. *EOS* **54**, 769–780 (*Trans. Am. Geophys. Union*).

Rosenberg, G. D., and Runcorn, S. K., eds. (1975). "The Growth Rhythms and the History of the Earth's Rotation." Wiley, New York.

Runcorn, S. K. (1975). Paleontological and astronomical observations on the rotational history of the earth and the moon. In Rosenberg and Runcorn (1975), pp. 285–291.

Russell, B. (1936). On order in time. *Proc. Cambridge Phil. Soc.* **32**(2), 216–228; also in Russell, B. (1971). "Logic and Knowledge." Capricorn, New York.

Rutman, J. (1974a). Relations between spectral purity and frequency stability. *Proc. ASFC*, *28th* pp. 160–165.

Rutman, J. (1974b). Characterization of frequency stability: A transfer function approach and its application to measurements via filtering of phase noise. *IEEE Trans. Instrum. Meas.* **23**, 40–48. [A bandpass variance is defined for the distinction between white and flicker phase noise.]

Saburi, Y., Yamamoto, M., and Harada, K. (1976). High precision time comparison via satellite and observed discontinuity etc. *IEEE Trans. Instr. Meas.* **25**(4), 473–477.

Salecker, H., and Wigner, E. P. (1958). Quantum limitations of the measurement of space-time distances. *Phys. Rev.* **109**(2), 571–577.
Saltzberg, G. R., and Zydney, H. M. (1975). Network synchronization. *BSTJ* **54**(5), 879–892.
Scheibe, A., and Adelsberger, U. (1936). Nachweis von Schwankungen der Astronomischen Tageslaenge im Jahre 1935 mittels Quarzuhren. *Phys. Z.* **37**(11), 415. [First real evidence of a periodic seasonal variation.]
Scheibe, A., and Adelsberger, U. (1950). Die Gangleistungen der PTB Quarzuhren und die jaehrliche Schwankung der astronomischen Tageslaenge. *Z. Phys.* **127**, 416–428.
Schild, A. (1959). The clock paradox in relativity theory. *Am. Math. Mon.* January, pp. 1–18.
Schroeder, R. (1974). Methode zur Verringerung der Phasendifferenz. *CIC* **A2**, 1–11.
Schroedinger, E. (1931). Spezielle Relativitaetstheorie und Quantummechanik. *Sitzung ber. Preuss. Akad. Wiss., Phys.-Math. Kl.* pp. 67–80. (See also Salecker and Wigner, 1958).
Sexl, R. U. (1976). Seasonal differences between clock rates. *Phys. Lett.* B **61**(1), 65.
Shapiro, I. I. (1976). Estimation of astrometric and geodetic parameters. In Meeks (1976), pp. 261–276.
Silverberg, E. C. (1974). *Appl. Opt.* **13**, 565.
Smith, H. M. (1953). Quartz clocks of the Greenwich Time Service. *Mon. Not. R. Astron. Soc.* **113**(1), 67–80.
Smith, H. M., and Tucker, R. H. (1953). The annual fluctuation in the rate of rotation of the earth. *Mon. Not. R. Astron. Soc* **113**(2), 251–257.
Spencer-Jones, Sir H. (1956). The rotation of the earth. *In* " Handbuch der Physik," (S. Flügge, ed.), Vol. 47, pp. 1–23. Springer-Verlag, Berlin and New York.
Stanley, J. T. (1972). The uses and limitations of HF standard broadcasts for time and frequency comparison. *Proc. PTTI Conf., 4th* pp. 249–258.
Stein, S. R., and Turneaure, J. P. (1975). *Proc. IEEE* **63**, 1249–1250.
Stein, S. R. (1975). Application of superconductivity to precision oscillators. *Proc. ASFC, 29th* pp. 321–327. (See also Stein and Turneaure, 1975).
Stone, R. R., Jr., Gattis, T. H., and Lieverman, T. N. (1972). Utilization of FSK communications for time. *Proc. PTTI Conf., 4th* pp. 213–224.
Stover, H. A. (1974). A time reference distribution concept. *Proc. PTTI Conf., 5th* pp. 505–527.
Stover, H. A. (1975). Applications of PTTI techniques in communications systems. *Proc. PTTI Conf., 7th* pp. 399–417.
Stoyko, A. (1970). La variation seculaire de la rotation de la terre et des problemes connexes. *Ann. Guebhard*, **46**, 293–316.
Stoyko, A., and Stoyko, N. (1974). La chronometrie et la variation de la rotation de la terre. *CIC* **A9**, 1–4.
Stoyko, M. N. (1936). Sur l'irregularité de la rotation de la terre. *C.R. Acad. Sci.* **203**, 39.
Stoyko, M. N. (1937). Sur la periodicité dans l'irregularité de la rotation de la terre. *C.R. Acad. Sci.* **205**, 79.
Swanson, E. R., and Kugel, C. P. (1973). A synoptic study of sudden phase anomalies (SPA's). *Proc. PTTI Conf., 5th* pp. 443–475.
Synge, J. L. (1960). " Relativity, The General Theory," p. 106. North-Holland Publ., Amsterdam.
Vanier, J., Larouche, R., and Audoin, C. (1975). The hydrogen maser wall shift problem. *Proc. ASFC, 29th* pp. 371–382.
Vessot, R. F. C. (1976). Frequency and time standards. In Meeks (1976), pp. 198–227.
Vessot, R. F. C., and Levine, M. W. (1970). A method for eliminating the wall shift in the atomic hydrogen maser. *Metrologia* **6**(4), 116–117.
Vessot, R., Mueller, L., and Vanier, J. (1966). The specification of oscillator characteristics from measurements made in the frequency domain. *Proc. IEEE* **54**(2), 199–207.

von Neumann, J., Kent, R. H., Bellinson, H. R., and Hart, B. I. (1941). The mean square successive difference. *Ann. Math. Stat.* **12**, 153–162.

Wait, J. R. (1969). On the optimum receiver bandwidth for propagated pulsed signals. *Proc. IEEE* **57**(10), 1784–1785.

Walls, F. L., and Wainwright, A. E. (1974). Measurements of the short-term stability of quartz crystal resonators—A window on future developments in crystal oscillators. *Proc. PTTI Conf., 6th* pp. 143–157. See also *IEEE Trans. Instrum. Meas.* **24**(1), 15–20.

Wells, J. W. (1963). Coral growth and geochronometry. *Nature (London)* **197**, 948–950.

Wesson, P. S. (1975). Gravity and the earth's rotation. In Rosenberg and Runcorn (1975), pp. 353–374.

Wiener, N. (1914). *Proc. Cambridge Philos. Soc.* **17**, 441–449.

Wineland, D. J., Allan, D. W., Glaze, D. J., Hellwig, H., and Jarvis, S., Jr. (1976). Results on limitations in primary cesium standard operation. *IEEE Trans. Instrum. Meas.* **25**(4), 453–458.

Winkler, G. M. R. (1972). Path delay, its variations etc. *Proc. IEEE* **60**(5), 522–529.

Winkler, G. M. R., and van Flandern, T. (1977). Ephemeris time, relativity, and the problem of uniform time in astronomy. *Astron. J.* **28**(1), 84–92.

Winkler, G. M. R., Hall, R. G., and Percival, D. B. (1970). The U.S. Naval Observatory clock time reference etc. *Metrologia* **6**(4), 126–134.

Woolard, E. W. (1953). Theory of the rotation of the earth around its center of mass. *Astron. Pap. Am. Ephemeris* **15**, Pt. I.

Woolard, E. W., and Clemence, G. M. (1966). "Spherical Astronomy." Academic Press, New York.

Yamamoto, M., Harada, K., and Saburi, Y. (1975). A time comparison experiment performed by SSRA system via ATS-1. *J. Radio Res. Lab.* **23**, 85–103.

Yoshimura, K. (1972). The generation of an accurate and uniform time scale with calibrations and predictions. *Nat. Bur. Stand. (U.S.), Tech. Note* No. 626.

Electrodynamic Concepts of Wave Interactions in Thin-Film Semiconductor Structures. I

A. A. BARYBIN

Department of Electron-Ion Processing of Solids
V. I. Ulyanov (Lenin) Electrical Engineering Institute
Leningrad, USSR

I. Introduction 99
II. Searches for Methods of Design and Analysis of Solid-State Traveling-Wave Amplifiers 100
    A. Solid-State Traveling-Wave Tube 100
    B. Acoustic-Wave Amplifier 102
    C. Magnetostatic and Spin Wave Amplification 103
    D. Traveling-Wave Amplifier Using Negative Differential Mobility Semiconductors 104
    E. Normal-Mode Theory As a Method of Wave Interaction Analysis ... 106
III. Generalized Theory of Normal-Mode Excitation in Active Polarized-Medium Waveguides by External Sources 110
    A. Active Polarized-Medium Models. General Equations and External Sources 110
    B. Small-Signal Power Theorem 121
    C. Reciprocity Theorem and Mode Orthogonality and Normalization Relations ... 125
    D. Normal-Mode Excitation by External Sources and Mutual Coupling among Modes 130
    References 135

I. INTRODUCTION

In recent years, there has been considerable interest among many scientists in studying wave interactions in solid-state plasmas, for two reasons: First, the investigation of wave propagation and instabilities in solid-state plasmas opens new possibilities for scientists to study the physical properties of solids. Second, wave interactions provide the basis for the operation of the traveling-wave amplifier (TWA) and make it attractive for engineers to realize a solid-state analog of the vacuum traveling-wave tube (TWT). The first, physical aspect of this problem has been discussed in detail by many authors in review papers and books (Bowers and Steele, 1964; Vedenov, 1964; Buchsbaum, 1965; Glicksman, 1965; Chynoweth and Buchsbaum, 1965; Kaner and Skobov, 1968; Steele and Vural, 1969; Hartnagel, 1969; Platzman and Wolff, 1973; Kaner and Yakovenko, 1975; Vladimirov, 1975). Here we shall give attention to the second aspect of the wave interaction problem in solid-state plasmas.

The known merits of TWT, such as the combination of unilateral amplification with high gain and broad bandwidth with relatively low noise, have excited the interest of scientists in searching for the ways and means of solid-state analog construction of the vacuum TWT. All the presently known principles of developing the solid TWA using the wave properties of nondegenerate semiconductor plasmas can be divided into two categories:

1. Amplification due to the interaction of drifting carriers in semiconductors with external, in respect to the carriers, systems supporting wave propagation (such as electromagnetic waves in slow-wave structures (SWS), acoustic waves in piezoelectric crystals, magnetostatic or spin waves in ferromagnetics).

2. Amplification of internal carrier waves in semiconductor plasmas due to the intrinsic properties of the electron subsystem in a crystal. Related to this category are (a) convective instability of traveling space-charge waves caused by a negative differential mobility (NDM) (for example, in transferred-electron materials such as n-GaAs, which exhibits the Gunn effect); and (b) two-stream instabilities in semiconductor plasmas. (The wave instability due to NDM in n-GaAs may be referred to the two-stream instabilities as a result of interaction between "light" and "heavy" electrons in two-valley semiconductors).

Strictly speaking, devices using the interactions of the first category are only direct analogs of the vacuum TWT, where the role of SWS is played by a corresponding external system (electromagnetic, acoustic, or spin).

Below we shall briefly consider the main directions and results of studying wave interactions and seeking ways to develop solid TWA in order to formulate basic electrodynamic problems, which should be analyzed within the scope of wave interactions in thin-film semiconductor structures.

II. Searches for Methods of Design and Analysis of Solid-State Traveling-Wave Amplifiers

A. Solid-State Traveling-Wave Tube

The idea of constructing a solid-state traveling-wave tube (STWT) as a direct analog of the vacuum TWT, where the electron beam is replaced by drifting carriers in a semiconductor, was first proposed by Pierce and Suhl (1955) and one year later by Peter (1956). After that proposal several investigators analyzed the interaction between slow electromagnetic waves and drifting carriers in a semiconductor.

A one-dimensional analysis of an n-type STWT was first published by Solymar and Ash (1966) following the well-known analysis for vacuum TWT of Pierce (1950). Sumi (1966, 1967) has published a two-dimensional analysis of the interaction between the surface TM waves on a semiinfinite semiconductor medium and the adjacent planar SWS as an infinitely thin current-sheet. An extension of Sumi's results was made by Steele and Vural (1969) to the case of an inductive–resistive plane as SWS.

Different aspects of interaction between drifting carriers in semiconductors and external SWS were discussed by other authors. Ettenberg and Nadan (1968, 1970), Fujisawa (1968a,b, 1970), Vikulov and Tager (1969), and Lefeuvre and Hanna (1973) have developed a one-dimensional model for the STWT. Ettenberg and Nadan (1970) have showed that the presence of two carrier species can essentially decrease the growth rate of STWT. Hammer (1967), following the treatment of Solymar and Ash and using a definition for the carrier mass, has discovered that the NDM, depending on the relation between a carrier drift velocity v_0 and an electromagnetic wave phase velocity v_{ph}, can either further (when $v_0 < v_{ph}$) or inhibit (when $v_0 > v_{ph}$) a wave amplification. Fujisawa (1968a,b, 1970) worked out a transmission-line analog and a kinetic power theorem for carrier waves in semiconductors. In papers of Fujisawa and Ichikawa (1969) and Lefeuvre and Hanna (1973) the coupled-mode theory for STWT was given. Gover et al. (1974) have taken into account the velocity distribution of drifting carriers in the collisionless regime by means of a one-dimensional analysis of the collisionless Boltzmann equation.

A two-dimensional analysis for the STWT was made by Steele and Vural (1969) and Thiennot (1972, 1975). As distinct from Sumi (1967), using the boundary condition on a semiconductor surface in the form of zero normal-current component $(j_{in} = 0)$, they take into account the effect of surface charge arising on the semiconductor surface in the zero diffusion limit. Thiennot (1975) infers that the thickness of a semiconductor plate and especially a dielectric gap between the SWS plane and semiconductor surface and also the surface charge may play an important part in the amplification mechanism of STWT, reducing its growth rate.

Hines (1969) and Swanenburg (1973) have proposed using mosaic and interdigital electrode structures as SWS instead of helical or meander-line ones. Analysis has showed the possibility of coupling these structures to space-charge waves in both bulk and thin-film semiconductors, for what is exhibited as a negative conductance of such structures. This fact has been confirmed experimentally (Swanenburg, 1972a,b).

The total feature of all the above theoretical analyses is an inference of the possibility of attaining high gains (about several hundreds of decibels per millimeter) in the STWT. However, the experiments performed by

Sumi and Suzuki (1968), Freeman et al. (1973), and Thiennot (1975) with n-InSb and Ge at 77 and 4.2°K, respectively, has confirmed only a principal possibility of amplification but has not given a net gain of electromagnetic waves. This fact testifies to more weak interaction between carriers and external SWS than that predicted by the above theories.

Discrepancy between theory and experiment can be related to either a simplification of theoretical models or an imperfection of experimental arrangements. Disadvantages of different theoretical models were noted by Ettenberg and Nadan (1970) and Thiennot (1975). Basic design and technological difficulties are related to the necessity to have a very high slowing of electromagnetic waves (a wave propagation velocity must be reduced by at least three orders of magnitude with respect to the velocity of light). The modern state of the technological art does not allow one to design SWS that provide for such extreme high slowing. Moreover, the insulator gap between SWS and a semiconductor surface determining an interaction efficiency must be utterly thin but endure high dc voltage gradients. To obtain such a thin dielectric of good quality is also a great technological problem at present. Apparently, direct solid-state analogs of the vacuum TWT will appear in the near future in connection with progress in molecular, electron-ion, and laser technology.

A further task of the STWT theory is a study of wave interactions in two- or three-dimensional models taking into consideration the properties of real crystal surfaces involving the surface states and charged layers.

B. *Acoustic-Wave Amplifier*

Considerably more progress has been achieved in the way of acoustic wave amplification in piezoelectric semiconductors with drifted carrier streams. Here the situation was the opposite to that which has taken place in searching for ways to design STWT. Hutson et al. (1961) first observed experimentally an effective interaction of ultrasonic waves with drifting carriers in a piezoelectric crystal CdS. The interaction result was displayed in terms of sound amplification when the drift velocity was in excess of the sound velocity. This confirmed the prediction of Weinreich (1956) about the change of an electronic attenuation sign of the sound in the presence of drifting carriers. A macroscopic theory of sound–carrier interactions in piezoelectric crystals was developed by Hutson and White (1962) and White (1962). Later Blötekjaer and Quate (1964) and Barybin (1968a,b) applied the coupled-mode theory to the study of wave interactions in the acoustic amplifier. Experimental results of Hutson et al. (1961) evoked a vast flow of both theoretical and experimental investigations, which showed good mutual agreement. The reader will find a detailed discussion of the state of the art of acoustic amplifiers and their theories in the reviews of McFee (1966), Spector (1966), Gurevich (1968), Pustovoit (1969), and in the special issue on ultrasonics of *Proceedings of IEEE* (Vol. 53, No. 10, 1965).

The bulk elastic wave's amplification restricts a class of materials suitable for acoustic amplifiers, with the crystals possessing simultaneously good piezoelectric and semiconductor properties. Crystals used in practice do not satisfy this requirement: CdS, CdSe, and ZnO are not good semiconductors, while GaAs and InSb have a weak piezoeffect. This fact resulted in the idea of using composite structures where the thin semiconductor film (Si, GaAs, or InSb) with drifting carriers is placed on a surface of strong piezoelectrics (for example, $LiNbO_3$) supporting Rayleigh surface wave propagation. The idea of surface wave amplification was first stated by Gulyaev and Pustovoit (1964) and Tien (1964). The subsequent experiments carried out by White and Voltmer (1966), Yoshida and Yamanishi (1968), Collins *et al.* (1968), Lakin *et al.* (1969), and Gulyaev *et al.* (1971) have totally confirmed this idea. A more detailed consideration of acoustic surface wave devices is given in a review of Gunshor (1975).

C. Magnetostatic and Spin Wave Amplification

The possibility of bulk spin wave amplification in ferromagnetic semiconductors with drifting carriers has been discussed by many authors (see, e.g., the books by Akhiezer *et al.*, 1968; Steele and Vural, 1969). An experimental check of this idea was impeded by the absence of materials having simultaneously good magnetic and semiconductive properties. For this reason interest arose in the study of interactions between surface magnetostatic or spin waves and drifting carriers in composite structures involving ferrite and semiconductor layers. The idea of using such multilayer structure for solid TWA was first stated by Chang and Matsuo (1968) and Schlömann (1969). The work of Robinson *et al.* (1970), Shapiro (1972), Lukomskii and Tsvirko (1973), Masuda *et al.* (1974), Chang and Matsuo (1974, 1975), Bespyatykh *et al.* (1975), and Tsvirko *et al.* (1975, 1976) carried out the detailed analysis of wave interactions in layered structures and the onset conditions for a convective instability of surface magnetostatic waves were established. The experiments carried out by Szustakowski and Wecki (1971, 1973), Vashkovsky *et al.* (1972), and Kawasaki *et al.* (1974) have confirmed the existence of interactions between surface magnetostatic waves and drifting carriers in layered structures. However, the interaction efficiency proved to be very low, so that an amplification effect was displayed only as a reducing magnetostatic wave attenuation at the presence of carrier drift. One possible reason for this may be a strong effect of surface charges on a semiconductor boundary (see, e.g., the papers of Kawasaki *et al.*, 1974; Tsvirko *et al.*, 1976). Further experimental progress in this direction can be expected in connection with the development of technological methods of multilayer heteroepitaxial compound structure growth.

D. Traveling-Wave Amplifier Using Negative Differential Mobility Semiconductors

From two above-mentioned mechanisms of internal carrier wave amplification in semiconductor plasmas we shall be interested only in the convective wave instability due to NDM, in particular, caused by electron transfer in two-valley semiconductors like n-type GaAs. Two-stream instabilities have been discussed in the literature (see, e.g., the books by Steele and Vural, 1969; Hartnagel, 1969). Their practical worth as a possible operation mechanism of solid TWA is, however, represented as doubtful owing to difficulties of a selective coupling to desirable carrier type in multicomponent plasmas.

The idea of using two-valley semiconductors like n-GaAs not only for generation but also for microwave amplification was first stated by McCumber and Chynoweth (1966) and experimentally realized by Thim and Barber (1966) in the form of a microwave one-port reflection amplifier. The subsequent work of numerous authors suggested and designed all kinds of reflection amplifiers using Gunn diodes and developed their theory. The reader having any interest in this problem should refer to the reviews of Narayan and Sterzer (1970) and Perlman et al. (1970) and also to the books by Bulman et al. (1972) and Levinshtein et al. (1975).

The principal disadvantage of all kinds of reflection amplifiers connected with a relatively narrow bandwidth and the absence of built-in isolation between input and output (which requires using an external circulator for separating input and output) makes more interesting and promising another type of amplifier: the two-port unilateral traveling-wave amplifier. In these amplifiers electromagnetic power fed into the input coupler (located near the cathode contact) excites carrier (space-charge) waves in a carrier stream drifted from the cathode to the anode. These waves grow, as a consequence of NDM in the transferred-electron semiconductor, when traveling toward the anode. There the amplified wave power is picked up by means of the output coupler. For this reason one keeps the term "traveling-wave amplifier" in respect to this type of amplifier although its operation mechanism differs from that of a TWT using a distributed-wave interaction.

To obtain stable amplification on n-type GaAs samples dc biased in the NDM region, it is necessary to stabilize them against the formation of moving high-field domains, which bring about spontaneous Gunn oscillations. As now known (see, e.g., McCumber and Chynoweth, 1966; Bulman *et* al., 1972; Levinshtein et al., 1975), the electric-field profile stabilization is achieved by reducing the donor density n_0 and the sample dimension either in the electron drift direction (length L) or transverse to it (thickness d). The stabilization parameters are $n_0 L$ or $n_0 d$, which have a critical value of 2×10^{11} to 5×10^{11} cm^{-2}. Below this critical value domain formation is suppressed. In the same manner as for stabilization one distinguishes the longitudinally stabilized amplifiers with $n_0 L < (n_0 L)_{crit}$ (so-called bulk TWA) and the transversely stabilized amplifiers with $n_0 d < (n_0 d)_{crit}$ (so-called thin-film TWA).

The first bulk TWA was designed and investigated by Robson et al. (1967). The couple electrodes were first made as two capacitative strip probes and then replaced by alloyed metal ohmic contacts to improve the coupling. The couplers were located on one lateral side of a sample near the cathode and anode contacts formed on its ends. The amplifier of Frey et al. (1971) was provided with both longitudinal ($n_0 L = 1.8 \times 10^{11}$ cm^{-2}) and transverse ($n_0 d = 1.2 \times 10^{11}$ cm^{-2}) stabilizations. The cathode and anode ohmic contacts have simultaneously played a part in terminal microwave couplers. Koyama et al. (1969), Kumabe and Kanbe (1970), and Kanbe et al. (1971) have proposed and designed a bulk GaAs TWA of planar construction, which is like an intermediate variation toward a thin-film type TWA. The sample stabilization was provided for reducing the donor density and the sample thickness so that $n_0 d = 0.5 \times 10^{11}$ to 1.2×10^{11} cm^{-2}. The cathode and anode ohmic contacts were formed on one lateral side of a sample. These two ohmic electrodes were also used for input and output terminals of microwave signals.

The measurements of the above-described bulk TWA's have shown linear gains about 10–30 dB at frequencies 1–4 GH$_z$ and saturation power about several milliwatts with a built-in isolation of 20 to 30 dB. All these results were obtained for pulsed bias conditions because the thick samples of compensated bulk-grown GaAs with a negative temperature coefficient of resistivity were used. For dc biasing and cw operation, one must have a positive temperature coefficient of resistivity, and this requires epitaxially grown material. In addition, thin-film structures allow one to provide for an effective heat sinking in the transverse direction from the active semiconductor layer to the heat-conducting substrate.

The thin-film TWA was first fabricated by Dean et al. (1970). Their design contains two large planar ohmic contacts serving as both dc bias electrodes (the cathode and anode) and microwave ground planes of a coplanar line. Between these contacts were input and output coupling electrodes in the form of narrow Schottky barrier contacts. Using the Schottky barrier contacts, rather than the ohmic contacts applied in all the previous bulk TWA, a more effective coupling between electromagnetic waves and carriers was obtained. The problems of theory and design of the thin-film TWA were discussed by Dean (1969, 1972), Dean and Matarese (1972), Bianco et al. (1973), and Dean and Robinson (1974). Experimental investigations of Dean et al. (1973; Dean, 1969), Dean and Matarese (1972), and Kanbe et al. (1973) showed the vitality and perspective of thin-film TWA as broad-band (up to an octave) and high-gain (up to 30 dB) microwave amplifiers with built-in isolation between input and output (about 20–30 dB), and capable of operating at very high frequencies. High measured noise figures (up to 25 dB) can be accounted for, on the one hand, by insufficient efficiency of couplers exciting undesirable modes in a carrier stream and, on the other hand, by an imperfection in the modern technology.

Interesting results were published by Fleming (1975) on the investigation of the new type of thin-film TWA with a wave propagation transverse to the carrier drift direction. The measurements showed a low noise figure (≤ 6 dB) with 15.6 dB amplification at 20 GHz.

E. *Normal-Mode Theory As a Method of Wave Interaction Analysis*

The above review of work devoted to the search for development methods of solid-state TWA allows one to single out two aspects of this problem: the design-technological aspect and calculation-theoretical aspect.

The general feature of all the proposed and fabricated solid TWA constructions capable of cw operation is the presence of a thin semiconductor layer with drifting carriers, together with other bulk or thin-film materials (metallic, dielectric, piezoelectric, ferromagnetic, etc.). We shall refer to such constructions as thin-film semiconductor structures (TFSS).

Experimental data of numerous authors mainly confirm the possibility of the principle of designing different kinds of solid TWA. The greatest progress in this direction was achieved for surface acoustic-wave amplifiers and thin-film TWA using NDM semiconductors. The principal design features of solid TWA give rise to technological requirements that are within or even beyond the possibilities of modern solid-state technology. At present, the technological difficulties represent just a basic obstacle in the way of technical embodiment of numerous ideas for designing solid TWA. Successes in modern solid-state technology, especially in the field of molecular, electron–ion, and laser processing of solids and epitaxial growth of multilayer heterostructures, allow one to hope in the near future to achieve considerable engineering progress in the field of solid TWA design.

We shall next turn our attention to the calculation-theoretical aspect of solid TWA problem using TFSS. Analysis of theoretical and experimental data published by different authors shows that there is essential disagreement among them that cannot be explained only by experimental design imperfections. Theoretical and calculation models, analysis methods, simplifications, approximations, and assumptions used by different authors are fairly various and not infrequently inconsistent with or contradictory to each other, and in some cases physically incorrect. This gives rise to the situation where, in spite of the commonality of all the models (related to the presence of semiconductor layers in all kinds of TFSS) establishing conformity between the results of different authors and understanding the general features peculiar to all the wave interactions is extremely difficult.

The most widely used method of wave analysis is so-called substitution analysis (Sturrock, 1958). This is the simplest and most straightforward technique for obtaining the total dispersion relation of a complicated system consisting of separate interacting subsystems. The basis of this method is the general dynamic equations for each of the subsystems, taking into account all the necessary interactions and boundary conditions. The substitution of a desirable dependence on time t and wave propagation coordinate z as a wave factor $\exp[i(\omega t - kz)]$ in these dynamic equations results in the total dispersion relation $D(\omega, k) = 0$. The dispersion equation gives the relationship between the frequency ω and the wave number k, each of which is generally complex (Sturrock, 1958). Substitution analysis is unable to give a physical interpretation of complex solutions of the dispersion equation. In the first place, this refers to the problem of distinguishing between absolute and convective instabilities (time instability with complex values of ω) and also amplifying and evanescent waves (space instability with complex values of k). In order to solve the distinguishing problem it is necessary to go beyond the scope of the substitution analysis by using the corresponding threshold criteria of instabilities (Akhiezer et al., 1974), for example, Briggs' criteria (Briggs, 1964). All the criteria reduce essentially to the study of the dispersion equation roots' behavior on the complex planes of ω and k. This requires the analytic solution of the dispersion equation $D(\omega, k) = 0$ in the form of both $\omega(k)$ and $k(\omega)$. To obtain such solutions in analytic forms is possible, as a rule, only for the simplest wave interaction systems (Steele and Vural, 1969). For complicated systems such as TFSS the dispersion equation contains transcendental functions and does not allow one to obtain the analytic solutions in the form of $\omega(k)$ and $k(\omega)$. Each author, proceeding by his own reasoning and physical intuition, assumes some simplifying approximations or introduces some restrictions of system characteristic parameters to simplify the dispersion equation. Therefore, it is not surprising that the results of different authors obtained for similar initial models differ due to discrepancies between assumptions and restrictions made on the final stage of the simplification of the total dispersion equation. Hence it follows that the mathematical description of wave processes in all kinds of TFSS from one point of view is necessary and very important. A theory allowing one to describe uniformly a physical picture of drifting carrier interactions in a semiconductor film with different "external" systems (electromagnetic, acoustic, spin, etc.) should make it possible to compare different kinds of TFSS with each other in order to estimate their applicability as an active structure of the solid TWA.

In general electromagnetic theory, the most fundamental method of solving boundary value problems is the normal-mode method (Marcuvitz and Schwinger, 1951; Collin, 1960; Ramo and Whinnery, 1965; Johnson, 1965). This method allows one to reduce vector Maxwell's equations to scalar differential equations by applying the normal-mode expansion of the total fields. It consists, in essence, in determining the mode amplitudes in the presence of external sources in a wave guiding (or oscillating) system.

The normal-mode method was used with great success in microwave electronics to analyze the processes in travelling- and backward-wave tubes (Pierce, 1950; Heffner, 1954; Chodorow and Susskind, 1964). The normal-mode conception was made the basis of coupled-mode theory, which played an important role in studying wave interactions and parametric oscillations in microwave vacuum electronics (Louisell, 1960). Application of waveguide excitation theory (Vainshtein, 1957) to the interaction between an electron beam and an electromagnetic field of SWS allowed us to obtain the coupled equations for normal-mode amplitudes of the electron beam and the SWS (Barybin and Ter-Martirosyan, 1969).

The normal-mode theory was recently generalized by Auld and Kino (Auld, 1969; Auld and Kino, 1971; Kino and Reeder, 1971) to solve some problems of acoustoelectronics, in particular, for the analysis of the interdigital transducer and the Rayleigh wave amplifier. The above papers together with those of Blötekjaer and Quate (1964) and Barybin (1968a) have demonstrated the fruitfulness of using normal- and coupled-mode theory in solid-state electronics. In our opinion, this theory can provide for a uniform mathematical consideration of wave interactions in different kinds of TFSS.

The basic idea of the normal-mode theory arose from the desire to use the entire body of available knowledge about properties of separate parts of a system without an interaction between them in order to extract new information about properties of the whole system in the presence of interactions. In practice, any complicated system can be represented as a number of simpler subsystems interacting with each other (for example, two coupled passive transmission lines, SWS and electron beam in TWT, piezoelectric crystal supporting surface acoustic wave, and semiconductor plate with drifting carriers in acoustic amplifier). The analysis of uncoupled subsystems gives a series of normal modes for each subsystem, characterized by the orthogonal eigenfunctions and corresponding eigenvalues. In the presence of a couple, the total values of physical quantities are represented as a sum of normal-mode eigenfunctions of all the sybsystems taking part in the interaction. The interacting subsystems are mutually external for each other. Each of them contains sources external for another subsystem, which excite the normal modes of the latter coupled to the former. The orthogonality of the eigenfunctions allows one to obtain the normal-mode excitation equations by external sources, which can be reduced to the form of the coupled equations by taking into account the mutual coupling between the modes. The application of the power theorem to eigenfunction normalization allows one to analyze the coupled modes in terms of power exchange between interacting modes (Louisell, 1960).

There is only a small number of very simple (mainly passive) systems for which the exact solution of the boundary value problem is possible. In practice, the analysis of any complicated system means an approximation to the exact solution. It refers also to the normal- and coupled-mode method, which is in fact a modification of perturbation theory (Louisell, 1960; Ter-Martirosyan, 1969). In this case, one such approximation is the weak-coupling approximation, when the normal modes of uncoupled subsystems are weakly perturbed at the expense of the coupling. To evaluate the validity of the weak-coupling approximation it is necessary in each particular case to estimate how the energy stored in a coupling element is small compared with the energy of each normal mode taking part in the interaction (Louisell, 1960). In many practical situations the weak-coupling approximation follows, however, from intuitive considerations on the basis, for instance, of geometric representations (filling the extent of a volume or a cross section by a perturbing body, the gap between interacting layers, etc.).

In our view, the normal-mode theory is the right mathematical instrument that allows one to study the wave interactions in different kinds of TFSS proceeding from a uniform position. In all kinds of TFSS the active role belongs to a thin semiconductor film. The knowledge of the normal-mode spectrum of this film is the basis for the analysis of different wave interactions. The elaboration of the normal- and coupled-mode theory for TFSS comprises the following points:

1. a choice and grounding of models for active media included in thin-film structures;
2. a derivation of the power theorem as the basis for the power normalization of modes;
3. a proof of the reciprocity theorem;
4. a derivation of the mode orthogonality relations;
5. a derivation of the normal-mode excitation equations by external sources, reducing them to the coupled-mode form.

In the following we shall adhere to the main results obtained by Auld and Kino (1971; Auld, 1969) and by the author (Barybin, 1974, 1975a,b).

III. Generalized Theory of Normal-Mode Excitation in Active Polarized-Medium Waveguides by External Sources

A. Active Polarized-Medium Models. General Equations and External Sources

We restrict our consideration to the macroscopic model of a medium. In the long-wavelength approximation (when excitation wavelengths are much greater than atomic distances) a crystal is represented as a continuum medium, the physical properties of which are characterized by pertinent phenomenological parameters.

In macroscopic electrodynamics, the electromagnetic properties of a medium are described by the electric and magnetic field intensity vectors $\mathbf{E}_1$ and $\mathbf{H}_1$, and also by the electric displacement vector $\mathbf{D}_1$ and the magnetic induction vector $\mathbf{B}_1$, which are related to each other via Maxwell's equations:

$$\nabla \times \mathbf{E}_1 = -\partial \mathbf{B}_1/\partial t \tag{1}$$

$$\nabla \times \mathbf{H}_1 = (\partial \mathbf{D}_1/\partial t) + \mathbf{j}_1 \tag{2}$$

$$\nabla \cdot \mathbf{D}_1 = \rho_1 \tag{3}$$

$$\nabla \cdot \mathbf{B}_1 = 0 \tag{4}$$

where ρ_1 and $\mathbf{j}_1$ are the charge and current densities of mobile carriers in a medium. Here and subsequently, the subscript 1 is ascribed to all the small-signal ac values, while all the unperturbed dc values will carry the subscript 0.

From a macroscopic point of view, the polarization properties of the medium are described by the polarization vectors: the electric polarization vector $\mathbf{P}_1$ for dielectric properties and the magnetic polarization (or magnetization) vector $\mathbf{M}_1$ for magnetic properties. The polarization vectors give the contribution to the vectors $\mathbf{D}_1$ and $\mathbf{B}_1$ in accordance with the relations (Johnson, 1965):

$$\mathbf{D}_1 = \varepsilon_0 \mathbf{E}_1 + \mathbf{P}_1 \tag{5}$$

$$\mathbf{B}_1 = \mu_0 (\mathbf{H}_1 + \mathbf{M}_1) \tag{6}$$

where ε_0 and μ_0 are the free-space values of the permittivity and permeability, respectively.

By an active polarized medium we mean a polarized medium that can support, besides the curl electromagnetic fields $\mathbf{E}_{1,c}$ and $\mathbf{H}_{1,c}$, also the propagation of potential waves with the potential field components $\mathbf{E}_{1,p} = -\nabla\varphi_1$ and $\mathbf{H}_{1,p} = -\nabla\psi_1$, so that the total electromagnetic fields are

$$\mathbf{E}_1 = \mathbf{E}_{1,c} - \nabla\varphi_1, \quad \mathbf{H}_1 = \mathbf{H}_{1,c} - \nabla\psi_1 \tag{7}$$

where

$$\nabla \cdot \mathbf{E}_{1,c} = 0, \quad \nabla \cdot \mathbf{H}_{1,c} = 0 \tag{8}$$

The potential wave propagation is due to the short-length interactions between adjacent polarization vectors (for example, spin waves in ferromagnetics or elastic piezoelectrically active waves in piezodielectrics) or the Coulomb interaction between the mobile charges. Therefore, besides dielectric and magnetic polarized media, we shall refer to the active polarized media also the drifting streams of mobile charge carriers in nondegenerate semiconductor plasmas, which in the polarization description are characterized by a polarization vector (see Section III,A,2).

For a uniaxial ferromagnet, which is uniformly magnetized by a dc magnetic field $\mathbf{H}_0 = \mathbf{e}_M H_0$ (directed along the easy magnetization axis $\mathbf{e}_M$) to the saturation magnetization $\mathbf{M}_0 = \mathbf{e}_M M_0$, the linearized equation of magnetization motion (neglecting magnetic losses) has the form (Steele and Vural, 1969; Akhiezer et al., 1968)

$$\partial \mathbf{M}_1/\partial t = -\gamma\mu_0[\mathbf{M}_1 \times \mathbf{H}_0 + \mathbf{M}_0 \times \mathbf{H}_1 + \alpha_e(\mathbf{M}_0 \times \nabla^2\mathbf{M}_1) - \beta_a(\mathbf{M}_0 \times \mathbf{M}_1)] \tag{9}$$

where γ is the gyromagnetic ratio, and α_e and β_a are the exchange and anisotropy constants, respectively.

The stressed state of an elastic medium is described by the stress and strain tensors $\mathbf{T}_1$ and $\mathbf{S}_1$. The dynamic equations giving the relationship between these tensors and the particle displacement velocity vector $\mathbf{V}_1$ have the form (Steele and Vural, 1969; Auld, 1969)

$$\rho_m \, \partial\mathbf{V}_1/\partial t = \nabla \cdot \mathbf{T}_1, \tag{10}$$

$$\partial \mathbf{S}_1/\partial t = \nabla_s \mathbf{V}_1 \tag{11}$$

where ρ_m is the mass density of the elastic medium. Here, instead of the conventional tensor notation, we have used the symbolic notation of Auld (1969) wherein $\nabla \cdot \mathbf{T}_1$ means the tensor divergence with the components $(\nabla \cdot \mathbf{T}_1)_i = \partial T_{1,ij}/\partial x_j$ and $\nabla_s \mathbf{V}_1$ means the symmetric part of the vector gradient with the components $(\nabla_s \mathbf{V}_1)_{ij} = \frac{1}{2}(\partial V_{1,i}/\partial x_j + \partial V_{1,j}/\partial x_i)$.

If the elastic dielectric medium possesses a piezoelectric property, the acoustic-wave propagation is accompanied by the appearance of an electric polarization $\mathbf{P}_1$. In turn, an electric field causes an elastic stress in the medium. In this case the constitutive relations (neglecting magnetostrictive effect) assume the conventional forms (Steele and Vural, 1969; Auld, 1969)

$$\mathbf{T}_1 = c : \mathbf{S}_1 - e \cdot \mathbf{E}_1 \qquad (12)$$

$$\mathbf{P}_1 = e : \mathbf{S}_1 + \varepsilon_0 \chi \cdot \mathbf{E}_1 \qquad (13)$$

where the single dot means the ordinary scalar product, so that $(e \cdot \mathbf{E}_1)_{ij} = e_{kij} E_{1,k}$, and the double dot indicates the double scalar product, so that $(c : \mathbf{S}_1) = c_{ijkl} S_{1,kl}$. In view of (13), relations (5) can be rewritten as

$$\mathbf{D}_1 = e : \mathbf{S}_1 + \varepsilon \cdot \mathbf{E}_1. \qquad (14)$$

In relations (12)–(14), χ and $\varepsilon = \varepsilon_0(1 + \chi)$ are the electric susceptibility tensor (χ_{ij}) and the permittivity tensor (ε_{ij}), e is the third-order piezoelectric stress tensor (e_{kij}), and c is the fourth-order stiffness tensor (c_{ijkl}).

1. Internal and External Sources of Electromagnetic Fields

The motion of the polarization vectors $\mathbf{M}_1$ and $\mathbf{P}_1$ is determined by the electromagnetic fields $\mathbf{E}_1$, $\mathbf{H}_1$, and on the other hand it produces these fields. In other words, the polarized media are sources of electromagnetic fields that should be taken into account in Maxwell's equations by means of the charge and current densities (in addition to those of mobile carriers in a medium). The polarized dielectric or magnetic finite media (for instance, the thin layers of TFSS) produce the *excess* (over surroundings) polarization currents defined for a monochromatic excitation with the frequency ω in the form (Vainshtein, 1957; Barybin, 1975b)

$$\mathbf{j}_1^p = i\omega(\mathbf{D}_1 - \varepsilon_0 \mathbf{E}_1) = i\omega \mathbf{P}_1 \qquad (15)$$

$$\mathbf{j}_1^m = i\omega(\mathbf{B}_1 - \mu_0 \mathbf{H}_1) = i\omega\mu_0 \mathbf{M}_1 \qquad (16)$$

Then Maxwell's equations (1) and (2), in view of (15) and (16), assume the form

$$\nabla \times \mathbf{E}_1 = -i\omega\mu_0 \mathbf{H}_1 - \mathbf{j}_1^m \qquad (17)$$

$$\nabla \times \mathbf{H}_1 = i\omega\varepsilon_0 \mathbf{E}_1 + \mathbf{j}_1^e \qquad (18)$$

with $\mathbf{j}_1^e = \mathbf{j}_1^p + \mathbf{j}_1$. If one formally introduces the polarization electric (ρ_1^p) and magnetic (ρ_1^m) charges expressed in terms of the currents $\mathbf{j}_1^p$ and $\mathbf{j}_1^m$ with the help of the continuity equations as (Vainshtein, 1957)

$$\rho_1^{p(m)} = -\frac{1}{i\omega} \nabla \cdot \mathbf{j}_1^{p(m)} \qquad (19)$$

then Eqs. (3) and (4) can be rewritten, by virtue of (7) and (8), in the form of

Poisson's equations:

$$\nabla^2 \varphi_1 = -\rho_1^e/\varepsilon_0 \qquad (20)$$

$$\nabla^2 \psi_1 = -\rho_1^m/\mu_0 \qquad (21)$$

We regard the active polarized finite media as peculiar waveguides that are able to support the propagation of either the curl electromagnetic waves ($\mathbf{E}_{1,c}$, $\mathbf{H}_{1,c}$) or the specific waves accompanied by the potential electric ($\mathbf{E}_{1,p} = -\nabla \varphi_1$) or magnetic ($\mathbf{H}_{1,p} = -\nabla \psi_1$) fields. These waves form the normal-modes set of the given system. Then the charges $\rho_1^e = \rho_1^p + \rho_1$, ρ_1^m, and the currents $\mathbf{j}_1^e = \mathbf{j}_1^p + \mathbf{j}_1$, $\mathbf{j}_1^m$ produced by the polarization vector's motion, are considered as the *internal* sources intrinsic to this polarized medium.

Besides polarized medium waveguides, we include electromagnetic waveguides as waveguiding structures containing metallic conducting surfaces, possibly filled (fully or partially) with a *passive* dielectric or magnetic medium. Such waveguides, in which are included SWS, can support only the curl electromagnetic-wave propagation. The electromagnetic processes in such waveguides will be governed by Eqs. (17) and (18) in which the electric $\mathbf{j}_1^e$ and magnetic $\mathbf{j}_1^m$ currents are now the *external* sources exciting the electromagnetic waveguide. From this view, the values of ε_0 and μ_0 appearing in expressions (15) and (16) for the *excess* polarization currents of an external medium and in Eqs. (17), (18), (20), and (21) can differ from their free-space values and be equal to the corresponding values for the passive medium filling the electromagnetic waveguide.

In consideration of a complex system consisting of several parts interacting with each other (for example, drifting carriers and electromagnetic, acoustic, or spin waves) we can always single out one of them, assuming the others to be external for the former. This medium under consideration possesses its own *internal* sources—electric ($\mathbf{j}_1^e$, ρ_1^e) or magnetic ($\mathbf{j}_1^m$, ρ_1^m)—due to the intrinsic processes in the medium. The influence of other parts of the complex system on the processes in the medium under consideration can be taken into account in terms of the *external bulk* sources—the electric and magnetic current and charge densities $\mathbf{j}_b^e$, $\mathbf{j}_b^m$, ρ_b^e, ρ_b^m. These external bulk sources will enter Maxwell's equations (17) and (18) in analogy with the internal sources of the medium, namely (Barybin, 1975b)

$$\nabla \times \mathbf{E}_1 = -i\omega\mu_0 \mathbf{H}_1 - \mathbf{j}_1^m - \mathbf{j}_b^m \qquad (22)$$

$$\nabla \times \mathbf{H}_1 = i\omega\varepsilon_0 \mathbf{E}_1 + \mathbf{j}_1^e + \mathbf{j}_b^e \qquad (23)$$

The external bulk charge densities ρ_b^e and ρ_b^m are related to the external bulk current densities $\mathbf{j}_b^e$ and $\mathbf{j}_b^m$ by the continuity equations analogous to (19) (Barybin, 1975b).

Together with the external bulk sources, there can be *external surface sources* in the form of the electric and magnetic current and charge densities j_s^e, j_s^m, ρ_s^e, ρ_s^m. The electric surface sources j_s^e and ρ_s^e appearing on a boundary of the drifting carrier streams were discussed in detail (Barybin, 1975a) and will be considered in Section IV. The magnetic surface sources j_s^m and ρ_s^m, formally introduced in analogy with the electric ones, are realized by means of slots and cuts on metallic surfaces (Vainshtein, 1957). The surface sources give the discontinuities of the electromagnetic field components at the surface on which these external sources are located, in the form of boundary conditions (Vainshtein, 1957; Barybin, 1975a):

$$\mathbf{n}_s^+ \times \mathbf{E}_1^+ + \mathbf{n}_s^- \times \mathbf{E}_1^- = -\mathbf{j}_s^m \qquad (24)$$

$$\mathbf{n}_s^+ \times \mathbf{H}_1^+ + \mathbf{n}_s^- \times \mathbf{H}_1^- = \mathbf{j}_s^e \qquad (25)$$

$$\mathbf{n}_s^+ \cdot \mathbf{D}_1^+ + \mathbf{n}_s^- \cdot \mathbf{D}_1^- = \rho_s^e \qquad (26)$$

$$\mathbf{n}_s^+ \cdot \mathbf{B}_1^+ + \mathbf{n}_s^- \cdot \mathbf{B}_1^- = \rho_s^m \qquad (27)$$

where the superscripts $\pm$ mean the field's values taken at points lying on different sides of the source surface, which are marked by the inward normal unit vectors $\mathbf{n}_s^\pm$ (see Fig. 2).

The external bulk and surface sources, introduced in such a way, will subsequently (see Section III,D) enter the excitation equations for the amplitudes of normal modes.

It should be emphasized that the one-dimensional analysis of the wave interaction of drifting carriers with external circuits usually consists of dividing the total field acting on a charge into two parts: the internal field related to the intrinsic carrier motion (the so-called space charge field) and the field of a system external to the carriers (Pierce, 1950; Solymar and Ash, 1966; Barybin, 1968a). This approach is valid and is the only possible approach in the one-dimensional analysis of infinite media. In boundary problem analysis it must be preceded by the consideration of an external field penetrating into the semiconductor. Thus in the subsequent normal-mode theory we shall regard the total field acting on the motion of the polarization vector as self-consistent (without dividing it into two parts): the field is produced by the intrinsic motion of polarization and in turn affects this motion. So the total field can be represented as a sum of the normal modes of the system under consideration. The influence of external sources (located inside or outside the system) manifests itself by changing the normal-mode amplitudes, which is just the interaction effect. In consideration of the interaction of two systems, we represent the physical quantities peculiar to each system by their own normal-mode expansions. Then they will mutually play the role of external sources exciting another system coupled to this one. This procedure will lead ultimately to the coupled-mode analysis, which allows one to single out among the whole set of modes the more strongly interacting (see Section III,D).

Let us turn our consideration to the polarization model of nondegenerate semiconductor plasmas.

2. Eulerian and Polarization Description of Nondegenerate Solid-State Plasmas

In microwave vacuum electronics the small-signal analysis of electron beams also uses, besides the well-known Eulerian and Lagrangian variables, the so-called polarization variables (Bobroff, 1959). The introduction of polarization variables has allowed to Bobroff et al. (1962) to ground rigorously the Hahn surface sources (Hahn, 1939), appearing with transverse displacements of a beam boundary, and to prove the small-signal power theorem, thereby clarifying the physical picture of power exchange between interacting modes in microwave electron beam tubes. The polarization variables were used by Durney et al. (1969) to analyze multistream interactions in a finite longitudinal magnetic field.

Later some authors (Hasegawa, 1965; Shin-ichi-Akai, 1966; Goodrich et al., 1974) automatically used the equations and results of the vacuum electron beam polarization description for the wave interaction analysis in semiconductor plasmas, neglecting those features related to the diffusion, collisions, and recombination of charge carriers. The successive introduction of polarization variables for nondegenerate solid-state plasmas was first done by the author (Barybin, 1970a). Below we shall briefly treat the main results of the above paper.

We shall use the hydrodynamic model of nondegenerate plasmas in which the drifting charge carriers are represented as a charged-fluid flow characterized by such macroscopic values as the mean particle density n (or the charge density $\rho = qn$), the mean particle velocity $\mathbf{v}$ (or the current density $\mathbf{j} = \rho\mathbf{v}$), and the pressure p. The microscopic processes of collisions, thermal motion (or diffusion), and recombination are taken into account in this model with the help of phenomenological parameters: the momentum relaxation time (or the mean time of free path) τ, the thermal velocity $v_T = (k_B T/m)^{1/2}$ (or the diffusion constant $\mathscr{D} = v_T^2 \tau$), and the mean lifetime τ'. When the intercarrier collisions are rather frequent, there is a thermal equilibrium inside the carrier ensemble with a temperature T different from the lattice temperature T_0 (in the case of high electric fields). Local equilibrium defines the relationship between pressure and the carrier temperature as $p = nk_B T$ (Steele and Vural, 1969; Akhiezer et al., 1974).

The macroscopic equations of the hydrodynamic model are derived from the kinetic Boltzmann equation by taking moments of the carrier velocity distribution (Akhiezer et al., 1974). In the approximation of local equilibrium (when $p = nk_B T$) and the uniform stationary heating of carriers by a high electric field (when heat flow and thermal ac perturbation in carrier ensemble are negligibly small) the macroscopic equations for the zeroth- and first-order moments of the carrier distribution function have the form (Steele and Vural, 1969; Akhiezer et al., 1974)

$$\frac{\partial \rho}{\partial t} + \nabla \cdot \mathbf{j} = 0, \tag{28}$$

$$\frac{\partial \mathbf{v}}{\partial t} + (\mathbf{v} \cdot \nabla)\mathbf{v} = \frac{q}{m}(\mathbf{E} + \mathbf{v} \times \mathbf{B}) - \frac{\nabla(nk_B T)}{mn} - \frac{\mathbf{v}}{\tau} \tag{29}$$

where q and m are the charge and the effective mass, respectively, of a carrier under consideration. In (28) we have neglected generation–recombination processes by the carrier.

The hydrodynamic description of plasmas with the help of Eqs. (28) and (29) is valid for two limiting cases (Akhiezer et al., 1974):

1. The *vacuum electronics limit* is realized when collisions are so infrequent that $\omega\tau \gg 1$ and a thermal spread of carrier velocities is negligibly small ($T \simeq 0$), i.e., the velocity distribution function has the δ function form. In this case the velocity of all the particles coincides with the hydrodynamic velocity of a medium $\mathbf{v}(\mathbf{r}, t)$, and Eq. (29) assumes the form of the equation of motion conventional for vacuum electronics.

2. The *semiconductor electronics limit* is realized when collisions are so frequent that $\omega\tau \ll 1$ and $\kappa l \ll 1$ (where $l = v_T \tau$ is the mean free path). In this case the collisions scatter a directed momentum of particles so intensively that one can neglect the inertial term on the left-hand side of (29) as compared with the collision term $\mathbf{v}/\tau$. Then, taking into consideration the dependence of τ on the carrier temperature T, Eq. (29) is reduced to the form

$$\mathbf{v} = \mu(\mathbf{E} - \alpha_T \nabla T) + \mathbf{v} \times \mu\mathbf{B} - \frac{1}{\rho}\nabla(\mathscr{D}\rho) \tag{30}$$

where $\mu(T) = (q/m)\tau(T)$ and $\mathscr{D}(T) = v_T^2 \tau(T)$ are, respectively, the mobility and diffusion constant, which depend in the general case on the carrier temperature;

$$\alpha_T = -\frac{k_B}{q}\frac{d \ln \mu(T)}{d \ln T}$$

is the thermal emf constant due to collisions. In the case of uniform carrier heating by high dc fields when $T \simeq \text{const} \, (T > T_0)$, the thermoelectric component of an electric field is absent, i.e., $\mathbf{E}_T = \alpha_T \nabla T \equiv 0$. Then (30) gives the conventional (for semiconductor electronics) expression of current density

$$\mathbf{j} = \sigma\mathbf{E} - \nabla(\mathscr{D}\rho) + \mathbf{j} \times \mu\mathbf{B} \tag{31}$$

where $\sigma(T) = qn\mu(T)$ is the conductivity of a semiconductor medium in consideration of a carrier heating by an electric field.

The analysis of an electric field heating of carriers requires an additional equation for the second-order moment of the distribution function, i.e., for the carrier temperature T. In the simplest case of the local thermal equilibrium this equation has the form of a balance relation between the power picked up by carriers from a field and the loss power (Conwell, 1967):

$$\mathbf{j} \cdot \mathbf{E} = \frac{3}{2} nk_B \frac{T - T_0}{\tau_e(T)} \tag{32}$$

where $\tau_e(T)$ is the energy relaxation time defining the rate of carrier temperature changes. Equation (32) holds true when a temperature has time for following ac changes of a field $\mathbf{E}(\mathbf{r}, t)$ and charges $\rho(\mathbf{r}, t)$ providing for local thermal equilibrium. The condition for this is $\tau_e \ll \tau_r$, where $\tau_r = \varepsilon/\sigma$ is the dielectric relaxation time determining the time scale of field and charge ac changes. With known functions $\tau(T)$ and $\tau_e(T)$ depending on a scattering mechanism of carriers (see Conwell, 1967), Eqs. (31) and (32), in principle, allow one to find a functional dependence $T(E)$, where E is the absolute value of a field in a given point. This permits the temperature dependences $\mu(T)$ and $\mathscr{D}(T)$ to be expressed in the form of the field dependences $\mu(E)$ and $\mathscr{D}(E)$. Thus the influence of high electric fields in Eqs. (30) and (31) can be taken into account in terms of a field dependence of phenomenological parameters $\mu(E)$ and $\mathscr{D}(E)$.

For subsequent derivation of the small-signal power theorem and the reciprocity theorem it is necessary to use not the particular cases of vacuum or semiconductor electronics but the general equation (29), wherein we assume the field dependence $\tau(E)$ to be phenomenological. The energy-power relations and the relations of reciprocity, orthogonality, and normalization of normal modes, obtained below using such an approach, will have a rather general character. Application of the above limiting operations to these relations yields the particular cases of vacuum and semiconductor electronics. Then the former allows one to evaluate the correctness of results obtained via a comparison of those relations with the analogous relations derived before in microwave vacuum electronics (Louisell, 1960; Bobroff et al., 1962; Chodorow and Susskind, 1964).

In the hydrodynamic model we associate with each space point a physically infinitesimal volume element chosen in the unperturbed or perturbed (by the ac signal effect) state of the charged fluid, depending on an appropriate system of variables. The *coordinates* of the volume element center are the *independent space* variables, while the *properties* of the charges (density, velocity, etc.) within the volume element are *dependent* variables (Bobroff, 1959; Barybin, 1970a).

In the Eulerian description the volume element (E element) is chosen in the *perturbed* state of the charged fluid, i.e., in the presence of an ac excitation. The coordinates of the E element center, characterized by the radius vector $\mathbf{r}$, are the Eulerian independent space variables. The position of the E element remains fixed, and so it contains entirely different groups of charges in the absence and in the presence of ac excitation. Consequently, the Eulerian small-signal ac velocity

$$\mathbf{u}_1(\mathbf{r}, t) = \mathbf{v}(\mathbf{r}, t) - \mathbf{v}_0(\mathbf{r}) \tag{33}$$

defined as a subtraction between the total instantaneous value $\mathbf{v}(\mathbf{r}, t)$ and the dc value $\mathbf{v}_0(\mathbf{r})$ taken at the same space point under condition $|\mathbf{u}_1| \ll |\mathbf{v}_0|$ is an abstract quantity and has no direct physical meaning. Indeed, it represents the ac variation of velocity not for a particular group of charges, but rather for different groups of charges, which occupy the chosen E element in the presence and in the absence of ac excitation. On the contrary, the Eulerian small-signal ac charge density

$$\rho_1(\mathbf{r}, t) = \rho(\mathbf{r}, t) - \rho_0(\mathbf{r}) \tag{34}$$

is a natural quantity and has a direct physical meaning as the ac variation of charge density at the given space point, caused by a small-signal excitation.

The linearization of Eqs. (28) and (29) gives the small-signal equations of continuity and motion in the Eulerian variables (Barybin, 1970a):

$$\frac{\partial \rho_1}{\partial t} + \mathbf{V} \cdot \mathbf{j}_1 = 0 \tag{35}$$

$$\frac{\partial \mathbf{u}_1}{\partial t} + (\mathbf{v}_0 \cdot \mathbf{V})\mathbf{u}_1 + (\mathbf{u}_1 \cdot \mathbf{V})\mathbf{v}_0 = \frac{q}{m}(\mathbf{E}_1 + \mathbf{v}_0 \times \mathbf{B}_1 + \mathbf{u}_1 \times \mathbf{B}_0)$$

$$- \frac{v_T^2}{\rho_0}\mathbf{V}\rho_1 + \frac{v_T^2}{\rho_0}\frac{\mathbf{V}\rho_0}{\rho_0}\rho_1 - \frac{\mathbf{u}_1}{\tau_0} + \frac{\mathbf{v}_0 \tau_1}{\tau_0 \tau_0} \tag{36}$$

with

$$\mathbf{j}_1 = \rho_0 \mathbf{u}_1 + \rho_1 \mathbf{v}_0 \tag{37}$$

Equation (36) takes into account, in the general case, a space nonuniformity of dc values ρ_0 and $\mathbf{v}_0$ caused by a nonuniform distribution of dc external forces or a nonuniform doping of a material. In other words, the dc value ρ_0 can differ from its equilibrium value. The field dependence $\tau(E)$, where under the linear approximation

$$E = [(\mathbf{E}_0 + \mathbf{E}_1)^2]^{1/2} \approx E_0 + (\mathbf{E}_0 \cdot \mathbf{E}_1)/E_0 \tag{38}$$

was represented as a sum $(\tau_0 + \tau_1)$ with

$$\tau_0 = \tau(E_0), \qquad \tau_1 = \left(\frac{d\tau}{dE}\right)_{E_0} \frac{\mathbf{E}_0 \cdot \mathbf{E}_1}{E_0} \tag{39}$$

The use of Eulerian description leads to a field treatment of the carrier stream problems wherein (as distinct from the Lagrangian or particle description) any information about individual properties of a particle is entirely lost. The polarization description takes an intermediate place between the Eulerian and Lagrangian. It supplements the Eulerian description rather than opposing itself to the latter.

In the polarization description the volume element (P element) is chosen in the *unperturbed* state of the charged fluid, i.e., in the absence of an ac excitation. The coordinates of the P element center characterized by the radius vector $\mathbf{r}_0$ are the polarization-independent space variables. Under the force of an ac excitation the charges that occupied the given $\tilde{P}$ element at its unperturbed position at the point $\mathbf{r}_0$ move to the new perturbed position at the point $\mathbf{r}$ corresponding to the E element. In this way the form and dimensions of the element can change. These two positions of one and the same charge group, arising as a consequence of an ac signal and taken at the same instant of time, are apart from each other by a distance of the small-signal displacement

$$\mathbf{r}_1(\mathbf{r}_0, t) = \mathbf{r}(t) - \mathbf{r}_0(t) \tag{40}$$

assumed as a function of the unperturbed radius vector $\mathbf{r}_0$. The displacement vector $\mathbf{r}_1$ (or the polarization vector $\mathbf{p}_1 = \rho_0 \mathbf{r}_1$) is the main dependent polarization variable.

The polarization small-signal velocity $\mathbf{v}_1(\mathbf{r}_0, t)$ is defined as the difference between the total instantaneous value $\mathbf{v}(\mathbf{r}, t)$ and the unperturbed dc value $\mathbf{v}_0(\mathbf{r}_0)$ of velocity for one and the same group of charges located for the given instant of time at the different space points $\mathbf{r}$ and $\mathbf{r}_0$ by virtue of an ac excitation effect, i.e.,

$$\mathbf{v}_1(\mathbf{r}_0, t) = \mathbf{v}(\mathbf{r}, t) - \mathbf{v}_0(\mathbf{r}_0) \tag{41}$$

This quantity has a real physical meaning as the ac change of velocity of a *particular* group of charges at a *given* instant of time caused by a small-signal excitation.

The Eulerian velocity (33) and the polarization velocity (41) taken at the same space point are related to each other by (Bobroff, 1959; Barybin, 1970a)

$$\mathbf{u}_1 = \mathbf{v}_1 - (\mathbf{r}_1 \cdot \nabla)\mathbf{v}_0 \tag{42}$$

The processes of diffusion, collision, and recombination in solid-state plasmas complicate the polarization description and do not permit the direct use of the corresponding results for electron beams. At first glance, these processes might seem inconsistent with the polarization description, because this description involves the examination of two positions (dc and ac) of the same group of charges, while the above elementary processes can prevent charges from collecting into one volume element. However, this reasoning is not true, as we consider the two positions (dc and ac) of a *particular* group of charges at the *same* instant of time. The sole difference between properties of the charges in these two positions is related only to the absence or presence of an ac excitation. We consider the ac excitation so small that it does not change the character of elementary processes in plasmas. In this case the polarization description of solid-state plasmas becomes completely correct.

The role of collisions and diffusion is that they further an additional change of form and dimensions of a volume element caused by the presence of an ac excitation. This has an effect on the Eulerian variables—the charge density ρ_1 and the current density $\mathbf{j}_1$ (Barybin, 1970a). These Eulerian variables, appearing in Maxwell's equations as the sources of electromagnetic fields, are related to the polarization vector $\mathbf{p}_1 = \rho_0 \mathbf{r}_1$ by (Bobroff, 1959; Barybin, 1967b, 1970a)

$$\rho_1 = -\nabla \cdot \mathbf{p}_1 \tag{43}$$

$$\mathbf{j}_1 = (\partial \mathbf{p}_1/\partial t) + \nabla \times (\mathbf{p}_1 \times \mathbf{v}_0) \tag{44}$$

As shown by Barybin (1970a), Eqs. (43) and (44) are valid for charge carrier streams in both vacuum and nondegenerate solid-state plasmas.

The equation of motion in the polarization variables has the form, taking into account collisions and diffusion (Barybin, 1970a),

$$\frac{\partial \mathbf{v}_1}{\partial t} + (\mathbf{v}_0 \cdot \nabla)\mathbf{v}_1 = \frac{q}{m}[\mathbf{E}_1 + (\mathbf{r}_1 \cdot \nabla)\mathbf{E}_0 + \mathbf{v}_1 \times \mathbf{B}_0 + \mathbf{v}_0 \times \mathbf{B}_1$$
$$+ \mathbf{v}_0 \times (\mathbf{r}_1 \cdot \nabla)\mathbf{B}_0] + \frac{v_T^2}{\rho_0}[\rho_0 \nabla(\nabla \cdot \mathbf{r}_1) + \nabla \mathbf{r}_1 \cdot \nabla \rho_0]$$
$$- \frac{\mathbf{v}_1}{\tau_0} + \frac{\mathbf{v}_0 \tau_1 + (\mathbf{r}_1 \cdot \nabla)\tau_0}{\tau_0} \tag{45}$$

The equivalent of the Eulerian continuity equation (35) in the polarization description is the relation between $\mathbf{v}_1$ and $\mathbf{r}_1$ in the form (Bobroff, 1959; Barybin, 1970a)

$$\mathbf{v}_1 = (\partial \mathbf{r}_1/\partial t) + (\mathbf{v}_0 \cdot \nabla)\mathbf{r}_1 \tag{46}$$

Equation (45) takes into consideration a space nonuniformity of dc values $\mathbf{E}_0$, $\mathbf{B}_0$, ρ_0, and τ_0 caused, for example, by nonuniform doping of a semiconductor medium. It is not difficult to check that Eqs. (44) and (45) can be obtained from the corresponding Eulerian equations (37) and (36) by means of the substitution of (42) and (43) into the latter.

Equations (42)–(45) give a connection in each space point between the Eulerian (ρ_1, $\mathbf{u}_1$, $\mathbf{j}_1$, $\mathbf{E}_1$, $\mathbf{B}_1$, τ_1) and polarization ($\mathbf{r}_1$, $\mathbf{v}_1$) variables. The values of these variables and their spatial derivatives are taken at the same point but correspond to different states of a carrier stream: the Eulerian values correspond to the perturbed state (in the presence of an ac signal), while the polarization values correspond to the unperturbed state (in the absence of an ac signal).

It should be noted that Maxwell's equations (1)–(4) contain as field sources the Eulerian variables ρ_1 and $\mathbf{j}_1$. Maxwell's equations with sources written in the form (43) and (44) represent the equations of electrodynamics of moving media in the Chu formulation (Fano et al., 1960; Penfield and Haus, 1967).

B. Small-Signal Power Theorem

In electrodynamics, Poynting's theorem is a useful tool for the study of power flow and energy balance relations (Vainshtein, 1957; Johnson, 1965; Ramo and Whinnery, 1965). In microwave vacuum electronics, the generalization of Poynting's theorem is the so-called kinetic power theorem, which was derived first by Chu (1951) for longitudinal oscillations in electron beams and generalized later by Bobroff et al. (1962) for the case of both longitudinal and transverse oscillations. This theorem was very useful for the understanding of wave interaction mechanisms and the analysis of electron beam noise properties (Haus, 1959; Lopukhin, 1961). The necessity of this theorem manifests itself especially in the energy consideration of coupled modes (Louisell, 1960), giving a clear physical picture of wave interactions from the point of view of power exchange between interacting modes.

The derivation of the small-signal power theorem using the Eulerian variables proved to be possible only for longitudinal electron oscillations (Chu, 1951) and for so-called irrotational electron beams (with the zero curl of the total generalized momentum) (Klüver, 1958). This is caused by the fact that the Eulerian variables, "working" with the fixed E element, cannot take into account a kinetic energy of transverse motion of electrons related, in particular, to their rotation in a magnetic field. Consideration of transverse electron motion in the Eulerian variables was done by Bobroff (1964) with the help of relation (44), expressing the Eulerian current density $\mathbf{j}_1$ via the polarization vector $\mathbf{p}_1$, which is in essence outside the pure Eulerian description. A rigorous and successive proof of the small-signal power theorem is possible only in the polarization variables, and was performed for electron beams by Bobroff et al. (1962).

Recently, different kinds of power approaches have been applied to study nondegenerate solid-state plasmas (Hasegawa, 1965; Shin-ichi-Akai, 1966; Barybin, 1967a; Vural and Bloom, 1967; Wessel-Berg, 1967; Ishii and Inuishi, 1968; Fujisawa, 1968a,b, 1970; Askne and Nilsson, 1969; Steele and Vural, 1969; Askne and Lind, 1970; Barybin, 1970b; Berger *et al.*, 1970; Meyer and Van Duzer, 1970; Lind and Askne, 1972; Freire, 1973; Askne, 1974). Some authors (Hasegawa, 1965; Shin-ichi-Akai, 1966; Ishii and Inuishi, 1968; Goodrich *et al.*, 1974) have transferred, with no arguments, the basic results obtained for vacuum electron beams, including the polarization variables, to the case of semiconductor plasmas with collisions and diffusion, thus causing errors and imprecision. Other authors (Shin-ichi-Akai, 1966; Barybin, 1967a; Vural and Bloom, 1967; Fujisawa, 1968a,b, 1970; Steele and Vural, 1969; Freire, 1973) have used the Eulerian equations with collision and diffusion phenomenological parameters to derive the small-signal power theorem for a one-dimensional approximation. A third group of authors (Wessel-Berg, 1967; Askne and Nilsson, 1969; Berger *et al.*, 1970; Askne and Lind, 1970; Meyer and Van Duzer, 1970; Lind and Askne, 1972; Askne, 1974) has developed their own energy approaches, giving up the kinetic power theorem.

The successive derivation of the small-signal power theorem for nondegenerate solid-state plasmas was carried out by the author (Barybin, 1970b) on the basis of the generalized polarization description (Barybin, 1970a). The model chosen was the hydrodynamic model of nondegenerate plasmas in a hypothetic semiconductive medium simultaneously possessing dielectric, magnetic, piezoelectric, and elastic properties. The initial system of equations describing wave processes in such a medium includes Eqs. (1), (2), (6), (9)–(12), (14), and (44)–(46). The small-signal power theorem derived for instantaneous values of physical quantities (Barybin, 1970b) expresses the small-signal power–energy conservation law. But in such a form it does not allow one to give an unambiguous interpretation of separate terms in the power theorem (Bobroff *et al.*, 1962; Barybin, 1970b). This is a fundamental characteristic of all conservation theorems written in terms of instantaneous values of physical quantities (Landau and Lifshitz, 1959; Bobroff *et al.*, 1962).

In all practical applications of the power theorem the quantities of physical interest are obtained from the corresponding instantaneous values by their time average and integration over the cross section of a system under consideration. After such a procedure the ambiguity of the power theorem disappears, and it assumes the form for monochromatic excitation with a frequency ω (Barybin, 1970b):

$$(dP/dz) + \Pi = 0 \qquad (47)$$

where z is the longitudinal axis of the system. As distinct from the theorem for instantaneous values, the term $\partial W/\partial t$ is not in (47) because in the pure monochromatic excitation case the time-averaged stored energy W does not change with time.

The quantity P in (47) represents the time-averaged power flow through the cross section S of a system (Barybin, 1970b):

$$P = P_{em} + P_{es} + P_{ms} + P_{ek} + P_{sp} + P_{ac} \tag{48}$$

where

$$P_{em} = \tfrac{1}{2} \operatorname{Re} \int_S (\mathbf{E}_{1,c} \times \mathbf{H}_{1,c}^*) \cdot \mathbf{e}_z \, dS \tag{49}$$

is the electromagnetic power flow carried by the curl fields $\mathbf{E}_{1,c}$ and $\mathbf{H}_{1,c}$;

$$P_{es} = \tfrac{1}{2} \operatorname{Re} \int_S \varphi_1 (i\omega \mathbf{D}_1)^* \cdot \mathbf{e}_z \, dS \tag{50}$$

is the power flow carried by the quasi-electrostatic (potential) fields $\mathbf{E}_{1,p} = -\nabla \varphi_1$;

$$P_{ms} = \tfrac{1}{2} \operatorname{Re} \int_S \psi_1 (i\omega \mathbf{B}_1)^* \cdot \mathbf{e}_z \, dS \tag{51}$$

is the power flow carried by the quasi-magnetostatic (potential) fields $\mathbf{H}_{1,p} = -\nabla \psi_1$;

$$P_{ek} = \frac{1}{2} \operatorname{Re} \int_S \left\{ -i\omega \frac{m}{q} \left[(\mathbf{p}_1^* \cdot \mathbf{v}_L) \mathbf{v}_0 - \frac{v_T^2}{\rho_0} (\nabla \cdot \mathbf{p}_1) \mathbf{p}_1^* \right] \right.$$
$$\left. + (\mathbf{p}_1^* \times \mathbf{v}_0) \times \mathbf{E}_{1,c} + \varphi_1 (i\omega \mathbf{p}_1)^* \right\} \cdot \mathbf{e}_z \, dS \tag{52}$$

is the electrokinetic power flow carried by mobile charge carriers as a result of their drift (with the velocity $\mathbf{v}_0$) and thermal motion (with the mean velocity v_T) and also at the expense of their coupling to the curl ($\mathbf{E}_{1,c}$) and potential ($\mathbf{E}_{1,p} = -\nabla \varphi_1$) electric fields [here we have denoted $\mathbf{v}_L = \mathbf{v}_1 - (q/2m)(\mathbf{r}_1 \times \mathbf{B}_0)$];

$$P_{sp} = \tfrac{1}{2} \operatorname{Re} \int_S \{i\omega \mu_0 \alpha_e (\nabla \mathbf{M}_1 \cdot \mathbf{M}_1^*)\} \cdot \mathbf{e}_z \, dS \tag{53}$$

is the power flow carried by spin waves;

$$P_{ac} = \tfrac{1}{2} \operatorname{Re} \int_S (-\mathbf{V}_1 \cdot \mathbf{T}_1^*) \cdot \mathbf{e}_z \, dS \tag{54}$$

is the power flow carried by acoustic waves.

Formulas (48)–(54) give the particular cases obtained previously by other authors:

(a) for electron beams in a vacuum (Bobroff et al., 1962),
$$P = P_{em} + P_{es} + P_{ek}$$
with $v_T \equiv 0$ in (52);

(b) for acoustic waves without charge carriers (Auld and Kino, 1971),
$$P = P_{es} + P_{ac};$$

(c) for longitudinal carrier oscillations in semiconductor plasmas (Barybin, 1967a; Vural and Bloom, 1967; Freire, 1973),
$$P = P_{em} + P_{es} + P_{ek} = P_{em} + (V_k + V_d)I_z$$
where $V_k = (m/q)v_0 v_{1z}$ is the kinetic potential, $V_d = (m/q)(v_T^2/\rho_0)\rho_1$ is the diffusion potential, and $I_z = i\omega p_{1z} S = (\rho_0 v_{1z} + \rho_1 v_0)S$ is the longitudinal current.

The quantity Π in (47) is given by (Barybin, 1970b)
$$\Pi = \frac{1}{2} \operatorname{Re} \int_S \left\{ -\frac{i\omega}{\tau_0} \frac{m}{q} \left[\mathbf{v}_1 - \mathbf{v}_0 \frac{\tau_1 + (\mathbf{r}_1 \cdot \mathbf{V})\tau_0}{\tau_0} \right] \cdot \mathbf{p}_1^* \right\} dS \quad (55)$$
and can be interpreted as the time-averaged loss power per unit length of a system due to the carrier collisions with lattice defects. Using (39) and the equality
$$(\mathbf{r}_1 \cdot \mathbf{V})E_0 = \frac{\mathbf{E}_0}{E_0} \cdot (\mathbf{r}_1 \cdot \mathbf{V})E_0$$

Eq. (55) can be rewritten in the form
$$\Pi = \frac{1}{2} \operatorname{Re} \int_S \left\{ -\frac{i\omega \mathbf{p}_1^*}{\mu_e} \cdot \left[\mathbf{v}_1 - \mathbf{v}_0 \left(1 - \frac{\mu_d}{\mu_e}\right) \frac{\mathbf{E}_0 \cdot (\mathbf{E}_1 + \mathbf{r}_1 \cdot \mathbf{V}\mathbf{E}_0)}{E_0^2} \right] \right\} dS \quad (56)$$
where we have introduced the notations $\mu_e = (q/m)\tau_0 = v(E_0)/E_0$ as the static mobility and $\mu_d = (dv/dE)_{E_0}$ as the differential mobility both determined by the form of the velocity field characteristic $v(E)$ of the semiconductor material.

Equation (47) shows that in a lossless system (for example, electron beams in a vacuum, wherein $\Pi = 0$) the total power flow is constant along the z axis. In this case a spatial change of one of the components (49)–(54) causes spatial changes of the others. This manifests itself in the form of either a passive power exchange between interacting modes or their active interaction, resulting in an amplification or a generation (convective or absolute instabilities). The latter takes place only in case at least one of the coupled waves carries negative power (Louisell, 1960; Briggs, 1964).

In a lossy system the situation can be substantially different. For the wave traveling along the z axis with a propagation constant $\gamma = \alpha + i\beta$ (the wave factor is $e^{(i\omega t - \gamma z)}$) we have $dP/dz = -2\alpha P$. Then (47) gives the expression for the attenuation constant

$$\alpha = \Pi/2P \tag{57}$$

It is evident from (57) that the wave carrying a positive power along the z axis $(P > 0)$ can have spatial growth in the direction of its propagation $(\alpha < 0)$ provided the energy dissipation is negative $(\Pi < 0)$. To understand the physical reasons for negative dissipation let us apply (56) to the longitudinal carrier oscillations of semiconductor plasmas in the zero diffusion limit $(v_T = 0)$. In this case we have (Barybin, 1971)

$$\rho_0 v_{1z} = (i\omega - \gamma v_0) p_{1z}, \quad j_{1z} = i\omega p_{1z} = \sigma_d E_{1z} - \gamma v_0 p_{1z} \tag{58}$$

Substituting (58) into (56), one can obtain

$$\Pi = \frac{\omega(\omega - \beta v_0)}{\sigma_d} |P_{1z}|^2 S \tag{59}$$

with $\sigma_d = \rho_0 \mu_d$. As seen from (59), negative dissipation arises in the two following cases:

(a) for a positive-differential conductivity medium $(\sigma_d > 0)$ if $\beta v_0 > \omega$ or $v_0 > v_{ph}$;

(b) for a negative-differential conductivity medium $(\sigma_d < 0)$ if $\beta v_0 < \omega$ or $v_0 < v_{ph}$.

The first case is realized in an acoustic amplifier wherein, as is known, amplification occurs when the carrier drift velocity exceeds the sound velocity. The acoustic-wave amplification mechanism has a substantially dissipative character in which losses play the principle role (Barybin, 1974).

The second case takes place in the case of a carrier (or space charge) wave amplification in the NDM medium. As known (Blötekjaer and Quate, 1964), the phase velocity of these waves always exceeds the carrier drift velocity $(v_{ph} \geq v_0)$. Then according to (59) we obtain $\Pi < 0$ with $\sigma_d < 0$. This mechanism is the basis of the operation of a thin-film semiconductor TWA and will be considered in Section V,C,3.

C. *Reciprocity Theorem and Mode Orthogonality and Normalization Relations*

In electrodynamics, the reciprocity theorem serves as a basis for the derivation of the normal-mode orthogonality relations and the excitation equations of normal modes by external sources (Vainshtein, 1957; Collin, 1960; Ramo and Whinnery, 1965). The acoustic counterpart of the reciprocity theorem was first obtained by Auld (1969). A generalization of this theorem for the case of nondegenerate plasmas in a hypotherical semiconductor possessing simultaneously piezodielectric, magnetic, and elastic properties was carried out by the author (Barybin, 1975b).

For the derivation of the reciprocity theorem one usually considers, besides the *original* problem (solutions of which carry the subscript 1), a *subsidiary* problem, which satisfies the same equations but with different external sources. In the complex form of the reciprocity theorem the complex-conjugate solutions of a subsidiary problem (with the subscript 2, and also having a small-signal meaning, as the subscript 1) are used. This procedure takes place in the case of lossless systems.

In our case the system can have losses due to collisions of mobile charge carriers with crystal defects in the semiconductor. In order to use the complex reciprocity theorem for lossy systems, we have to apply an artificial method: we must change the sign of the loss parameters [in the given case, of τ_0 in Eq. (45) for carrier motion] in the equations of the subsidiary problem and take their complex conjugates. Such a subsidiary problem with the sign of the loss parameter changed in the absence of external sources is referred to as the *associated* problem.

As was shown previously (Barybin, 1975b) the normal-mode spectra of the original and associated problems coincide in the following sense. Each normal n mode of the original problem with the propagation constant $\gamma_n = \alpha_n + i\beta_n$ corresponds to a normal $\tilde{n}$ mode of the associated problem (indicated by a tilde) with propagation constant $\gamma_{\tilde{n}} = \alpha_{\tilde{n}} + i\beta_{\tilde{n}}$ such that $\alpha_{\tilde{n}} = -\alpha_n$ and $\beta_{\tilde{n}} = \beta_n$, i.e., $\gamma_{\tilde{n}}^* = -\gamma_n$.

Consequently, attenuated modes of the original problem are accompanied by growing modes of the associated problem. This spatial growth has no physical meaning and is a consequence of the artificial change of sign of the loss parameter τ_0. If for the n mode the distribution of physical quantities over a cross section of a system is described by the *original eigenfunction* $\hat{A}_n(x, y)$, then for the $\tilde{n}$ mode of the associated problem there is a corresponding *complex-associated eigenfunction* $\hat{A}_{\tilde{n}}^*(x, y)$ obtained from $\hat{A}_n(x, y)$ by the conventional complex-conjugate operation (indicated by an asterisk) with simultaneous change of sign of the loss parameter (indicated by a tilde). The complex-associated eigenfunctions $\hat{A}_{\tilde{n}}^*(x, y)$ introduced in such a way will be helpful later to obtain the mode orthogonality relations for lossy systems. It is easy to see that in the absence of losses the complex-associated eigenfunctions $\hat{A}_{\tilde{n}}^*$ coincide with the corresponding complex-conjugate eigenfunctions $\hat{A}_n^*$. Here and below, the caret denotes a transverse distribution of physical quantities over a cross section.

It should be noted that for obtaining the eigenvalues γ_n^* and eigenfunctions $\hat{A}_n^*(x, y)$ it is not necessary to solve again the boundary-value complex-associated problem. For this purpose one should apply the above procedure to the corresponding eigenvalues and eigenfunctions.

The generalized reciprocity theorem is derived on the basis of Eqs. (9)–(13), (22), (23), and (43)–(46) and corresponding complex-conjugate equations of a subsidiary problem (indicated by the subscript 2). In the final form, the reciprocity theorem appears as (Barybin, 1975b)

$$\nabla \cdot \mathbf{G}_{12} = F_{12}^{(b)} \tag{60}$$

where

$$\begin{aligned}
\mathbf{G}_{12} = {} & [\mathbf{E}_{1,c} \times (\mathbf{H}_{2,c} - \mathbf{p}_2 \times \mathbf{v}_0)^* + \mathbf{E}_{2,c}^* \times (\mathbf{H}_{1,c} - \mathbf{p}_1 \times \mathbf{v}_0)] \\
& + [\varphi_1(i\omega\mathbf{D}_2 + i\omega\mathbf{p}_2)^* + \varphi_2^*(i\omega\mathbf{D}_1 + i\omega\mathbf{p}_1)] \\
& + [\psi_1(i\omega\mathbf{B}_2)^* + \psi_2^*(i\omega\mathbf{B}_1)] \\
& + \frac{m}{q}\mathbf{v}_0[(i\omega\mathbf{p}_1) \cdot \mathbf{v}_{L,2}^* + (i\omega\mathbf{p}_2)^* \cdot \mathbf{v}_{L,1}] \\
& - \frac{m\, v_T^2}{q\, \rho_0}[(i\omega\mathbf{p}_1)(\nabla \cdot \mathbf{p}_2^*) + (i\omega\mathbf{p}_2)^*(\nabla \cdot \mathbf{p}_1)] \\
& - \frac{i}{\omega\tau_0}\frac{m\,\mathbf{v}_0}{q\,\rho_0}(i\omega\mathbf{p}_1) \cdot (i\omega\mathbf{p}_2)^* \\
& - \alpha_c[\nabla\mathbf{M}_1 \cdot (i\omega\mu_0\,\mathbf{M}_2)^* + \nabla\mathbf{M}_2^* \cdot (i\omega\mu_0\,\mathbf{M}_1)] \\
& - (\mathbf{V}_1 \cdot \mathbf{T}_2^* + \mathbf{V}_2^* \cdot \mathbf{T}_1) \tag{61}
\end{aligned}$$

$$\begin{aligned}
F_{12}^{(b)} = {} & -(\mathbf{j}_{b,1}^e \cdot \mathbf{E}_{2,c}^* + \mathbf{j}_{b,2}^{e*} \cdot \mathbf{E}_{1,c}) \\
& - (\mathbf{j}_{b,1}^m \cdot \mathbf{H}_{2,c}^* + \mathbf{j}_{b,2}^{m*} \cdot \mathbf{H}_{1,c}) \\
& + [(i\omega\rho_{b,1}^e)\varphi_2^* + (i\omega\rho_{b,2}^e)^*\varphi_1] \\
& + [(i\omega\rho_{b,1}^m)\psi_2^* + (i\omega\rho_{b,2}^m)^*\psi_1] \tag{62}
\end{aligned}$$

and the notation

$$\mathbf{v}_{L,1(2)} = \mathbf{v}_{1(2)} - \frac{q}{2m}(\mathbf{r}_{1(2)} \times \mathbf{B}_0)$$

has been used.

The different particular cases can be obtained from (61) and (62) by assuming the following simplifications:

1. electromagnetic waveguide (with no active polarized media):

$$\mathbf{p}_1 = \mathbf{p}_2 \equiv 0, \quad \mathbf{M}_1 = \mathbf{M}_2 \equiv 0, \quad \mathbf{V}_1 = \mathbf{V}_2 \equiv 0,$$
$$\varphi_1 = \varphi_2 \equiv 0, \quad \psi_1 = \psi_2 \equiv 0 \tag{63}$$

2. ferromagnetic medium (with no conductivity and elasticity):

$$\mathbf{p}_1 = \mathbf{p}_2 \equiv 0, \qquad \mathbf{V}_1 = \mathbf{V}_2 \equiv 0, \qquad \varphi_1 = \varphi_2 \equiv 0 \tag{64}$$

3. elastic piezodielectric medium (with no magnetic properties):

$$\mathbf{p}_1 = \mathbf{p}_2 \equiv 0, \qquad \mathbf{M}_1 = \mathbf{M}_2 \equiv 0, \qquad \psi_1 = \psi_2 \equiv 0 \tag{65}$$

4. charge carrier stream in a semiconductive medium (with no elasticity and magnetic properties):

$$\mathbf{V}_1 = \mathbf{V}_2 \equiv 0, \qquad \mathbf{M}_1 = \mathbf{M}_2 \equiv 0, \qquad \psi_1 = \psi_2 \equiv 0 \tag{66}$$

Let us apply the reciprocity theorem (60) to a generalized waveguide in the form of the finite polarized medium surrounded by a grounded electric wall (Fig. 1). The normal modes of the original problem are assumed to be

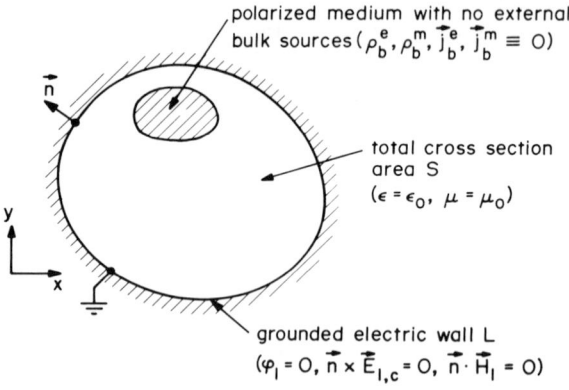

FIG. 1. A schematic illustration of the cross section of a polarized-medium waveguide with no external sources. From Barybin (1975b).

known. In other words, we know the sets of γ_n and $\hat{A}_n(x, y)$ $(n = 1, 2, \ldots)$ obtained as a result of solving the original boundary value problem with no external sources. According to the above, we thus know the set of the complex-associated eigenfunctions $\hat{A}_n^*(x, y)$ that correspond to the propagation constants $\gamma_n^* = -\gamma_n$.

In the absence of external sources the right-hand side of (60) is equal to zero. Thus after integration of (60) over the total cross-section area S of the polarized-medium waveguide (Fig. 1) we obtain (Barybin, 1975b)

$$\frac{\partial}{\partial z} \int_S \mathbf{G}_{12} \cdot \mathbf{e}_z \, dS = 0 \tag{67}$$

Let the solution marked by the subscript 1 be the m mode of the *original* problem and the solution marked by the subscript 2 be the $\tilde{n}$ mode of the *complex-associated* problem. Because all the physical quantities in the m mode have a z dependence of the form $\exp(-\gamma_m z)$ and those in the $\tilde{n}$ mode have the z dependence $\exp(-\gamma_{\tilde{n}}^* z)$, Eqs. (61) and (67) yield

$$(\gamma_m + \gamma_{\tilde{n}}^*) \mathcal{N}_{m\tilde{n}} \exp[-(\gamma_m + \gamma_{\tilde{n}}^*)z] = 0 \tag{68}$$

where

$$\mathcal{N}_{m\tilde{n}} = \int_S \hat{\mathbf{G}}_{m\tilde{n}} \cdot \mathbf{e}_z \, dS \tag{69}$$

with $\hat{\mathbf{G}}_{m\tilde{n}}$ obtained from (61) by substituting the subscript m for 1 and the subscript $\tilde{n}$ for 2 and by writing the caret above all the ac values of physical quantities (the latter means the absence of a dependence on the longitudinal coordinate z).

Equation (68) gives, on the one hand, the generalized mode orthogonality relation when $m \neq n$:

$$\mathcal{N}_{m\tilde{n}} = \int_S \hat{\mathbf{G}}_{m\tilde{n}} \cdot \mathbf{e}_z \, dS = 0, \quad \text{with} \quad \gamma_m \neq -\gamma_{\tilde{n}}^* \tag{70}$$

and, on the other hand, the normalization condition of the n mode when $m = n$:

$$\mathcal{N}_n = \mathcal{N}_{n\tilde{n}} = \int_S \hat{\mathbf{G}}_{n\tilde{n}} \cdot \mathbf{e}_z \, dS \neq 0, \quad \text{with} \quad \gamma_n = -\gamma_{\tilde{n}}^* \tag{71}$$

The mode orthogonality relation (70) and expression (71) for the norm N_n of the n mode of the original problem in view of (61) are the generalizations of the conventional expressions for lossless system (Vainshtein, 1957; Johnson, 1965; Auld and Kino, 1971) to the case of lossy systems.

In a lossless system the given n mode is orthogonal to all the other modes ($m \neq n$) obtained by solving the original problem. For this case the norms of the n mode are simplified to (Barybin, 1975b)

$$\mathcal{N}_n = 2 \operatorname{Re} \int_S \left\{ [\hat{\mathbf{E}}_{c,n} \times (\hat{\mathbf{H}}_{c,n} - \hat{\mathbf{p}}_n \times \mathbf{v}_0)^*] + \hat{\varphi}_n (i\omega \hat{\mathbf{D}}_n + i\omega \hat{\mathbf{p}}_n)^* \right.$$

$$+ \hat{\psi}_n (i\omega \hat{\mathbf{B}}_n)^* + i\omega \frac{m}{q} \left[(\hat{\mathbf{p}}_n \cdot \hat{\mathbf{v}}_{L,n}^*) \mathbf{v}_0 - \frac{v_T^2}{\rho_0} \widehat{(\nabla \cdot \mathbf{p}_n^*)} \hat{\mathbf{p}}_n \right]$$

$$+ i\omega \mu_0 \alpha_e \widehat{(\nabla \mathbf{M}_n)} \cdot \hat{\mathbf{M}}_n^* + (-\hat{\mathbf{V}}_n \cdot \hat{\mathbf{T}}_n^*) \bigg\} \cdot \mathbf{e}_z \, dS \tag{72}$$

If one compares (72), written for the particular cases (63)–(66), with the corresponding expressions (49)–(54) for the time-averaged power flow carried by the n mode through the cross section of a generalized waveguide, then one can see that the norm of a mode is equal to four times the power flow carried by this mode, i.e., for a lossless system we have always

$$N_n = 4P_n \tag{73}$$

The presence of losses has required an extension of the concept of mode orthogonality. In a lossy system the given n mode of the original problem is orthogonal to all the modes of the associated problem except the corresponding complex-associated $\tilde{n}$ mode. The eigenfunctions of the n mode and $\tilde{n}$ mode form the norm N_n of the n mode. Then Eq. (73) fails and the norm loses its power meaning. This must be expected because in a lossy system the separate modes cannot carry power independently of each other and, in addition to self-powers, they carry also a cross power, provided there is a simultaneous excitation of a few modes (Wessel-Berg, 1967; Barybin, 1971).

D. *Normal-Mode Excitation by External Sources and Mutual Coupling among Modes*

To derive the normal-mode excitation equations one should consider the polarized-medium waveguide, shown in Fig. 1, which now contains the bulk $(\rho_b^e, \rho_b^m, \vec{j}_b^e, \vec{j}_b^m)$ and surface $(\rho_s^e, \rho_s^m, \vec{j}_s^e, \vec{j}_s^m)$ external sources as depicted in Fig. 2. The surface sources are located along the contour L_S and lead to the discontinuity of electromagnetic field components, according to (24)–(27). The bulk external sources can be located both inside the polarized medium under consideration (for example, other kinds of mobile charge carriers, acoustic or spin waves for a drifting carrier stream in the same semiconductor crystal) and outside it (for example, other crystals adjacent to the given crystal).

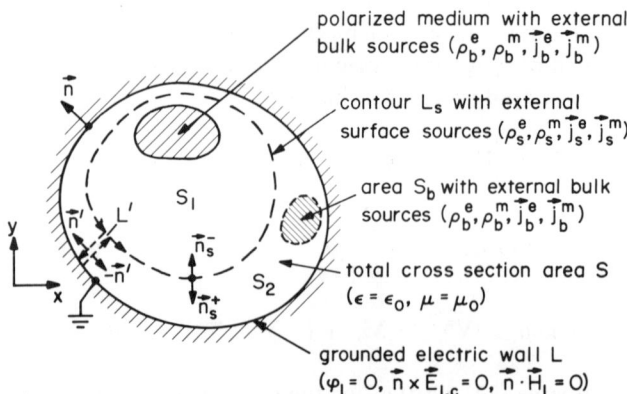

FIG. 2. A schematic illustration of the cross section of a polarized-medium waveguide with external bulk and surface sources. From Barybin (1975b).

Let us again apply the reciprocity theorem (60) to the polarized-medium waveguide with sources shown in Fig. 2. However, at this point we assume that the solution with the subscript 1 corresponds to the *total* solution ($\mathbf{E}_1$, $\mathbf{H}_1$, $\mathbf{p}_1$, $\mathbf{M}_1$, $\mathbf{T}_1$, etc.) of the original problem in the presence of the bulk and surface external sources, while the solution with the subscript 2 means, as before, the $\tilde{n}$ mode of the complex-associated problem with no external sources. The total values of *all* the physical quantities concerning the original problem with external sources can be represented in the form of the expansion over mode eigenfunctions of this problem, for example,

$$\mathbf{E}_{1,c}(x, y, z) = \sum_m c_m(z) \hat{\mathbf{E}}_{c,m}(x, y) e^{-\gamma_m z}$$

$$\mathbf{H}_{1,c}(x, y, z) = \sum_m c_m(z) \hat{\mathbf{H}}_{c,m}(x, y) e^{-\gamma_m z}$$

$$\varphi_1(x, y, z) = \sum_m c_m(z) \hat{\varphi}_m(x, y) e^{-\gamma_m z}$$

$$\psi_1(x, y, z) = \sum_m c_m(z) \hat{\psi}_m(x, y) e^{-\gamma_m z}$$

$$\mathbf{p}_1(x, y, z) = \sum_m c_m(z) \hat{\mathbf{p}}_m(x, y) e^{-\gamma_m z} \tag{74}$$

The coefficients $c_m(z)$ in Eqs. (74) are unknown mode amplitudes associated with each of the eigenfunctions ($\hat{\mathbf{E}}_{c,m}$, $\hat{\mathbf{H}}_{c,m}$, $\hat{\varphi}_m$, $\hat{\mathbf{p}}_m$, etc., $m = 1, 2, \ldots$). The mode amplitudes $c_m(z)$ contain a z dependence due to the driving effect of external sources, while the wave dependence on z is taken into account by the wave factor $\exp(-\gamma_m z)$. It is convenient to combine these two z factors and write a new amplitude of the m mode as

$$a_m(z) = c_m(z) e^{-\gamma_m z}$$

which now contains all dependences of physical quantities on z.

The excitation equations for mode amplitudes are derived by means of the substitution of the series expansions (74) into the reciprocity relation (60), and the subsequent integration over the waveguide cross section S. After using the boundary conditions (24)–(27) and the orthogonality relation (70), the amplitude excitation equation for any n mode assumes the form (Barybin, 1975b)

$$\frac{da_n(z)}{dz} + \gamma_n a_n(z) = \frac{1}{\mathcal{N}_n} \int_S \hat{F}_{1\tilde{n}}^{(b)} dS + \frac{1}{\mathcal{N}_n} \int_{L_S} \hat{F}_{1\tilde{n}}^{(s)} dl \tag{75}$$

The exciting terms $\hat{F}_{1\tilde{n}}^{(b)}$ and $\hat{F}_{1\tilde{n}}^{(s)}$ are related, respectively, to the bulk and surface external sources as (Barybin, 1975b)

$$\hat{F}_{1\tilde{n}}^{(b)} = -(\mathbf{j}_b^e \cdot \hat{\mathbf{E}}_{c,\tilde{n}}^* + \mathbf{j}_b^m \cdot \hat{\mathbf{H}}_{c,\tilde{n}}^*) + (i\omega \rho_b^e \hat{\varphi}_{\tilde{n}}^* + i\omega \rho_b^m \hat{\psi}_{\tilde{n}}^*) \tag{76}$$

$$\hat{F}_{1\tilde{n}}^{(s)} = -(\mathbf{j}_s^e \cdot \hat{\mathbf{E}}_{c,\tilde{n}}^* + \mathbf{j}_s^m \cdot \hat{\mathbf{H}}_{c,\tilde{n}}^*) + (i\omega \rho_s^e \hat{\varphi}_{\tilde{n}}^* + i\omega \rho_s^m \hat{\psi}_{\tilde{n}}^*) \tag{77}$$

The excitation equations (75) with the exciting terms (76) and (77) can be written for each particular case (63)–(66), which is done in Barybin 1975b. In particular, the given excitation equations coincide with those previously obtained for electromagnetic waveguides (Vainshtein, 1957) and acoustic waveguides (Barybin, 1968b; Auld and Kino, 1971).

As seen from (76) and (77), the external currents $\mathbf{j}_{b(s)}^e$ and $\mathbf{j}_{b(s)}^m$ excite a system coupled to them, as a result of their interaction with the curl electromagnetic fields $\mathbf{E}_{c,\hat{n}}$ and $\mathbf{H}_{c,\hat{n}}$ supported by the system. The external charges $\rho_{b(s)}^e$ and $\rho_{b(s)}^m$ provide the system excitation via their interaction with the potential fields $\mathbf{E}_{p,\hat{n}} = -\nabla\varphi_{\hat{n}}$ and $\mathbf{H}_{p,\hat{n}} = -\nabla\psi_{\hat{n}}$ supported by the system. Any electromagnetic structure supporting only the curl fields, for example, SWS, with respect to external systems (carrier stream in a semiconductor film, ferromagnetic film, etc.), can be uniquely represented with the help of the charge and current distributions on metallic surfaces of the structure. These charges and currents are the exciting sources for systems external with respect to SWS. Therefore, when SWS are coupled to the potential fields of an external system, the exciting sources for the system are surface charges on metallic parts of SWS. If SWS are coupled to an external system through the curl electromagnetic fields of the system, then surface currents on metallic parts of SWS are the exciting sources for the system.

The excitation equations thus obtained allow one to extend the general boundary value problem to coupled-mode analysis. For this purpose we divide the complicated system under consideration into two (or more) subsystems and investigate their normal modes in the absence of coupling between them. In the presence of coupling these subsystems become mutually external with respect to each other, and their bulk and surface charges and currents determine the external sources in the excitation equations of another subsystem.

We consider the case of interaction between a polarized medium (ferromagnetic or elastic piezodielectric medium or carrier stream in a semiconductor) and an electromagnetic waveguiding metallic structure (for example, SWS), which are two subsystems of the total system schematically shown in Fig. 3. The normal-mode problems for the electromagnetic structure and the polarized-medium waveguide in the absence of coupling between them (i.e., for regions I and II in Fig. 3) are assumed to be solved. In other words, we know sets of propagation constants and eigenfunctions for the electromagnetic waveguide supporting only the curl fields ($\gamma_w = i\beta_w$, $\hat{\mathbf{E}}_{c,w}$ and $\hat{\mathbf{H}}_{c,w}$, denoted by the subscript $w = 1, 2, \ldots$) and for the polarized medium waveguide ($\gamma_m = \alpha_m + i\beta_m$, $\hat{\mathbf{E}}_{c,m}$, $\hat{\mathbf{H}}_{c,m}$, $\hat{\varphi}_m$, $\hat{\psi}_m$, $\hat{\mathbf{p}}_m$, etc., denoted by the subscript $m = 1, 2, \ldots$). In the interaction region (see Fig. 3) all the physical quantities including the bulk and surface charges and currents of each subsystem are represented in the form of linear combinations of known eigenfunctions with unknown amplitude factors that should be determined.

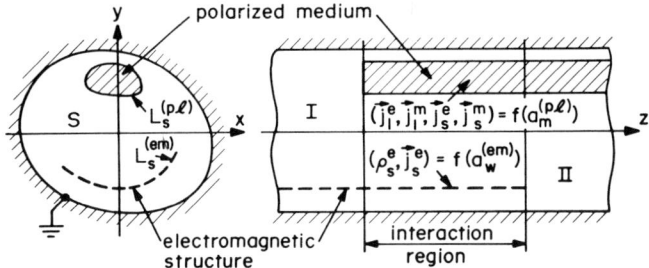

FIG. 3. The coupling of a polarized medium with an electromagnetic waveguide structure. From Barybin (1975b)

For the electromagnetic structure we have

$$\mathbf{E}_{1,c}^{(em)} = \sum_w a_w^{(em)}(z)\hat{\mathbf{E}}_{c,w}(x, y)$$

$$\mathbf{H}_{1,c}^{(em)} = \sum_w a_w^{(em)}(z)\hat{\mathbf{H}}_{c,w}(x, y) \qquad (78)$$

The surface electric charge $\rho_s^{e(em)}$ and electric current $\mathbf{j}_s^{e(em)}$ densities on metallic surfaces, defined according to (25) and (26) as the discontinuities of corresponding components of $\mathbf{E}_{1,c}^{(em)}$ and $\mathbf{H}_{1,c}^{(em)}$, can be uniquely expressed in terms of the amplitudes $a_w^{(em)}(z)$ with the help of (78):

$$\rho_s^{e(em)} = \sum_w a_w^{(em)}(z)\hat{\rho}_{s,w}^e(x, y) \qquad (79)$$

$$\mathbf{j}_s^{e(em)} = \sum_w a_w^{(em)}(z)\hat{\mathbf{j}}_{s,w}^e(x, y) \qquad (80)$$

where the values of $\hat{\rho}_{s,w}^e(x, y)$ and $\hat{\mathbf{j}}_{s,w}^e(x, y)$ are taken at the points of the contour $L_s^{(em)}$ of the electromagnetic structure (see Fig. 3). Equations (79) and (80) give the exciting sources for the polarized medium.

For the polarized-medium waveguide, in turn we have

$$\mathbf{E}_1^{(pl)} = \sum_m a_m^{(pl)}(z)\hat{\mathbf{E}}_{c,m}(x, y)$$

$$\mathbf{H}_{1,c}^{(pl)} = \sum_m a_m^{(pl)}(z)\hat{\mathbf{H}}_{c,m}(x, y)$$

$$\varphi_1^{(pl)} = \sum_m a_m^{(pl)}(z)\hat{\varphi}_m(x, y)$$

$$\psi_1^{(pl)} = \sum_m a_m^{(pl)}(z)\hat{\psi}_m(x, y)$$

$$\mathbf{p}_1^{(pl)} = \sum_m a_m^{(pl)}(z)\hat{\mathbf{p}}_m(x, y) \qquad (81)$$

etc. All the bulk ($\mathbf{j}_1^{e(pl)}$, $\mathbf{j}_1^{m(pl)}$) and surface ($\mathbf{j}_s^{e(pl)}$, $\mathbf{j}_s^{m(pl)}$) current densities of the polarized medium, exciting the electromagnetic waveguide (charges in the polarized medium do not excite the SWS because the latter does not support potential fields), can be uniquely expressed in terms of the amplitudes $a_m^{(pl)}(z)$ with the help of (81):

$$\mathbf{j}_1^{e(pl)} = \sum_m a_m^{(pl)}(z) \hat{\mathbf{j}}_{1,m}^e(x, y) \tag{82}$$

$$\mathbf{j}_1^{m(pl)} = \sum_m a_m^{(pl)}(z) \hat{\mathbf{j}}_{1,m}^m(x, y) \tag{83}$$

$$\mathbf{j}_s^{e(pl)} = \sum_m a_m^{(pl)}(z) \hat{\mathbf{j}}_{s,m}^e(x, y) \tag{84}$$

$$\mathbf{j}_s^{m(pl)} = \sum_m a_m^{(pl)}(z) \hat{\mathbf{j}}_{s,m}^m(x, y) \tag{85}$$

where the values $\hat{\mathbf{j}}_{s,m}^e(x, y)$ and $\hat{\mathbf{j}}_{s,m}^m(x, y)$ are taken at the points of the countour $L_s^{(pl)}$ of the polarized-medium boundary (see Fig. 3).

The polarized medium as an external subsystem for the electromagnetic waveguide is completely equivalent to the bulk and surface currents (82)–(85), which are linear functions of the mode amplitudes $a_m^{(pl)}$. These currents play the role of external sources appearing in the excitation equations (75)–(77) for the SWS, i.e.,

$$\mathbf{j}_b^e \equiv \mathbf{j}_1^{e(pl)}, \qquad \mathbf{j}_b^m \equiv \mathbf{j}_1^{m(pl)}, \qquad \mathbf{j}_s^e \equiv \mathbf{j}_s^{e(pl)}, \qquad \mathbf{j}_s^m \equiv \mathbf{j}_s^{m(pl)} \tag{86}$$

Then the excitation equations of the electromagnetic waveguide by the polarized medium assume the form

$$(da_w^{(em)}(z)/dz) + i\beta_w a_w^{(em)}(z) = \sum_m C_{wm} a_m^{(pl)}(z), \qquad w = 1, 2, \dots \tag{87}$$

where the coupling coefficients C_{wm}, by virtue of (75)–(77) and (82)–(86), are

$$C_{wm} = -\frac{1}{\mathcal{N}_w^{(em)}} \int_S (\hat{\mathbf{j}}_{1,m}^e \cdot \hat{\mathbf{E}}_{c,w}^* + \hat{\mathbf{j}}_{1,m}^m \cdot \hat{\mathbf{H}}_{c,w}^*) \, dS$$

$$- \frac{1}{\mathcal{N}_w^{(em)}} \int_{L_S^{(pl)}} (\hat{\mathbf{j}}_{s,m}^e \cdot \hat{\mathbf{E}}_{c,w}^* + \hat{\mathbf{j}}_{s,m}^m \cdot \hat{\mathbf{H}}_{c,w}^*) \, dl \tag{88}$$

with the norm of the w mode of the electromagnetic waveguide $\mathcal{N}_w^{(em)}$.

On the other hand, the electromagnetic waveguide structure as an external subsystem for the polarized-medium waveguide is completely equivalent to the surface electric charge (79) and current (80), which are linear functions of the mode amplitudes $a_w^{(em)}$. These charges and currents play the role of the external surface sources in the excitation equations (75)–(77) for the polarized medium, i.e.,

$$\rho_s^e \equiv \rho_s^{e(em)}, \qquad \mathbf{j}_s^e \equiv \mathbf{j}_s^{e(em)} \tag{89}$$

while the surface magnetic sources and all the bulk sources in the case of an electromagnetic waveguide as an external subsystem are equal to zero.

Then the excitation equations of the polarized medium waveguide by SWS assume the form

$$(da_m^{(pl)}(z)/dz) + \gamma_m a_m^{(pl)}(z) = \sum_w C_{mw} a_w^{(em)}(z), \quad m = 1, 2, \ldots \quad (90)$$

where the coupling coefficients C_{mw} by virtue of (75)–(77), (79), (80), and (89), are

$$C_{mw} = -\frac{1}{\mathcal{N}_m^{(pl)}} \int_{L_S^{(pl)}} (\hat{\mathbf{j}}_{s,w}^e \cdot \hat{\mathbf{E}}_{c,\hat{m}}^* - i\omega \hat{\rho}_{s,w}^c \hat{\varphi}_{\hat{m}}^*) \, dl \quad (91)$$

with the norm of the m mode of the polarized-medium waveguide $\mathcal{N}_m^{(pl)}$.

Equations (87) and (90) represent the system of coupled equations. The coupling coefficients (88) and (91) are calculated from the known distributions of eigenfunctions over a waveguide cross section for each of the interacting subsystems. By calculating and evaluating the coupling coefficients and applying a conventional procedure of the coupled-mode theory (Louisell, 1960; Ter-Martirosyan, 1969) one can simplify the analysis by neglecting weakly coupled modes in the first approximation.

The generalized theory of normal-mode excitation in active polarized-medium waveguides by external sources treated in this section allows one to analyze wave interactions in different kinds of thin-film semiconductor structures. Problems concerning wave propagation in thin semiconductor films (such as boundary conditions on carrier stream surfaces or normal-mode spectrum) will be considered in Part II of this review.

References

Askne, J. (1974). *Int. J. Electron.* **36**, 543.
Askne, J., and Lind, B. (1970). *Phys. Rev. A* **2**, 2335.
Askne, J., and Nilsson, O. (1969). *Proc. IEEE* **57**, 1423.
Akhiezer, A. I., Bar'yakhter, V. G., and Peletminskii, S. V. (1968). "Spin Waves." North-Holland Publ., Amsterdam.
Akhiezer, A. I., Akhiezer, I. A., Polovin, R. V., Sitenko, A. G., and Stepanov, K. N. (1974). "Electrodynamics of Plasmas." Nauka, Moscow.
Auld, B. A. (1969). *IEEE Trans. Microwave Theory Tech.* **MTT-17**, 800.
Auld, B. A., and Kino, G. S. (1971). *IEEE Trans. Electron Devices* **ED-18**, 898.
Barybin, A. A. (1967a). *Radio Eng. Electron. Phys. (USSR)* **12**, 1729.
Barybin, A. A. (1967b). *Radio Eng. Electron. Phys. (USSR)* **12**, 2086.
Barybin, A. A. (1968a). *Radio Eng. Electron. Phys. (USSR)* **13**, 1783.

Barybin, A. A. (1968b). *Izv. LETI* **64**, 111.
Barybin, A. A. (1970a). *Radiotekh. Elektron.* **15**, 1556.
Barybin, A. A. (1970b). *Radiotekh. Elektron.* **15**, 2205.
Barybin, A. A. (1971). *Radiotekh. Elektron.* **16**, 897.
Barybin, A. A. (1974). *IEEE Trans. Electron Devices* **ED-21**, 516.
Barybin, A. A. (1975a). *J. Appl. Phys.* **46**, 1684.
Barybin, A. A. (1975b). *J. Appl. Phys.* **46**, 1707.
Barybin, A. A., and Ter-Martirosyan, L. T. (1969). *Radio Eng. Electron. Phys. (USSR)* **14**, 237.
Berger, H., Harrison, R. I., and Denker, S. P. (1970). *IEEE Trans. Microwave Theory Tech.* **MTT-18**, 105.
Bespyatykh, Y. I., Vashkovsky, A. V., and Zubkov, V. I. (1975). *Radiotekh. Elektron.* **20**, 1003.
Bianco, B., Chiabrera, A., and Ridella, S. (1973). *Alta Freq.* **42**, 181.
Blötekjaer, K., and Quate, C. F. (1964). *Proc. IEEE* **52**, 360.
Bobroff, D. L. (1959). *IRE Trans. Electron Devices* **ED-6**, 68.
Bobroff, D. L. (1964). *J. Appl. Phys.* **35**, 3044.
Bobroff, D. L., Haus, H. A., and Klüver, J. W. (1962). *J. Appl. Phys.* **33**, 2932.
Bowers, R., and Steele, M. C. (1964). *Proc. IEEE* **52**, 1105.
Briggs, R. J. (1964). "Electron-Stream Interaction with Plasmas." MIT Press, Cambridge, Massachusetts.
Buchsbaum, S. J. (1965). *In* "Plasma Effects in Solids," p. 3. Dunod, Paris.
Bulman, P. J., Hobson, G. S., and Taylor, B. C. (1972). "Transferred Electron Devices." Academic Press, New York.
Chang, N. S., and Matsuo, Y. (1968). *Proc. IEEE* **56**, 765.
Chang, N. S., and Matsuo, Y. (1974). *Trans. IECE Jpn.* **57-B**, 474.
Chang, N. S., and Matsuo, Y. (1975). *Trans. IECE Jpn.* **58-B**, 315.
Chodorow, M., and Susskind, C. (1964). "Fundamentals of Microwave Electronics." McGraw-Hill, New York.
Chu, L. J. (1951). *IRE Conf. Electron. Devices, Durham, New Hampshire* p. 83.
Chynoweth, A. G., and Buchsbaum, S. J. (1965). *Phys. Today* **11**, 26.
Collin, R. E. (1960). "Field Theory of Guided Waves." McGraw-Hill, New York.
Collins, J. H., Lakin, K. M., Quate, C. F., and Shaw, H. J. (1968). *Appl. Phys. Lett.* **13**, 314.
Conwell, E. M. (1967). "High Field Transport in Semiconductors." Academic Press, New York.
Dean, R. H. (1969). *Proc. IEEE* **57**, 1327.
Dean, R. H. (1972). *IEEE Trans. Electron Devices* **ED-19**, 1148.
Dean, R. H., and Matarese, R. J. (1972). *Proc. IEEE* **60**, 1486.
Dean, R. H., and Robinson, B. B. (1974). *IEEE Trans. Electron Devices* **ED-21**, 61.
Dean, R. H., Dreeben, A. B., Kaminski, J. F., and Triano, A. (1970). *Electron. Lett.* **6**, 775.
Dean, R. H., Dreeben, A. B., Hughes, J. J., Matarese, R. .'., and Napoli, L. S. (1973). *IEEE Trans. Microwave Theory Tech.* **MTT-21**, 805.
Durney, C. H., Christensen, D. A., and Grow, R. W. (1969). *IEEE Trans. Electron Devices* **ED-16**, 609.
Ettenberg, M., and Nadan, J. S. (1968). *Proc. IEEE* **56**, 741.
Ettenberg, M., and Nadan, J. S. (1970). *IEEE Trans. Electron Devices* **ED-17**, 219.
Fano, R. M., Chu, L. J., and Adler, R. B. (1960). "Electromagnetic Fields, Energy and Forces." Wiley, New York.
Fleming, P. L. (1975). *Proc. IEEE* **63**, 1253.
Freeman, J. C., Newhouse, V. L., and Gunshor, R. L. (1973). *Appl. Phys. Lett.* **22**, 641.
Freire, G. F. (1973). *Proc. IEEE* **61**, 1368.
Frey, W., Engelmann, R. W. H., and Bosch, B. G. (1971). *Arch. Elektron. Übertragung.* **25**, 1.
Fujisawa, K. (1968a). *Electron. Commun. Jpn.* **51-C**, 120.

Fujisawa, K. (1968b). *Electron. Commun. Jpn.* **51-C**, 128.
Fujisawa, K. (1970). *Electron. Lett.* **6**, 754.
Fujisawa, K., and Ichikawa, H. (1969). *Electron. Commun. Jpn.* **52-C**, 122.
Glicksman, M. (1965). *In* "Plasma Effects in Solids," p. 149. Dunod, Paris.
Goodrich, L. C., Durney, C. H., and Grow, R. W. (1974). *J. Appl. Phys.* **45**, 357.
Gover, A., Burrell, K. H., and Yariv, A. (1974). *J. Appl. Phys.* **45**, 4847.
Gulyaev, Y. V., and Pustovoit, V. I. (1964). *Zh. Eksp. Teor. Fiz.* **47**, 2251.
Gulyaev, Y. V., Kmita, A. M., Kotelyansky, I. M., Medved', A. V., and Tursunov, S. S. (1971). *Fiz. Tekh. Poluprovodn.* **5**, 80.
Gunshor, R. L. (1975). *Solid-State Electron.* **18**, 1089.
Gurevich, V. L. (1968). *Fiz. Tekh. Poluprovodn.* **2**, 1557.
Hahn, W. C. (1939). *Gen. Elec. Rev.* **42**, 258.
Hammer, J. M. (1967). *Appl. Phys. Lett.* **10**, 358.
Hartnagel, H. L. (1969). "Semiconductor Plasma Instabilities." Amer. Elsevier, New York.
Hasegawa, A. (1965). *J. Phys. Soc. Jpn.* **20**, 1072.
Haus, H. A. (1959). *In* "Noise in Electron Devices" (L. D. Smullin and H. A. Haus, eds.), p. 104. Wiley, New York.
Heffner, H. (1954). *Proc. IRE* **42**, 930.
Hines, M. E. (1969). *IEEE Trans. Electron Devices* **ED-16**, 88.
Hutson, A. R., and White, D. L. (1962). *J. Appl. Phys.* **33**, 40.
Hutson, A. R., McFee, J. H., and White, D. L. (1961). *Phys. Rev. Lett.* **7**, 237.
Ishii, T., and Inuishi, Y. (1968). *J. Phys. Soc. Jpn.* **25**, 1406.
Johnson, C. C. (1965). "Field and Wave Electrodynamics." McGraw-Hill, New York.
Kanbe, H., Kumabe, K., and Nii, R. (1971). *Rev. Electron. Commun. Lab.* **19**, 917.
Kanbe, H., Shimizu, N., and Kumabe, K. (1973). *Electron. Lett.* **9**, 29.
Kaner, E. A., and Skobov, V. G. (1968). *Adv. Phys.* **17**, 605.
Kaner, E. A., and Yakovenko, V. M. (1975). *Usp. Fiz. Nauk* **115**, 41.
Kawasaki, K., Takagi, H., and Umeno, M. (1974). *IEEE Trans. Microwave Theory Tech.* **MTT-22**, 918, 924.
Kino, G. S., and Reeder, T. M. (1971). *IEEE Trans. Electron Devices* **ED-18**, 909.
Klüver, J. W. (1958). *J. Appl. Phys.* **29**, 618.
Koyama, J., Ohara, S., Kawazura, K., and Kumabe, K. (1969). *Rev. Electron. Commun. Lab.* **17**, 1102.
Kumabe, K., and Kanbe, H. (1970). *Rev. Electron. Commun. Lab.* **18**, 913.
Lakin, K. M., Collins, J. H., and Hagon, P. J. (1969). *Proc. IEEE* **57**, 740.
Landau, L. D., and Lifshitz, E. M. (1959). "The Classical Theory of Fields." Addison-Wesley, Reading, Massachusetts.
Lefeuvre, S., and Hanna, V. F. (1973). *Int. J. Electron.* **35**, 145, 163.
Levinshtein, M. E., Pozhela, Y. K., and Shur, M. S. (1975). "Gunn Effect." Sov. Radio, Moscow.
Lind, B. I., and Askne, J. I. H. (1972). *IEEE Trans. Electron Devices* **ED-19**, 239.
Lopukhin, B. M. (1961). *Radiotekh. Elektron.* **6**, 683.
Louisell, W. H. (1960). "Coupled Mode and Parametric Electronics." Wiley, New York.
Lukomskii, V. P., and Tsvirko, Y. A. (1973). *Fiz. Tverd. Tela (Leningrad)* **15**, 700.
McCumber, D. E., and Chynoweth, A. G. (1966). *IEEE Trans. Electron Devices* **ED-13**, 4.
McFee, J. H. (1966). *In* "Physical Acoustics" (W. P. Mason, ed.), Vol. 4, Part A, p. 3. Academic Press, New York.
Marcuvitz, N., and Schwinger, J. (1951). *J. Appl. Phys.* **22**, 808.
Masuda, M., Chang, N. S., and Matsuo, Y. (1974). *IEEE Trans. Microwave Theory Tech.* **MTT-22**, 132.
Meyer, M., and Van Duzer, T. (1970). *IEEE Trans. Electron Devices* **ED-17**, 193.

Narayan, S. V., and Sterzer, F. (1970). *IEEE Trans. Microwave Theory Tech.* **MTT-18**, 773.
Penfield, P., Jr., and Haus, H. A. (1967). "Electrodynamics of Moving Media." MIT Press, Cambridge, Massachusetts.
Perlman, B. S., Upadhyayula, C. L., and Marx, R. E. (1970). *IEEE Trans. Microwave Theory Tech.* **MTT-18**, 911.
Perlman, B. S., Upadhyayula, C. L., and Siekanowicz, W. W. (1971). *Proc. IEEE* **59**, 1229.
Peter, R. (1956). U.S. Patent 2,760,013.
Pierce, J. R. (1950). "Travelling Wave Tubes." Van Nostrand, Princeton, New Jersey.
Pierce, J. R., and Suhl, P. (1955). U.S. Patent 2,743,322.
Platzman, P. M., and Wolff, P. A. (1973). *Solid State Phys.* **13**, Suppl.
Pustovoit, V. I. (1969). *Usp. Fiz. Nauk* **97**, 257.
Ramo, S., and Whinnery, J. R. (1965). "Fields and Waves in Communication Electronics." Wiley, New York.
Robinson, B. B., Vural, B., and Parekh, J. P. (1970). *IEEE Trans. Electron Devices* **ED-17**, 224.
Robson, P. N., Kino, G. S., and Fay, B. (1967). *IEEE Trans. Electron Devices* **ED-14**, 612.
Schlömann, E. (1969). *J. Appl. Phys.* **40**, 1422.
Shapiro, R. K. (1972). *Fiz. Tverd. Tela (Leningrad)* **14**, 3209.
Shin-ichi-Akai (1966). *Jpn. J. Appl. Phys.* **5**, 1227.
Solymar, L., and Ash, E. A. (1966). *Int. J. Electron.* **20**, 127.
Spector, H. N. (1966). *Solid State Phys.* **19**, 291.
Steele, M. C., and Vural, B. (1969). "Wave Interactions in Solid State Plasmas." McGraw-Hill, New York.
Sturrock, P. A. (1958). *Phys. Rev.* **112**, 1488.
Sumi, M. (1966). *Appl. Phys. Lett.* **9**, 251.
Sumi, M. (1967). *Jpn. J. Appl. Phys.* **6**, 688.
Sumi, M., and Suzuki, T. (1968). *Appl. Phys. Lett.* **13**, 326.
Swanenburg, T. J. B. (1972a). *Phys. Lett. A* **38**, 311.
Swanenburg, T. J. B. (1972b). *Electron. Lett.* **8**, 351.
Swanenburg, T. J. B. (1973). *IEEE Trans. Electron Devices* **ED-20**, 630.
Szustakowski, M., and Wecki, B. (1971). *Bull. Acad. Pol. Sci., Ser. Sci. Tech.* **19**, 231.
Szustakowski, M., and Wecki, B. (1973). *Proc. Vibr. Probl.* **14**, 155.
Ter-Martirosyan, L. T. (1969). *Radio Eng. Electron. Phys. (USSR)* **14**, 1269.
Thiennot, J. (1972). *J. Phys. (Paris)* **33**, 219, 781.
Thiennot, J. (1975). *J. Appl. Phys.* **46**, 3925.
Thim, H. W., and Barber, M. R. (1966). *IEEE Trans. Electron Devices* **ED-13**, 110.
Tien, P. K. (1964). U.S. Patent 3,158,819.
Tsvirko, Y. A., Lukomskii, V. P., and Chovnyuk, Y. B. (1975). *Fiz. Tverd. Tela (Leningrad)* **17**, 1646.
Tsvirko, Y. A., Lukomskii, V. P., and Chovnyuk, Y. B. (1976). *Fiz. Tverd. Tela (Leningrad)* **18**, 1077.
Vainshtein, L. A. (1957). "Electromagnetic Waves." Sov. Radio, Moscow.
Vashkovsky, A. V., Zubkov, V. I., Kildishev, B. N., and Murmuzhev, B. A. (1972). *Sov. Phys.— JETP Lett.* **16**, 4.
Vedenov, A. A. (1964). *Usp. Fiz. Nauk* **84**, 533.
Vikulov, I. K., and Tager, A. S. (1969). *Radio Eng. Electron. Phys. (USSR)* **14**, 1711.
Vladmirov, V. V. (1975). *Usp. Fiz. Nauk* **115**, 73.
Vural, B., and Bloom, S. (1967). *IEEE Trans. Electron Devices* **ED-14**, 345.
Weinreich, G. (1956). *Phys. Rev.* **104**, 321.

Wessel-Berg, T. (1967). "Electromagnetic Properties of Drifted Semiconductor Plasmas," AE-Rep. 4. Norwegian Inst. Technol. Trondheim.
White, D. L. (1962). *J. Appl. Phys.* **33**, 2547.
White, R. M., and Voltmer, F. M. (1966). *Appl. Phys. Lett.* **8**, 40.
Yoshida, K., and Yamanishi, M. (1968). *Jpn. J. Appl. Phys.* **7**, 1143.

Microwave Power Semiconductor Devices. II*

Critical Review

S. TESZNER

Centre National d'Etudes des Télécommunications, France

AND

J. L. TESZNER

Direction des Recherches et Moyens d'Essais, Paris, France

Three-Terminal Devices 141
  I. Bipolar Transistors 141
    A. Basic Considerations 141
    B. Device Design and Fabrication 151
    C. Electrical Characteristics and Performances 158
    D. General Considerations and Conclusions 170
    References for Section I 172
  II. Field-Effect Transistors 174
    A. Basic Considerations 174
    B. Device Design and Fabrication 193
    C. Electrical Characteristics and Performances 203
    D. General Considerations and Conclusions 212
    References for Section II 213
Present Trends and Future Development Prospects for Microwave Power Semiconductor Devices. 216
  I. Present Trends 216
  II. Future Development Prospects 217

Three-Terminal Devices

I. Bipolar Transistors

A. Basic Considerations

1. General Theory

The first three-terminal solid-state device adaptable to the amplification of oscillating currents in radio communication was proposed by Lilienfeld in 1925 (*1.1*). However, it was only in 1948 that an operational device was

---

* For part I see Vol. 39, p. 291.

presented by Bardeen and Brattain *(1.2)* under the name of "transistor," the structure of which was quite different from the one described by Lilienfeld. Then in 1949, Shockley proposed a junction bipolar transistor and gave the basic theory of this fundamental structure *(1.3)*.

This theory as well as its numerous complements and extensions are too well known to be recalled in detail here, and so the discussion shall be restricted to microwave power bipolar transistor problems. In this domain, Early *(1.4)* and Pritchard *(1.5)* are the pioneers, particularly in the study of high-frequency effects, and Ebers, Moll *(1.6)*, Fletcher *(1.7)*, as well as Emeis *et al.* *(1.8)* in the study of high-power effects on transistor operation.

Some papers have been devoted to the analysis of the mutual dependence of these two effects *(1.9–1.11)*. On the other hand, some more recent general studies have been published concerning the theory and characterization of microwave bipolar transistors *(1.12–1.16)*. We shall consider briefly now these theoretical bases, in particular with a view to specifying the parameter trade-offs unavoidable in the design of microwave power bipolar transistors.

Figure 1 shows an *n–p–n* bipolar transistor in symbolic presentation;

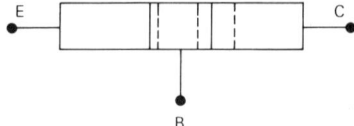

FIG. 1. Bipolar transistor symbolic presentation. E, emitter; C, collector; B, base.

Fig. 2 gives its simplified high-frequency equivalent circuit. Regarding the operating frequency limit, it is usual at the present time, to determine the frequency characteristic values of the cut-off frequency f_t and the maximum oscillation frequency f_{max}. f_t is determined by the emitter–collector signal delay time τ_{ec} and is given by

$$f_t = \frac{1}{2\pi\tau_{ec}} \quad (1)$$

where

$$\tau_{ec} = \tau_e + \tau_b + \tau_d + \tau_c \quad (2)$$

with τ_e the emitter–base junction charging time, τ_b the base transit time, τ_d the collector depletion layer transit time, and τ_c the collector depletion layer charging time.

In compliance with the equivalent circuits in Fig. 2,

$$\tau_e \cong r_e C_e \quad (3)$$

Since r_e is relatively small, its delaying effect should be neglected;

$$\tau_b = W_B^2/\eta D_B \quad (4)$$

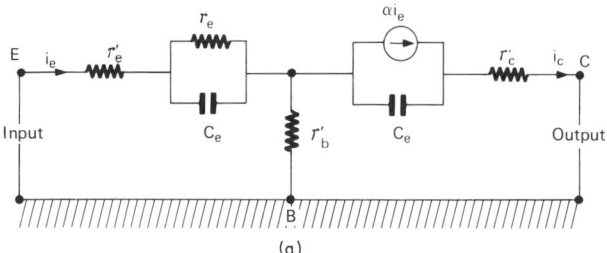

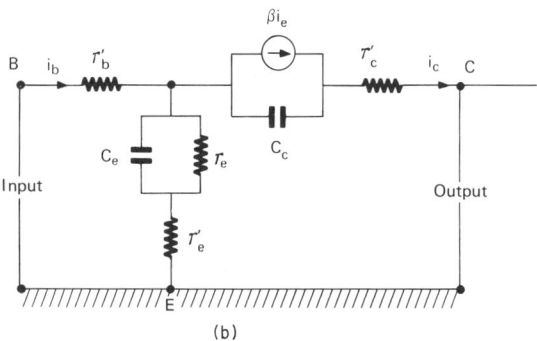

FIG. 2. Simplified high-frequency transistor circuit for (a) common-base and (b) common-emitter configurations. E, emitter; C, collector; B, base; r_e, emitter space–charge resistance; r'_e, r'_b, r'_c, respectively, emitter, base, and collector spreading resistances; C_e, emitter space–charge capacitance (usually under forward bias); C_c, collector transition capacitance (usually under reverse bias); α and β, short-circuit current gain:

$$\alpha = \partial I_c/\partial I_e; \quad \beta = \partial I_c/\partial I_b = \alpha/(1 - \alpha).$$

where W_B is the base width, D_B the minority carrier diffusion coefficient, and η a factor related to the drift field in the base region associated with impurity gradient;*

$$\tau_d = X_{cd}/nv \qquad (5)$$

where X_{cd} is the base–collector depletion layer width, v the carrier velocity within this layer, and n a factor that depends on the layer response time relation. The electrical field within the depletion layer being generally $> 10^4$ V/cm, v is almost equal to the scattering limiting velocity v_{sl}; on the

---

* In microwave transistors η can vary from ~ 2 for a uniform impurity profile to ~ 3 for a double-diffused and ~ 4 for a diffused-base alloyed-emitter structure.

other hand, $n = 2$ if the sinusoidal response of the depletion layer is required, and relation (5) becomes

$$\tau_d = X_{cd}/2v_{sl} \qquad (6)$$

For the last τ_{ec} component,

$$\tau_c = r'_c C_c \qquad (7)$$

Thus

$$\tau_{ec} \cong r_e C_e + \frac{W_B^2}{\eta D_B} + \frac{X_{cd}}{2v_{sl}} + r'_c C_c \qquad (8)$$

The cut-off frequency f_t is defined (1.5) as the frequency at which the common-emitter short-circuit current gain β ($\equiv \partial I_c/\partial I_\beta$) = 1 [the common-base current gain $\alpha = \beta/(1 + \beta)$ being then $\frac{1}{2}$]. However, in our opinion it does not seem evident, at least for microwave transistors, that the f_t value as determined above effectively corresponds to that definition.

In the earliest bipolar transistor devices, where the base-spreading resistance could be neglected in comparison with the emitter space–charge resistance, the τ_{ec} defined through (8) could profitably be considered as the total delay time of the transistor operation. At the present time, however, and especially for microwave devices characterized—as we shall see later—by an extremely narrow base width, this simplification no longer seems to be acceptable. Thus a fifth elementary time should be added to τ_{ec}. This time τ_{be} will be defined as

$$\tau_{be} = r'_b C_e \qquad (9)$$

Therefore,

$$\tau_{bec} = \tau_{ec} + \tau_{be} \qquad (10)$$

and

$$f'_t = 1/\tau_{bec} < f_t \qquad (11)$$

In the calculation commonly performed, the base-spreading resistance is introduced into the definition of f_{max}, the maximum oscillation frequency. f_{max} is extrapolated from the expression of the unilateral power gain U, through the statement $U = 1$ at $f = f_{max}$. It should be recalled that the unilateral gain is obtained when a device is unilateralized with a lossless reciprocal feedback network and matched at both ports.

For the microwave region, U is usually given by the simplified expression

$$U = \frac{f_t}{8\pi f^2 r'_b C_c} \qquad (12)$$

where f is the operating frequency.

Now according to the previous remarks, it seems preferable to replace f_t by f'_t, as defined above (11).
Then

$$U = \frac{f'_t}{8\pi f^2 r_b C_c} \tag{13}$$

and the f_{max}, for $U = 1$, is given by

$$f_{max} = \tfrac{1}{2}(f'_t/2\pi C_c)^{1/2} \tag{14}$$

In order to increase f'_t and likewise f_{max} as much as possible, from (8)–(11) and (14), one consequently must reduce r_e, r_b, r'_c, C_e, C_c, W_B, and X_{cd}, and on the other hand increase ηD_B. We shall discuss these different proposals briefly.

To reduce r_e one should make the emitter–base junction as narrow as possible and the emitter cross section as large as possible; however, C_e would be simultaneously increased. Then one must seek a trade-off; the best one would be to increase the ratio of the effective area in the total emitter area.

A similar conclusion is reached for the base-spreading lateral resistance r_b decreasing. However, on the other hand it is necessary to try to reduce the base width W_B, the consequence being an increase in r_b. To compensate this detrimental effect it would be adequate to simultaneously reduce the lateral length L of the base area.

An analogous problem arises concerning the collector depletion layer. In order to reduce the transit time τ_d, it would be convenient to reduce X_{cd}, but in that case the capacitance C_c would be proportionately increased. In order to prevent τ_c from increasing, it is necessary to minimize the thickness of the collector high-resistivity region in series with the depletion layer.

Finally, concerning ηD_B, the value of which should be increased, this effect could be obtained through grading the impurities in the base, which will produce within an electric field. The minority carriers are then moving through the base by a combination of the diffusion and the drift caused by this electric field, resulting in an important reduction of the transit time τ_b.

As a necessary complement of the above considerations, which are particularly valid for the small-signal operation, specific large-signal operational problems should be discussed briefly.

One fundamental difference between large- and small-signal operations results from their operating domains. In order to explain this, Fig. 3 gives the transistor dc characteristics for a common-emitter configuration showing three regions of operation. The small-signal operation is focused in the narrow part of the active region, guaranteeing the highest current gain. On the contrary, the large-signal operation passes successively from saturation to the active and the off regions in order to obtain the highest power output

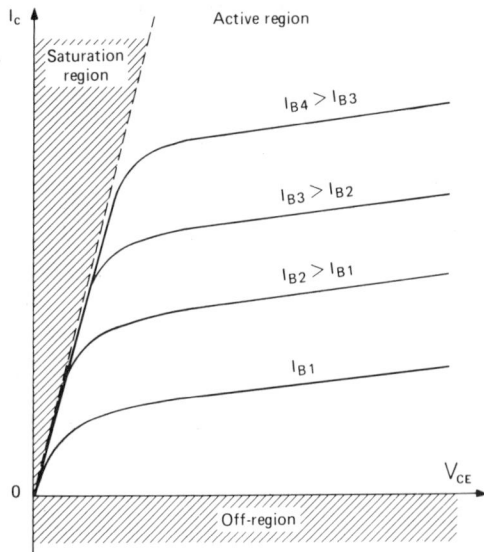

FIG. 3. Transistor dc characteristics for common-emitter circuit configuration.

with the greatest possible power gain. Frequently, moreover, the transistor is predominantly in the saturation or off regions and passes only briefly through the active region. Then obviously the amplification factor and therefore the power gain are reduced, as are f_t and f_{max}. As the load conductance is higher, this decrease becomes more and more effective.

Another problem concerns the effective emitter area of a power transistor that is only a part of the total emitter area, namely, the outer rim area (1.7, 1.8). This is a consequence of the lateral voltage drop relating to the base current, which in turn reduces the voltage drop of the emitter–base junction when the distance from the base contact increases. It is particularly important for the microwave power transistors to maintain the level of the emitter current base area ratio as high as possible. Therefore, the effective area ratio of the emitter structure and then the ratio of the emitter periphery to the base area (EP/BA) should be as large as possible. This conclusion also remains valid for reducing r_e and r_b as explained above.

We shall bear in mind the general considerations outlined above, in the following discussion of the physical bases of transistor structure-type geometries.*

---

* Regarding the output power increasing (and therefore the power–frequency product) its limitation stems from the practical limits (a) of the device's effective current density and cross section, (b) of the device's cells coupling in parallel configuration with forced-emitter ballasting (through series resistances), (c) of the circuit output and input matching. We shall come to these problems as well as to those of power efficiency in Section I,C,1.

2. Structure-Type Geometries

We shall discuss successively the emitter lateral geometry, the base layer width, and the collector depletion layer width.

a. *Emitter lateral geometry.* The figure of merit of this geometry is given by the aspect ratio $(EP/BA) = A$ defined above. Thus for a given power output level, the higher the aspect ratio, the lower the base area. The base lateral resistance r'_b is then reduced.

Different types of geometry have been performed up to now in the three fundamental types of structure: (1) interdigitated, (2) overlay, (3) mesh.

The interdigitated structure *(1.7, 1.17, 1.18)* contains a large number of emitter strips alternating with base strips (Fig. 4a), both being metallized.

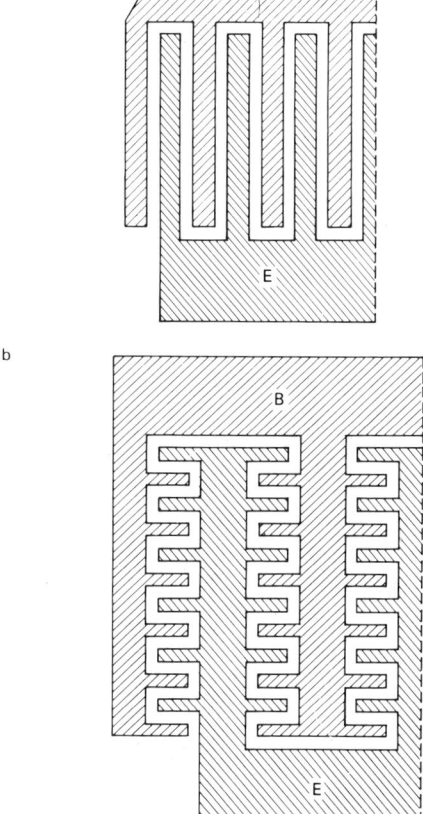

FIG. 4. (a) Early interdigitated and (b) "fishbone" emitter–base elementary geometries. E, emitter; B, base.

The aspect ratio essentially depends on the strips width. Then, in order to obtain high aspect ratios, very fine geometries attaining the limits of the processing technologies are required.

In a recently improved interdigitated structure called "fishbone," the emitter strips are perpendicularly elongated (Fig. 4b) in order to increase their internal periphery/area ratio also. This structure gives a higher aspect ratio without apparently increasing the definition geometry rate.

The overlay structure (*1.19–1.21*) contains a large number of segmented emitters overlaid through a plurality of wide metal strips (Fig. 5a).

In the early overlay structures the metal base strips had been replaced by

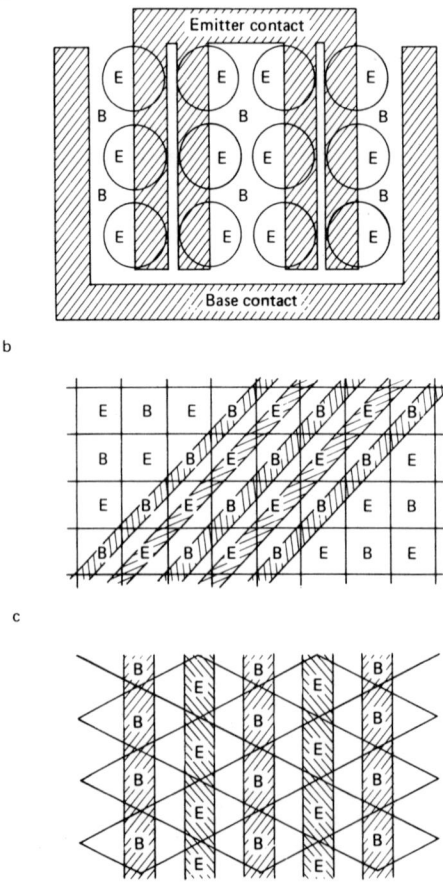

FIG. 5. (a) Early overlay, (b) improved checkerboard, and (c) diamond emitter–base elementary geometries.

a P^{2+} diffusion. Now, in recent overlay structures two improvements have been implemented (*1.24*). A two-layer metal system is used, the P^{2+} contacts being supplied through base strips of metal silicide and wide metal emitter strips being overlaid and insulated from the base by an oxide layer. On the other hand, the additional emitters have been inserted in checkerboard fashion.

If the segmented emitters are square, the checkerboard structure is as presented in Fig. 5b, the interdigitated emitter and base strips running at 45° angles. If the geometry of the emitter is then improved by increasing its internal periphery/area ratio through supplying the square by the diamond (*1.24*) as shown in Fig. 5c, the structure improvement is still more enhanced.

Mesh structure (also called matrix or grid-emitter) contains the emitter forming a grid, the base filling the meshes of this grid with a P^{2+} contact area in the middle of each mesh. Figure 6 gives an elementary geometry of this structure, assuming that the meshes are square.

Obviously, as for the structures considered above, more sophisticated emitter geometries are possible with a view to enhancing the emitter grid periphery for the same grid area.

However, it does not seem—and this remark is valid for the three structure types—that such shape complications could be implemented without any change of the geometry definition rate.

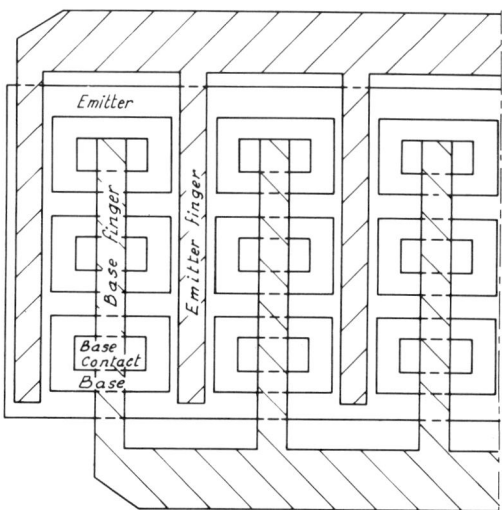

FIG. 6. Elementary mesh emitter–base geometry. Hatched areas, emitter and base metallizations.

With the present geometry definition maximum usually achievable for the power microwave transistors (without electron beam or X-ray exposure), i.e., $\sim 1~\mu$m linewidth, an aspect ratio of $\sim 20:1$ would be theoretically attainable with these three structure types. The practical limit, however, seems to be rather near to $10:1$.

Another structure geometry characterization is the current density I_0 in the metal emitter strips. Generally $I > 2 \times 10^5$ A/cm^2. The higher I_0, the more marked is the need for gold metallizing. We will come back to this item in Section I,B,3.

b. *Base layer width.* The base layer width W_B must be sufficiently narrow to reduce the time constant τ_b [Eq. (4)] up to ~ 0.1 to $0.2 \times (2\pi f'_T)^{-1}$, where f'_T is the required cut-off frequency.

On the other hand, however, making W_B narrow will increase the base lateral resistance r'_b, therefore raising the time τ_{be} [Eq. (9)] and f'_T [Eq. (11)] as well as the product $r'_b C_c$ and f_{max} [Eq. (14)].

This detrimental effect should be offset as far as possible through a parallel reduction of the emitter and the base finger length and consequently by increasing their number connected in parallel. On the other hand, the base impurity concentration should be increased up to the highest technological limit.

Finally, a compromise must be sought for the choice of the W_B value as dictated by the specified contradictory conditions. Currently the lowest W_B value is $\sim 0.1~\mu$m.

c. *Collector depletion layer width.* The conditions to fulfil are likewise opposite in this case. In fact, the depletion layer width should be as narrow as possible in order to reduce the time constant τ_d [Eq. (5)]. On the contrary, this depletion layer should be as large as possible in order to reduce the charging time constant τ_c [Eq. (7)] as well as the charging time $r'_b C_c$.

A balance should obviously be sought between these opposite requirements. However, in this case a compromise seems easier to define. In fact, since the carrier drift velocity is here attaining v_{nsl}, it is possible to have a collector depletion layer width much larger than the base width by decreasing its impurity concentration in order to reduce C_c and nevertheless to obtain a quite acceptable $\tau_d \sim 0.2(2\pi f'_T)^{-1}$.

On the other hand, in order to minimize τ_c, r'_c is limited by the restriction of the weakly doped portion of the collector layer thickness, which can be swept out, to X_{cd} width (5). On the contrary, the other part of that layer will be highly overdoped.

The foregoing outline of the theoretical discussion on the transistor structure characteristics provides an appropriate basis for device design and fabrication examination.

B. Device Design and Fabrication

The problems regarding the design and fabrication of microwave bipolar transistors have already been treated in many articles (*1.12, 1.13, 1.17–1.44*). Transistor design depends strictly on emitter geometry, as explained above. However, the transistor vertical cross section is roughly similar for the three fundamental types of this geometry, the only differences being minor. It would thus be in order to discuss the transistor structure and fabrication process using only one, i.e., Fig. 7, which gives a vertical cross section of an

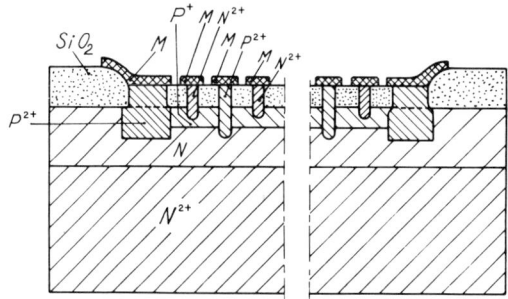

FIG. 7. Vertical cross section of a double-diffused, epitaxial planar interdigitated n–p–n transistor (schematic presentation). N^{2+}, substrate layer and emitter fingers; N, collector epitaxial layer; P^{2+}, base guard-ring and base fingers; P^+, active base; M, metallizations.

interdigitated transistor. Nevertheless, we shall take care in this discussion to specify briefly the design and fabrication differences between these three structure types.

On the other hand, we shall consider as semiconductor material only silicon, since it is practically the only material presently used for the fabrication of the microwave power bipolar transistors. It is obviously quite possible that in the future another semiconductor, particularly a compound semiconductor, could be adapted for this purpose.* For the moment it is not. In fact there are two types of difficulties in the way of this adaptation: (1) the technological processes available for realizing a microwave bipolar transistor structure are practically inapplicable with compound semiconductors; (2) currently the minority carrier lifetime into these semiconductors is too low, almost 100 times into GaAs, for instance.

Thus, Fig. 7 shows the fabrication with silicon integrated-circuit technology commonly used at the present time, namely, one epitaxial deposition

---

* Thus, with experimental types of heterojunction transistors containing such compound semiconductors, some promising performances have been already obtained [see, e.g., Konagai and Takahashi (*1.43*)].

and double (or triple) diffusion. However, we shall also give in Fig. 9 an experimental structure example in which diffusion has been replaced through ion implantation.

1. *Usual Technology*

The structure in Fig. 7 contains a substrate N^{2+} (with resistivity $\rho \cong 0.01\ \Omega$ cm) on which an epitaxial layer is grown, with a doping concentration N generally of $\sim 3 \times 10^{15}$ to 3×10^{16} cm^{-3} and a thickness of ~ 1.5 to 2.5 μm, comprising the emitter and base widths and the collector depletion layer. They are successively diffused through appropriate photoengraved oxide masks:

1. the base guard ring and contact fingers P^{2+} (segmented base contacts in the case of the mesh structure);
2. the base layer P^+; and
3. the emitter fingers (segmented emitters or emitter grid, respectively, in the case of an overlay or a mesh structure).

The patterns of those masks are specified by the base and emitter geometries relating to the different types of structure as explained above. The strip width is generally ~ 1–3 μm and the length to width ratio ~ 10–20; the emitter geometry can be calculated for 1 to 1.5 mA/linear mil.

The impurity diffused for the base fingers and layer formation is boron with a doping concentration giving at the surface, respectively, $> 10^{20}$ and 2×10^{19} cm^{-3}. The P^{2+} finger (or segmented contact) depth is generally 0.6–1.5 μm and the P^+ depth ~ 0.4–1 μm, these depths being somewhat increased during the following N^+ diffusion.

The depth of that diffusion is generally ~ 0.3–0.6 μm, the doping concentration at the surface being generally 4×10^{20} to 1×10^{21} cm^{-3}. During this fabrication step one has to deal with the "cooperative diffusion" effect, usually called "emitter-push" or "emitter-dip" effect (*1.25*). This phenomenon, illustrated in Fig. 8, results in an enhanced diffusion of the diffused base layer through a penetration in the silicon of the high-concentration diffused emitter layer. This effect may be explained as an interaction between

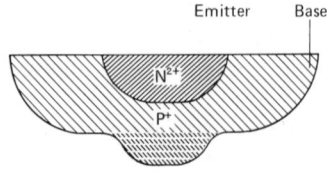

FIG. 8. Enhanced base layer diffusion through the emitter-push effect (indicated with dashed line).

boron ions and the lattice disorder created by the phosphorous diffusion, the base diffusion being enhanced along lattice dislocations.

This phenomenon would introduce a very serious limitation in the fabrication of microwave transistors with very narrow base width. Fortunately now it seems to be possible:

1. to attenuate this effect by using a solid–solid diffusion technique within which the lattice disorder formation is very much less than in the case of classical diffusion techniques (*1.26, 1.29*); and

2. moreover, almost to eliminate it through simultaneous reduction of the phosphorous diffusion temperature to 900°C (*1.12*), or by replacing the phosphorous by arsenic as the impurity diffused (*1.27–1.30*). However in the latter case, there are some ensuing inconveniences, namely that the emitter surface concentration is somewhat reduced and the emitter diffusion time is greatly increased;* during this time extension the normal diffusion of the base is enhanced.

Nevertheless, in any case, it seems quite feasible at the present time to obtain the final base width of ~ 0.1 μm.

This fabrication difficulty, as well as some other problems concerning the realization and the reproducibility of the structures with very fine geometries, seemed to be avoidable by using ion implantation technology (*1.33–1.37*).

2. *Ion Implantation Application*

The most important predicted advantage of the implantation technology for the bipolar transistor fabrication was that the concentration profile could be reproduced exactly, since the implantation dose and uniformity could be accurately regulated. The operating temperature of the ion implantation being relatively low, the redistribution of impurities during the successive operations would be almost suppressed. However, it is well known today that ion implantation introduces a lattice disorder and the carrier lifetime is thus very low. On the other hand, only a very small fraction of the implanted ions is located on the substitutional sites and their electrical activity is likewise very low. Now, to anneal the lattice disorder and improve the ion electrical activity, proper thermal treatment must be performed. The operating temperature and duration depend on the implanted ion, species, energy, dose, and dose rate.

---

* Although at high arsenic concentrations, "anomalous" diffusion with higher diffusion constant has been observed (*1.30*).

In the case of the microwave bipolar transistor one should implant

a. for base formation, a P$^+$ boron layer of moderate depth (~ 0.5–0.6 μm), with a moderate dose and dose rate (2×10^{12}–10^{13} cm^{-2}), the implant energy being 150–300 keV;

b. For emitter formation, a shallow (~ 0.2–0.4 μm) N^{2+} layer with a high dose and, particularly, a high dose rate (5×10^{15}–2×10^{16} cm^{-2}) of arsenic,* the implant energy being ~ 25–100 keV.

Figure 9 shows a vertical cross section of an experimental structure with implanted emitter and active base; the nonactive base and base contact are diffused. Note that it is quite possible to fabricate also the fully implanted structure (1.37); however, it does not seem to be preferable to the structure presented above.

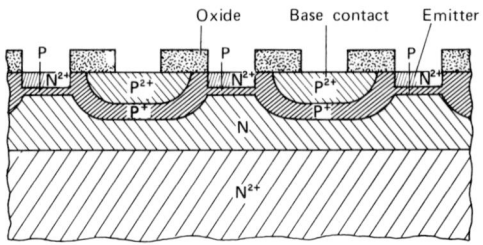

FIG. 9. Schematic presentation of a vertical cross section of a bipolar transistor experimental structure partially implanted. P, active base; P$^+$, nonactive base; N, collector layer.

The annealing process of the base layer needs a temperature of ~ 850°C applied for 20 min. This operation can be performed almost without base redistribution. On the contrary, the annealing process of the emitter layer needs a temperature > 900°C, preferably close to 1000°C, applied for 15 to 30 min. In this case some emitter redistribution takes place, especially for the highest temperature and duration of the operation.

Now, the problem is to maintain the carrier concentration profiles for emitter and base within the geometry compatible for use in the fabrication of microwave bipolar transistors. Two fabrication processes have been tried, differing in the emitter implantation and, in particular, in the annealing conditions. In the first case (1.34, 1.35), the emitter being implanted with a dose rate of 1×10^{16} cm^{-2}, the annealing temperature and duration were 900°C and 15 min, this treatment being the only anneal stage the wafer underwent. In the other case (1.33, 1.36, 1.37), with a dose rate of the same

---

* The arsenic seems to be definitely preferable to phosphorus here since it gives a better carrier concentration profile (without a pronounced tail).

order of magnitude, $\sim 4 \times 10^{15}$–2×10^{16} cm^{-2}, the annealing temperature was 1000°C and the treatment duration $\sim$ 20–30 min. An appreciable carrier redistribution is then forthcoming; consequently, the emitter profile is driven past the tail region and the emitter–base junction is located in damage-free material.

The experimental results known to date are not sufficient to provide a complete picture of each of these processes. However, it seems very probable that the average carrier electrical activity and the carrier lifetime should be higher after the annealing process of higher temperatures. Now, if the base were implanted before the emitter annealing treatment, a diffusional broadening of the base would occur during this operation, with the risk of transistor geometry alteration. In order to avoid this risk, one can implant the base after the emitter thermal treatment (*1.33*, *1.37*); however, in that case a second heat treatment is required after the base implantation.

On the other hand, if a high temperature regime is adopted the fundamental advantages of the implantation versus diffusion technology are somewhat reduced.

Regarding the wafer masking, many common films such as metals, oxides, or nitrides can be employed as masking layers. The implantation depth being relatively moderate in microwave bipolar transistor technology, it is possible in this case to use common oxide films but with a slightly increased thickness ($>$ 5000 Å), compared to that used in the diffusion process, in order to obtain a convenient ion stopping power.

Summing up, it seems today that the implantation process can improve industrial transistor characteristics, such as reproductibility and reliability, but it does not seem that the physical characteristics of transistors can be likewise substantially improved.

3. *Proton-Enhanced Diffusion*

Another technology presently under experimentation, namely proton-enhanced diffusion using the relatively small nuclei of the hydrogen atom to generate vacancies in the crystal, consists essentially in the introduction of a proton bombardment at a temperature of $\sim$ 500 to 700°C following the deposition of the impurity on the surface of the wafer. These impurity atoms then quickly diffuse into the crystal to a depth accurately determined by the energy of the protons. In addition, a new phenomenon has been discovered by using a high-temperature (850–900°C) proton irradiation after boron diffusion (*1.44*). In effect, "anomalous" redistribution profiles of boron in silicon, containing one or more concentration peaks within the wafer, have been observed. Investigations concerning this interesting discovery and its practical applying are being pursued at the present time.

4. Complementary Processes and Mask Creation

Whatever technology is adopted, the last step of the fabrication process is necessarily to provide the base, emitter, and collector with ohmic contacts (a) for base and emitter contacts, by deposing a metallic layer and cutting it out in an appropriate way, and (b) for collector contact, by evaporating and alloying a metallic film onto the back surface of the wafer.

This processing step is particularly important in view of long-term device reliability. Two flaw types should be noted: (1) metal microcracking and (2) failure modes due to the electromigration effects.

Microcracking occurs more frequently where the narrow metal fingers cross over an oxide step to make contact with the emitter–base junction. It may result from poor step coverage, from a thermal mismatch between silicon and metal layer, or from the finger etching process. The void formation through microcracking risks breaking the continuity of the metal fingers. This risk depends chiefly on the oxide step steepness and on the metal characteristics.

Electromigration (*1.38*, *1.39*) is a term applied to mass transport by momentum exchange between conducting electrons and metal ions. This phenomenon now depends essentially on the electron flow concentration and thus is enhanced through a highcurrent density, particularly in the emitter metal fingers. Two failure modes might be consequences of electromigration:

1. The void formation in a metal finger through the condensation of the vacancies in the metal, resulting in breaking the continuity of this finger, as in the microcracking case.

2. The growth of etch pits into the silicon, where electrons leave the silicon and enter the metal, and then by solid-state dissolution of silicon into metal and the transport of the solute ions down the metal conductor and away from the silicon–metal interface. The process will be pursued normally until the depth of an etch pit becomes sufficient to short an underlying junction.

The risk level of these failures depends on the current density in the metal and, as for microcracking, on the metal characteristics.

At the present time the metal most commonly used in bipolar transistors for the constitution of metal fingers on the silicon layer is aluminum. It is true that this metal is rather easy to use for this purpose, since it can be deposited directly over the silicon through the relatively high Si–Al eutectic formation temperature. However, it seems currently that it is not the best choice for narrow-finger constitution from the point of view of both microcracking and electromigration. Concerning microcracking, there is some thermal mismatch between silicon and aluminum; with electromigration, the drawback of aluminum is its relatively low level of activation energy. This is a limiting factor of the current density [a limit between 5×10^4 A cm^{-2} and 2×10^5 A cm^{-2} (*1.38, 1.39*) has been suggested] and of the working temperature (150°C).

The simplest way of decreasing this drawback is aluminum passivation by forming an aluminum compound (Al–Si, Al–Cu, or Al–Cu–Si alloys), aluminum oxide, or passivated aluminum glass. A more sophisticated solution consists in replacing the aluminum by gold. However, as it is not possible to place gold directly over silicon because of the relatively low Au–Si eutectic formation temperature (~ 380°C), it is necessary to interpose a thin layer of a refractory metal, the eutectic formation temperature of which with silicon is substantially higher. This is the case of molybdenum, for example. Finally, in order to improve the electrical contact of this refractory metal with silicon, a shallow platinum–silicide layer is formed.

Several such multimetal systems have been developed, some of which are currently used in microwave power transistor fabrication. In fact, the trend toward ever higher structure geometry resolution is leading to increasingly higher current densities in the metal fingers. These multimetal layers are now providing this possibility without the risk of decreasing the transistor reliability.

From that point of view, the structure geometries providing the same transistor electrical performance with relatively wider metal fingers, i.e., with a lower finger current density, seem to be preferable. This is, for instance, the case of the geometries presented in Figs. 4b and 5c; but the structure of Fig. 6 also allows for a similar improvement.

On the other hand, the transistor ultimate performance level obviously depends upon the geometry definition rate of the emitter and base structures. Then the resolution of these geometries is usually limited by the processes used to generate the optical masks corresponding to these structures, by the mask resolution itself, and by the accuracy with which the mask is located with respect to the structure formed during previous operations. Now, in solely optical processes, accuracy is limited by the fixed wavelength of light and the associated diffraction problems. The highest resolution pattern, at the present time, consists of 1 μm wide lines, which seems to benear the practical limit of this technique.

However, mask creation through recently developed electron beam technology (*1.40, 1.41*) makes possible much greater improvement in pattern resolution. Theoretically a resolution comprising a linewidth of ~ 0.1 μm could be obtained; practically, a linewidth of ~ 0.3 μm can be expected. In fact, a 0.5 μm resolution for emitter fingers has just been attained for experimental devices (*1.42*). This improvement might also affect successive pattern interregistration.

The consequence of each mask resolution improvement is obviously an increase in the aspect ratio and therefore in the emitter effective current density, expressed by the emitter current/base area ratio. Then in order to ensure a similar increase in heat dissipation without thermal runaway it becomes necessary to break down the power transistor structure, i.e., the base area, into some smaller cells. These cells will be spaced apart in order to eliminate high heat concentration and combined one by one or in appropriate groups. This technique permits multiplying the transistor package leads in parallel and thus results in a reduction of the package base and emitter inductances. Moreover, by introducing as a package lead a hybrid equivalent of a quarter line it becomes feasible to bring the input and output impedance levels at the package terminals to the desired value for a desired frequency, realizing internally matched transistors, called chip carriers.

Finally, the internal emitter ballasting becomes much easier. It consists, as discussed in Section I,C,2, of the insertion of an appropriate resistance in series with the emitter (*1.42, 1.45–1.48*) in order to limit the collector current variations, with respect to temperature, with forward biasing of the emitter–base junction or with output mismatching. These resistances are currently inserted inside the transistor package. They are realized by sputtering a short narrow line of a high-resistivity compound [for example, nichrome (*1.42*), tungsten, tantalum nitride (*1.48*), or polycrystalline silicon] in series preferably with each cell, i.e., each emitter, or even with each emitter finger.

C. Electrical Characteristics and Performances

In this section we shall consider successively (1) electrical characteristics and frequency–power transistor performances in amplifier, oscillator, and subsidiary switching operating modes, (2) transistor thermal properties and reliability problems, (3) transistor distortion and noise characteristics in amplifier and oscillator operations. All of these problems have been treated in part in many articles: some of the latest publications are cited in the reference list (*1.11, 1.13, 1.24, 1.33, 1.36, 1.45–1.78*).

1. Electrical Characteristics and Frequency–Power Performances

Electrical characteristics and performances depend first on the transistor operation mode (amplifier, oscillator, or switching) and second on the operation class (generally C or A, in the amplifier and oscillator cases). On the other hand, frequency–power performances are influenced through the transistor packaging conditions. Thus the transistor chip dynamic characteristics can be seriously degraded through the package inductances and capacitances, making external input and output matching for microwave operating frequencies difficult or even impossible.

The first attempt to eliminate this drawback was to reduce these inductances and capacitances as much as possible. Two more sophisticated solutions have been developed recently. The first, explained in Section I,B, consists of input and output matching within the transistor package: the so-called chip carriers; the second does away with the package and introduces transistors directly in a microwave integrated circuit called MIC (1.48–1.50). The first type of device is already commercially available; the second is currently under commercial development. In the tables that follow only the first solution is included. However, we shall come back in this section to the MIC possibilities.

Concerning the operation class, the bipolar transistors have been developed primarily for class C applications, particularly because in this operation class power efficiency attains its highest values, while the output power is also the highest possible. However, in recent years the field of application of bipolar transistors has been extended to class A, with a view to reducing the nonlinearity and noise figures, and then replacing traveling-wave tubes for microwave communications, especially in the low-noise area. The output power of the class A transistors is characterized otherwise by a

TABLE I.1

ELECTRICAL CHARACTERISTICS OF MICROWAVE POWER BIPOLAR TRANSISTORS (COMMERCIALLY AVAILABLE) FOR CLASS C AMPLIFIER AND OSCILLATOR OPERATIONS

	Frequency range (GHz) 1–4.5
Collector–base breakdown voltage, $BV_{CBO}(V)$	35–70
Collector–emitter breakdown voltage, with shunted BE junction,[a] $BV_{CER}(V)$	35–60
Collector–emitter breakdown voltage (with opened BE junction), $BV_{CEO}(V)$	15–40
Emitter–base breakdown voltage, $BV_{EBO}(V)$	2–4
Collector–emitter continuous voltage, $V_{CE}(V)$	12–28[b]
Emitter metal finger average current density, $I_{Emfa}(A\ cm^{-2})$	$\sim 5 \times 10^4$–5×10^5

[a] The shunting must be of sufficiently low resistance to limit the minority carrier injection as much as possible.

[b] For cw operation; in pulsed-wave operation, V_{EC} is generally increased, up to 40 V.

large bandwidth and gain flatness. Their characteristics will be discussed apart from those of transistors designed for class C operation.

Table I.1 gives the electrical characteristics of transistors for class C amplifier and oscillator operation. Note that the upper and lower parameter ranges correspond, respectively, to relatively high and relatively low power and frequency devices.

The performances of bipolar transistors operating in the class C microwave power amplifier are given in summary form in Table I.2.

Similarly, Table I.3 sets out electrical characteristics of transistors working in class A amplifier and oscillator modes, while Table I.4 gives transistor performances in the class A amplifier operation mode.

The oscillator performances are somewhat higher as far as operating frequency is concerned, which could approach the f_{max} value more closely. However, they are somewhat lower regarding output power, as a result of the input power subtraction. The relations between the respective f_0^- and P_{out} values depend on the particular case, i.e., on the specific structure and packaging of the transistor. Often, for instance, one can obtain $f_{0\,max}$

TABLE I.2

ELECTRICAL PERFORMANCES OF MICROWAVE POWER BIPOLAR TRANSISTORS, WITH OR WITHOUT INTERNAL MATCHING, FOR CLASS C AMPLIFIER OPERATION[a]

	Frequency range (GHz)								
	1	2	4	4	4.2	4.5	6	8	10
Amplifier transistor chip Output power in cw operation[b] $P_{out\,max}$(W)	35	20	5	7[f]	3[d]	1[e]	1.5[g]	1.2[h]	1[h]
Amplifier power added efficiency[c] $\eta\%$	60	50	30	33[f]	40[d]	25[e]	36[g]	24[h]	20[h]
Power gain PG (dB)	10	10	5	7[f]	4[d]	4[e]	4.5[g]	~5[h]	4.4[h]

[a] The data given up to 4 GHz concern commercially available transistors, of both types, with or without internal matching.

[b] In pulsed-wave operation, an output power level of 150 W (duty cycle ~ 1%) at 1 GHz with $PG = 10$ dB is obtained with commercially available transistors (V_{CE} being increased to 40 V).

[c] The η values indicated here correspond to the $P_{out\,max}$; however, some higher values have been obtained for lower P_{out}: in particular, at 1 GHz, η goes up to 65% and at 2 GHz, up to 60%.

[d] Internally matched devices.

[e] Devices without internal matching.

[f] Experimental (1.54).

[g] Experimental (1.55).

[h] Experimental (1.56).

TABLE I.3

ELECTRICAL CHARACTERISTICS OF MICROWAVE POWER BIPOLAR TRANSISTORS (COMMERCIALLY AVAILABLE) FOR CLASS A AMPLIFIER AND OSCILLATOR OPERATIONS

	Frequency range (GHz) 1–4
Collector–base breakdown voltage, $BV_{CBO}(V)$	≥ 40
Collector–emitter breakdown voltage, $BV_{CEO}(V)$	$\geq 15^a$
Emitter–base breakdown voltage, $BV_{EBO}(V)$	≥ 3
Collector–emitter continuous voltage, $V_{CC}(V)$	≥ 15
Collector–emitter continuous current, $I_{CC}(mA)$	≥ 50

[a] For BV_{CER}, the values approximate BV_{CBO}.

TABLE I.4

ELECTRICAL PERFORMANCES OF MICROWAVE POWER BIPOLAR TRANSISTORS, WITHOUT INTERNAL MATCHING, FOR CLASS A[a] AMPLIFIER OPERATION[b]

	Frequency range (GHz)				
	1	2	3	4	6
Amplifier transistor chip output power in cw operation, $P_{out\,max}(W)$	6	3	1.2	0.8	0.8
Amplifier power efficiency $\eta\%$[c]	~30	~25	~20	~15	17
Power gain PG (dB)[c]	10	10	8	6	4

[a] A linear amplifier operating in class B or AB with active broadband bias circuit has been developed in the laboratory (1.58). Compared with class A amplifiers it would provide for identical I/C (intermodulation ratio) = -20 dB, an efficiency ~2 times higher (15.5 vs 8.5%) with $PG = 10$ dB, $P_{out} = 0.4$ W at 4 GHz. This efficiency ratio seems to be increased for lower I/C; thus it becomes ~5 for $I/C = -30$ dB, but with η decreasing to 12 and 2.2%, respectively, and proportionately P_{out}. However, all the efficiencies quoted remain low compared with the efficiencies of commercially available devices given above.

[b] The data given up to 4 GHz apply to commercially available transistors. The data given in the last column are for an experimental model, under laboratory development (1.55). In both cases the output power is obtained at 1 dB gain compression.

[c] The η and PG values correspond to $P_{out\,max}$.

(oscillator) $= 1.2 f_{0\,\text{max}}$ (amplifier) and P_{out} (oscillator) $= 0.7$–$0.8 P_{\text{out}}$ (amplifier).

We turn now to the switching operation (subsidiary in the power microwave region), including turn-on and turn-off periods. The turn-on time can be approximated by $2\tau_{\text{bec}}$, where τ_{bec} is the base–emitter–collector signal delay time defined (10) in Section I,A. Thus it depends strictly on the frequency–power transistor capability specified above. Moreover, the turn-off time is a function of the so-called storage time, corresponding to the evacuation of the carriers stored within the transistor structure during the current pulse. However, with the very thin epitaxial layers used presently in the fabrication of the microwave transistors, the stored charge per unit area is extremely reduced, as is the storage time. To enhance this charge evacuation and thus to minimize this storage time, a relatively high reverse base-emitter bias voltage should be recommended. Then one should increase the emitter–base breakdown voltage. Thus switching turn-on and turn-off times of the order of 100 to 1000 psec can currently be obtained. Finally, for relatively high-power switching a relatively high recovery voltage should be sustained; BV_{CEO} should then be as close as possible to BV_{CBO}. For this purpose, the collector–base junction reverse current should first be lowered. On the other hand, the minority carrier injection by the emitter–base junction should be somewhat delayed; it could be obtained, for example, by widening the emitter–base junction, by reverse bias current injection, and/or by enhancing the carrier recombination.

In succession we shall discuss, in particular, the amplifier performances and their dependence on the transistor structure and packaging parameters. In order to do that, we shall refer to the equivalent transistor circuit of the intrinsic elements of Fig. 2 and shall consider Fig. 10 as giving this equivalent circuit in the common-base configuration completed by the parameters of the typical microwave stripline package.

The intrinsic transistor parameter influence on f_T or f_{max}, as well as the structure geometry problems in the microwave power device, have been already discussed in Section I,A. It is noteworthy that the recent very interesting performances at 8 and 10 GHz specified in Table I.2 have been reached through substantially improving the emitter geometry resolution by introducing electron-beam technology, the emitter finger width being reduced to 0.5 μm and the aspect ratio being increased to 11/1 (*1.42, 1.56*).

The required power output obtainable with acceptable power efficiency in a wide microwave region is, moreover, important in maintaining a suitable parameter value: for class C with a large current range at a convenient operating voltage, and for class A with large current and voltage excursion ranges. The current amplitude is limited first by the current gain and then by αi_E (see Fig. 10) vanishing: the problem is to maintain a suitable α with i_E as

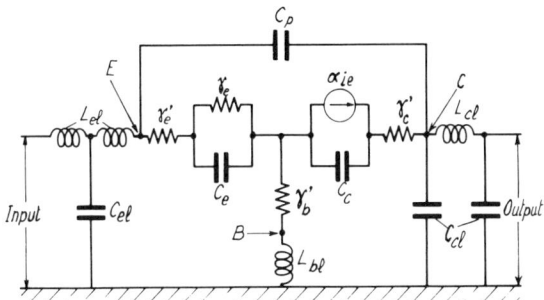

FIG. 10. Equivalent common-base transistor circuit of Fig. 2a completed by the package parameters. E, emitter; C, collector; B, base; r_e, emitter space-charge resistance; r'_e, r'_b, r'_c, emitter, base, and collector spreading resistances, respectively; C_e, emitter space-charge capacitance; C_c, collector transition capacitance; L_{e1}, L_{b1}, L_{c1}, packaging inductances from emitter, base, and collector sides, respectively; C_{e1}, C_{c1}, packaging capacitances from emitter and collector sides, respectively.

high as possible. A second limitation stems from the device's thermal behavior (it is necessary to avoid thermal runaway; we shall return to this problem later). The voltage excursion is limited at the low edge, mostly through an excessive increase in the collector capacitance C_c; consequently f_{max} decreases (14). At the high excursion edge, voltage is limited through the BV_{CE} value under the operating conditions: thus it would be desirable that this BV_{CE} value approach the BV_{CBO} value as closely as possible.

However, these considerations concern only the intrinsic chip parameters: their interaction with the packaging parameters, shown in Fig. 10, could result in severe degradation of the transistor performances. In effect, the most important flaw risk is the resonance possibility in the series $L_{b1} - C_e - L_{e1}$ at the input side or in the series $L_{b1} - C_c - L_{c1}$ at the output side. In any case, an appreciable reduction of input and output admittances occurs. Another drawback is the increase of the equivalent input and output capacitances through adjunction of C_{e1} and C_{c1}.

The first idea for lessening these drawbacks is to minimize the lead inductances, first by reducing the package area, and second, as indicated in Section I,B, by breaking down the transistor structure into some cells working in parallel. However, one becomes rapidly limited in this attempt at improvement by an increase of the lead capacitances, through narrowing between leads, up to the same order as those of the transistor chips. In that case the transistor performances would also be severely degraded.

A complementary improvement could then be introduced by placing the matched capacitances in series with the leads (possibly outside the package) in order to almost compensate the lead inductances (1.57). This cunning approach is particularly efficient in the base connection and makes it possible to obtain a substantial improvement of frequency–power performances without notable bandwidth narrowing.

To achieve maximum microwave transistor performances, appropriate input and output matching networks must be placed as close as possible to the transistor package (Fig. 11a). These networks are prepared with a view to transforming the input and output impedances to the nominal impedance level of the amplifier circuit (usually 50 Ω). Generally these input and output networks are formed by an approximate quarter-wave adapted by the user for obtaining the most convenient performances for operating frequency, bandwidth, power output, and power gain.

However, for a specific standard frequency, bandwidth, and input–output impedance levels, a transistor matching within the package (Fig. 11b) may provide somewhat higher performances. As shown in Table I.2, the power efficiency and power input are improved. Finally, in the most sophisticated solution, the MIC, which leads to suppression of the transistor package and integration of the device chip into a thin-film circuit, the parasitic elements associated with packaging are minimized. For example, the output (collector) lead may be eliminated, the collector constituting part of a microstrip transmission line. Consequently the transistor frequency–power performances could approach those of the chip itself. Thus MIC seems to be, in principle, the best solution for obtaining the highest possible operating

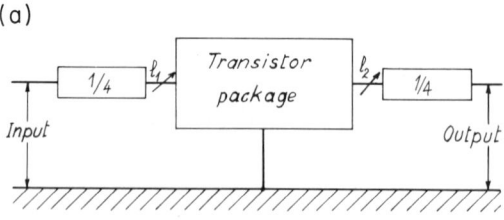

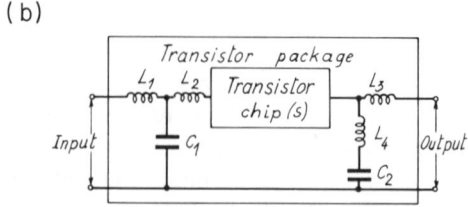

FIG. 11. (a) Transistor external input and output matching. $\lambda/4$, quarter-wave matching network; l_1, l_2, connections adjustable for minor tuning. (b) Transistor internal matching. C_1, L_1, L_2, and C_2, L_3, L_4, input and output matching equivalent circuits, respectively.

frequency for a given transistor structure. Concerning the power output, MIC amplifiers providing $P_{out} \sim 100$ W at 1 GHz have already been constructed (*1.51*). In fact, MIC power amplifiers are being used currently, notably in communication, telemetry, and phased array radar.

Another important problem lies in the choice of the operation class, essentially class C or A. When the major requirement is the highest possible power output and efficiency, class C is preferable, as can be seen by a comparison of the data in Tables I.2 and I.4. On the contrary, when the major requirement is above all to obtain as large an operating frequency bandwidth as possible and the best possible output signal linearity, class A should be chosen. In effect, in this operation class an octave bandwidth and simultaneously a relative signal linearity $I/C \cong -20$ dB are obtainable. We shall come back again to this subject in Section I,C,3.

The main reasons for the decrease of the power efficiency in the class A operation are obviously the relatively high continuous collector bias current, and moreover, the necessity for emitter finger ballasting through relatively high resistances to approximately equalize the current distribution, thereby reducing the risk of thermal runaway. Obviously these ballasting resistances cause an appreciable loss of output power, gain, and power efficiency.

However, from the reliability point of view the use of emitter ballasting resistors seems to be likewise necessary in class C operation. We shall now examine this important problem.

2. *Thermal Properties and Reliability Problems*

The thermal properties of a transistor are of considerable importance as far as its reliability is concerned. Consequently they have been considered in many publications (e.g., *1.24, 1.60–1.66*). The dynamic thermal behavior of a microwave power bipolar transistor is defined by (a) the transistor structure thermal resistance and response delay, and (b) the current–temperature relation and the current density distribution.

In thermal resistance one should also take into account, over and above the internal chip resistance, the external spreading resistances, all the more so as the chip size is reduced, whence the interest of dividing the device into separate cells forming chips of smaller and smaller size. No chip is really isothermal. Within reasonable limits, however, dividing the device into separate chips is beneficial as shown in Fig. 12. The thermal response delay is much higher than the electrical response delay. The former may be estimated at 50 to 100 nsec, the latter at some 100 psec. Thus, the chip temperature and its distribution remain almost without change during cw operation. However, the steady-state operation is only assumed if the external temperature and collector current average value remain constant.

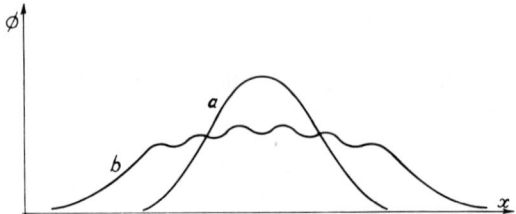

FIG. 12. Schematic diagram illustrating the thermal effect of dividing a transistor into an array of separate cells. Temperature (ϕ) distribution as function of lateral distance x. (a) Single-chip transistor; (b) transistor divided into six separate cells.

On the other hand, if the external temperature or the collector current increases, the risks of the stability breaking down will appear through the positive current–temperature coefficient common to bipolar devices. In effect, when temperature increases, the collector current likewise increases, enhancing the temperature rise. A steady state will be recovered at temperature and current levels relatively higher than the level corresponding only to the temperature increase. The process is quite similar if the current rises while the external temperature remains constant.

Current distribution uniformity and stability is another important problem. It concerns, in particular, the distribution within the emitter fingers and between fingers working in parallel. The first item has already been discussed in Section I,A,2: the current being concentrated at the edge of the emitter periphery, the practical solution lies in extending this periphery as much as possible without modifying the base area. However, the current density being thus increased, some improvement in thermal characteristics must be simultaneously incorporated: such an improvement may consist in decreasing each transistor chip area. As far as the second item is concerned, it seems to be practically impossible to assume an equal distribution among the emitter fingers without external regulation. This regulation can be realized through implementation of the emitter finger ballast resistors. This emitter ballasting is especially important not only because it assumes an almost regular current distribution, but it likewise reduces the risk of thermal runaway.

In effect, the reciprocal actions of the current and temperature increases explained above are, in principle, capable of triggering thermal runaway (*1.65*, *1.66*), if the circuit is not provided with any elements to limit the current increase. Now, the ballasting resistors included in series with each emitter, or preferably with each emitter finger, constitute such current-limiting elements. Then this ballasting seems to be absolutely necessary to guarantee acceptable short-term reliability (*1.61*, *1.62*), regardless of the transistor operation class.

The ballast resistance is obviously limited by the drop in power efficiency and output power (1.47): in effect an appropriate trade-off should be chosen between the contradictory conditions relating to reliability and power efficiency. If the resistor is formed with an alloy the resistance of which increases relatively rapidly with temperature, then this trade-off is easier to obtain.* Whatever trade-off will be admitted, it seems impossible that this technique could entirely eliminate the risk of thermal runaway. In particular, development of hot spots through local current concentrations in some technological structure defects in the transistor active area (1.64) remains possible. On the other hand, a severe circuit degrading, for instance, through output shorting would cause, despite ballasting, an high overcurrent and consequently excessive overheating. In both cases thermal runaway could result (for the second case, however, only from the highly ballasted class transistors).

Thermal runaway is the main basis for so-called second breakdown, which seems to be a peculiarity of bipolar devices. The start of this phenomenon, well known today, consists of an abrupt change in the dV/dI value, from almost zero during prime breakdown to a negative value in the second breakdown, as soon as a certain current density is reached (Fig. 13). While the former phenomenon does not overthrow the transistor, the second is

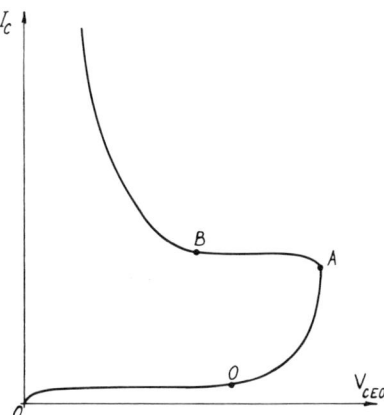

FIG. 13. Transistor I_c-V_{CEO} characteristic showing second breakdown transition; OA, prime breakdown avalanche conduction; A, starting second breakdown transition; B, second breakdown conduction.

* A noteworthy recent proposition lies in replacing the series emitter ballast resistor by a "temperature-sensing semiconductor shunt resistor" (1.63). This solution consists of shunting the base–emitter with a resistor, the resistance of which is falling as temperature increases. Theoretically, such an arrangement seems to be attractive; from the practical viewpoint, it seems to be less adaptable to the distributed emitter structure. For proper evaluation, experimental data would be necessary.

generally destructive. Thus, the most important short-term reliability problem is to prevent this phenomenon from occurring.

For this purpose, the most efficient external means seems to be to limit the increase in the amplitude and rate of the current density. In the best case the amplitude would be limited below point A of Fig. 13. In fact, however, such a perfect reliability solution would necessitate a ballast resistance value that would be excessive from the transistor operation viewpoint. This is particularly evident in the case of microwave power transistors, by reason of their operational high current density limit, relatively near the prime breakdown starting current density. Another reason for an especially critical situation in this case is the very fast rate of increase of temperature with an increase in transistor current. Consequently, the current amplitude of critical point A will be relatively lower for the microwave power transistors than for devices with smaller operation current density.

Nevertheless, the emitter ballasting can be efficient even if the starting point of the second breakdown process is passed, since the emitter resistors delay this process and introduce a limitation on the final current. Thus the second breakdown transition could inherently be nondestructive and reliability would then be improved.

The problem is different when this breakdown process results from carrier injection from the collector into the base region due to local thermal generation, impact ionization (1.67), avalanche injection (1.68), and collector current filament nucleation (1.69). In those cases the emitter ballast resistors are obviously ineffectual. Preventing such eventualities from arising, as far as possible, is a fabrication problem that concerns mainly the structure and process (discussed in Section I,B) regularity. Only with this proviso and with the transistor attached to an adequate heat-sink can acceptable short-term reliability be expected.

Regarding long-term reliability, the problem is all the more difficult as the structure geometry resolution and the current density become higher (overall and local, in particular). Thus, the difficulty is necessarily maximized for microwave power devices. The problem becomes critical if the finest structure geometry lies on the surface of the device, because of the passivation difficulty of surfaces designed in this way.

Finally a fair maintenance of the contact metallizations is likewise of prime importance from a long-term reliability viewpoint; this problem has been already discussed in Section I,B,4.

3. *Distortion and Noise Characteristics*

Since the relationship I_c (V_{BE}) and I_c (I_B) is essentially exponential, the relation P_{out}/P_{in} is nonlinear. However, in the class A operation with a properly limited voltage and current excursion, the amplified signal distortion may be acceptable for TWT replacement, in the "linear" applications mentioned above.

For the current excursion, the limitation is more stringent than that discussed before, in particular, the highest admissible value must be substantially reduced. The grounds for this reduction are (1) the rapid falloff of transistor gain and cut-off frequency at high collector currents (cf. *1.70–1.72*), and (2) the enhancement of the current instability beyond a certain limit of current density, the hotter areas tending to have higher gain than the cooler areas; in turn, through current filament concentration these hot zones become even hotter.

With this current and output power reduction significant results have been obtained as far as reduction of distortion is concerned, defined by the intermodulation ration limitation. However, the data concerning commercially available devices are generally restricted to the third-order intermodulation ratio, giving $I/C \leq -30$ dB. This indication is effectively quite important but it would also be appreciable for overall distortion estimation to know the second-order intermodulation ratio. In effect, the laboratory research results presented very recently (*1.58*) and given already above (p. 161) show, for a comparable power efficiency, $I/C = -20$ dB. $I/C = -30$ dB then corresponds in that case to a substantially lower power efficiency.

With regard to the noise characteristics (*1.12, 1.73–1.78*), there are three sources of noise to be considered in the case of microwave transistors: (1) thermal noise generally caused by the random collision of carriers with the lattice, (2) shot noise caused by the random emission of electrons and random passage of carriers across potential barriers, and (3) as a subsidiary, generation–recombination noise (GR noise) caused by the random generation and recombination of hole–electron pairs, carrier generation from traps, and carrier recombination with empty traps. Flicker noise ($1/f$ noise) may be practically neglected for microwave frequency bands, except in the case of the emitter–base avalanching (*1.76*).

The transistor common-base equivalent noise circuit (*1.73*) is presented in Fig. 14, where R_s is the source resistance, r_e the emitter resistance, r'_b the base resistance, and the corresponding thermal noise emf's are $(4kTR_s \Delta f)^{1/2}$, $(2kTr_e \Delta f)^{1/2}$, and $(4kTr'_b \Delta f)^{1/2}$; i is the noise current generator. In the microwave region the noise figure's minimum value will be expressed by

$$F_{\min} \simeq 1 + 2g_n r'_b + 2(g_n r'_b + g_n^2 r'^2_b)^{1/2} \qquad (15)$$

where g_n (noise conductance) $= \overline{i^2}/|\alpha|^2 4kT_0 \Delta f$, the noise minimization being obtained through an R_s value optimization:

$$(R_s)_{\mathrm{opt}} = (r'_b/g_n)(1 + r'_b g_n)$$

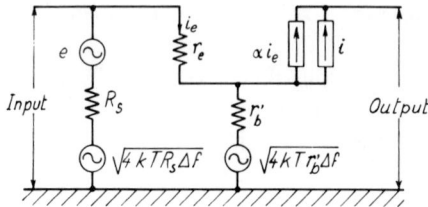

FIG. 14. Transistor equivalent noise circuit for common-base configuration. After Van der Ziel (1.74).

The noise figure then essentially depends on the product $r'_b g_n$. In the microwave devices g_n is given by an approximated expression:

$$g_n \simeq \frac{1}{2\alpha_0 r_e}\left(1 - \alpha_0 + \frac{f^2}{f_\alpha^2}\right) \tag{16}$$

where f_α is the alpha cutoff frequency, $f_\alpha = 1/2\pi\tau_b$ [τ_b is the base transit time, defined by Eq. (4)], and α_0 is the low-frequency common-base current gain.

Therefore, to reduce the noise level it would be recommended to reduce r'_b and to increase f_α. In that case, the operating frequency would be increased at the same time. Theoretically, it seems to be possible. In fact, however, this is not the case, since with f_α increasing the base width will be reduced and r'_b will be increased (1.77), despite the improvement of emitter geometry resolution. On the other hand, r_e would be reduced through a reduction of device's cross section. Consequently, the noise level usually rises with operating frequency increasing in the microwave region, for a given type of device.

The high-level injection effects in the base–collector region (1.73), which becomes space-charge limited, provide a considerable drop in f_α, with a subsequent rise of F_{min}. Otherwise, any structural homogeneity defect results in the noise level increasing, in particular, the generation-recombination noise.

Finally, regarding a transistor operating as an oscillator, the fm noise resulting from parameter instability must be considered. For this application it is then generally required to insert a properly tuned cavity resonator into the oscillator circuit.

The present values of microwave power transistor noise level lie between 10 and 15 dB (as typical values) for the devices cited in Table I.4.

D. General Considerations and Conclusions

Summing up, we define the bipolar transistor's place in the microwave power semiconductor device family. First let us comment on their characteristics.

Their advantages may be recapitulated as follows:

1. Due to their three-terminal configuration, their application, particularly as amplifiers or switching devices, is much easier and the corresponding circuits much simpler than for two-terminal devices.
2. Due to their operation with both majority and minority carrier types, very high local current density can be reached, much higher than in the majority carrier devices.
3. The output powers attained for different microwave frequencies are relatively high, particularly in the low bands of the microwave region.
4. Their operating power efficiency is high, particularly for class C amplifiers.
5. Their operational bandwidth is large, particularly in class A amplifiers.
6. Their power gain in amplifier operation is relatively high.
7. Signal distortion is lower than in two-terminal devices. Noise level is lower than in the avalanche diodes; it is of the same order of magnitude as in TED and in BARITT diodes.

Their disadvantages, on the other hand, seem to be as follows:

1. Their structure is more complex than that of two-terminal devices.
2. Since current distribution is essentially nonhomogeneous, quite sophisticated structures must be prepared in order to obtain an adequate overall current density.
3. Through their bipolar operation, the current–temperature coefficient being positive, each positive variation of current or temperature involves device operation instability with some risk of thermal runaway and second breakdown. These risks would almost be obviated by emitter ballasting with an appropriate resistor in series with each emitter finger; however, some input power is therefore dissipated and the power efficiency and gain are reduced. In return, the current distribution between emitter fingers working in parallel, is improved and the risks resulting from output mismatching be also prevented.
4. Signal distortion and noise level seem to be, in principle and in fact, higher than for unipolar transistors (which will be discussed in the next section).
5. Finally, it is noteworthy that in special applications, like nuclear and space systems, the immunity of bipolar devices to neutrons and ionizing radiations is lower than that of unipolar devices.

We will now try to give the prospects for microwave power bipolar transistor development versus that of microwave diodes, discussed in Part I. This comparison will first be limited to the microwave region presently accessible to bipolar transistors, at least under laboratory development, i.e., a region extending up to and including the X band. Later, however, we will likewise consider the coupling transistor–varactor diode limited only by the varactor possibilities.

If the transistor operates without any coupling device, its performance will be better than that of IMPATT or TRAPATT diodes or TED, which are the only diodes whose performance can be compared. In effect, compared to the IMPATT diode or TED for similar output power, transistor operating power efficiency is higher, up to 4 GHz in the oscillator and amplifier operation; moreover, in the amplifier case the power gain is also higher and with lower signal distortion. Then, in comparison with TRAPATT diodes, the transistor superiority lies essentially in much lower noise and, in the amplifier operation, lower signal distortion. On the contrary, its power output and efficiency in the oscillator operation are somewhat lower.

Regarding the coupling transistor–varactor, recent advances in the techniques of both components are such that this coupling seems to be currently capable of competing with the IMPATT diode or TED, up to the millimeter wave region and output power of ~ 1 W. The choice between these opposite solutions will depend on the particular case that arises. If the problem of power output, gain, and efficiency is primordial, then either IMPATT or TED could be preferable; on the contrary, if relatively high signal purity (low distortion and low noise level) is required, the choice of transistor–varactor coupling would be more appropriate.

However, the use of a transistor alone or a transistor coupled with varactor is not only available with bipolar transistors but also with unipolar transistors. The elements of appropriate choice, depending on the particular case, between these two transistor types will be proposed at the end of the next section, which is devoted to the unipolar, field-effect microwave power transistor examination. Finally, in this review's concluding section, our personal views on the present and future trends in the microwave power semiconductor device field will be presented.

References for Section I

1.1. J. E. Lilienfeld, U.S. Patent 1,745,175 (Appl. in Canada, October 22, 1925).
1.2. J. Bardeen and W. H. Brattain, *Phys. Rev.* **74**, 230 (1948).
1.3. W. Shockley, *Bell Syst. Tech. J.* **28**, 435 (1949).
1.4. J. M. Early, *Bell Syst. Tech. J.* **32**, 1271 (1953).
1.5. R. L. Pritchard, *Proc. IRE* **42**, 786 (1954).
1.6. J. J. Ebers and J. L. Moll, *Proc. IRE* **42**, 1761 (1954).
1.7. N. H. Fletcher, *Proc. IRE* **43**, 551 (1955).
1.8. R. Emeis, A. Herlett, and E. Spenke, *Proc. IRE* **46**, 1220 (1958).
1.9. E. O. Johnson, *RCA Rev.* **26**, 163 (1965).

1.10. R. J. Whittier and D. A. Tremere, *IEEE Trans. Electron Devices* **16**, 39 (1969).
1.11. R. L. Bailey, *IEEE Trans. Electron Devices* **17**, 108 (1970).
1.12. H. F. Cooke, *Proc. IEEE* **59**, 1163 (1971).
1.13. M. H. White and M. O. Thurston, *Solid-State Electron.* **13**, 523 (1970).
1.14. H. M. Rein, T. Schad, and R. Zühlke, *Solid-State Electron.* **15**, 481 (1972).
1.15. A. J. Wahl, *IEEE Trans. Electron Devices* **21**, 40 (1974).
1.16. P. W. Shackle, *IEEE Trans. Electron Devices* **21**, 32 (1974).
1.17. A. Goetzberger, M. Zetterquist, and R. M. Scarlett, *IEEE Int. Conv. Rec.* Pt. 3, p. 57 (1963).
1.18. J. G. Tatum, *Electronics* **41** (Feb. 19), 93 (1968).
1.19. D. R. Carley, *Int. Electron Devices Meet., Washington, D.C., 1963*.
1.20. D. R. Carley, P. L. McGeough, and J. F. O'Brien, *Electronics* **38** (Aug. 23), 70 (1965).
1.21. D. R. Carley, *Electronics* **41** (Feb. 19), 98 (1968).
1.22. M. Fukuta, H. Kisaki, and S. Maekawa, *Proc. IEEE* **56**, 743 (1968). (Letters.)
1.23. J. Andeweg and T. H. J. van den Hurck, *IEEE Trans. Electron Devices* **17**, 717 (1970).
1.24. J. A. Benjamin, *Microwave J.* October, p. 39 (1972).
1.25. K. H. Nicholas, *Solid-State Electron.* **9**, 35 (1966).
1.26. E. V. C. Rao and P. Tronc, *Conf. Solid-State Devices, 2nd, Manchester, 1968*.
1.27. R. B. Fair, *J. Electrochem. Soc.* **119**, 1389 (1972).
1.28. K. Reindl, *Solid-State Electron.* **16**, 18 (1973).
1.29. P. C. Parekh and K. Kolmann, *Solid-State Electron.* **17**, 3 (1974).
1.30. R. L. Kronquist, J. P. Soula, and M. E. Brilman, *Solid-State Electron.* **16**, 1159 (1973).
1.31. J. A. Kerr and F. Berz, *IEEE Trans. Electron Devices*, **22**, 15 (1975).
1.32. S. Dash and M. L. Joshi, *IBM J. Res. Dev.* **14**, 453 (1971).
1.33. V. G. K. Reddi and A. Y. C. Yu, *Solid State Technol.* **15** (Oct.), 35 (1972).
1.34. M. K. Barnoski and D. D. Lofer, *Solid-State Electron.* **16**, 433 (1973).
1.35. M. K. Barnoski and D. D. Lofer, *Solid-State Electron.* **16**, 441 (1973).
1.36. B. Pruniaux, J. L. Assemat, M. Delandre, A. Dumetz, P. Gabillet, and J. M. Lagorsse, *Proc. Eur. Microwave Conf., Brussels* A. 115 (1973).
1.37. R. S. Payne, R. J. Scavuzzo, K. H. Olson, J. M. Nacci, and R. A. Moline, *IEEE Trans. Electron Devices* **21**, 273 (1974).
1.38. J. R. Black, *IEEE Trans. Electron Devices* **16**, 338 (1969).
1.39. J. R. Black, *Proc. IEEE* **57**, 1587 (1969).
1.40. R. F. M. Thornley, M. Hatzakis, and V. A. Dhaka, *IEEE Trans. Electron Devices* **17**, 961 (1970).
1.41. G. R. Brewer, *Solid State Technol.* **15**(July), 36 (1972); **15**(Aug.), 42.
1.42. J. B. Kruger, You-Sun Wu, and Han-Tzong-Yuan, *Symp. Electron, Ion Laser Beam Technol., 13th, Colorado Springs, 1975*.
1.43. M. Konagai and K. Takahashi, *J. Appl. Phys.* **46**, 2120 (1975).
1.44. P. Baruch, J. Monnier, B. Blanchard, and C. Castaing, personal communication (1975).
1.45. G. Fodor and J. Causse, *Onde Electr.* **51**, 57 (1971).
1.46. R. Arnold and D. S. Zoroglu, *IEEE Trans. Electron Devices* **21**, 385 (1974).
1.47. M. M. Sayed, J. T. C. Chen, and S. Kakihana, *Int. Electron Devices Meet., Washington, D.C.* 13.6 (1974).
1.48. A. Presser and E. F. Belohoubek, *RCA Rev.* **33**, 737 (972).
1.49. K. Hartmann, W. Kotyczka, and M. J. O. Strutt, *Proc. IEEE* **59**, 1720 (1971). (Letters.)
1.50. P. T. Chen, *Int. Solid-State Circuits Conf., Philadelphia* p. 76 (1973).
1.51. W. E. Poole and D. Renkowitz, *Microwave J.* October, p. 23 (1972).
1.52. H. F. Cooke and F. E. Emery, *Microwave J.* November, p. 47 (1973).
1.53. D. Renkowitz, *Natl. Telecommun. Conf., Atlanta* 23 C-1 (1973).

1.54. H. T. Yuan and D. W. Mueller, personal communication (1976).
1.55. J. T. C. Chen and K. Verma, *Int. Electron Devices Meet., Washington, D.C.* 13.5 (1974).
1.56. J. M. Pankratz, J. B. Kruger, You-Sun Wu, and Han-Tzong-Yuan, *Cornell Electr. Eng. High Freq. Generation Amplif. Conf., Ithaca, N.Y., 1975.*
1.57. S. Teszner, P. Durand, and J. Laplanche, Fr. Patent 2,122,777 (1971).
1.58. F. N. Sechi, *Cornell Electr. Eng. High Freq. Generation Amplif. Conf., Ithaca, N.Y., 1975.*
1.59. H. Bernard and G. Orthras, *Proc. Eur. Microwave Conf., Hamburg* C. 10.7 (1975).
1.60. R. C. Joy and E. S. Schlig, *IEEE Trans. Electron Devices* **17**, 586 (1970).
1.61. F. Bergmann and D. Gerstner, *IEEE Trans. Electron Devices* **13**, 630 (1966).
1.62. D. Navon and R. E. Lee, *Solid-State Electron.* **13**, 981 (1970).
1.63. D. Navon, *IEEE Trans. Electron Devices* **20**, 907 (1973). (Correspondence.)
1.64. E. M. Juleff, *Solid-State Electron.* **16**, 1173 (1973).
1.65. H. A. Schafft, *Proc. IEEE* **55**, 1272 (1967).
1.66. C. Popescu, *Solid-State Electron.* **13**, 887 (1970).
1.67. H. C. Josephs, *IEEE Trans. Electron Devices* **13**, 778 (1966).
1.68. P. L. Hower and V. G. K. Reddi, *IEEE Trans. Electron Devices* **17**, 320 (1970).
1.69. W. B. Smith, D. H. Pontius, and P. P. Budenstein, *IEEE Trans. Electron Devices* **20**, 731 (1973).
1.70. D. L. Bowler and F. A. Lindholm, *IEEE Trans. Electron Devices* **20**, 257 (1973).
1.71. A. Van der Ziel and D. Agouridis, *Proc. IEEE* **54**, 411 (1966). (Letters.)
1.72. R. J. Whittier and D. A. Tremere, *IEEE Trans. Electron Devices* **16**, 39 (1969).
1.73. D. C. Agouridis and A. Van der Ziel, *IEEE Trans. Electron Devices* **14**, 808 (1967).
1.74. A. Van der Ziel, *Proc. IEEE* **58**, 1178 (1970).
1.75. H. Fukui, *IEEE Trans. Electron Devices* **13**, 329 (1966).
1.76. B. A. McDonald, *IEEE Trans. Electron Devices* **17**, 134 (1970).
1.77. J. A. Archer, *Solid-State Electron.* **17**, 387 (1974).
1.78. K. F. Knott, *Solid-State Electron.* **16**, 1429 (1973).

II. Field-Effect Transistors

A. Basic Considerations

1. General Theory

The basic idea of field-effect transistors (FET) consists in using an electric field effect in order to control the current flow cross section. As with bipolar transistors, it was again Lilienfeld who proposed in 1928 (*2.1*) a device using this idea for obtaining electric power amplification. Then in 1934, Heil proposed (*2.2*) a device structure in which such an effect occurs through air insulation. In 1952, Shockley (*2.3*) and Prim and Shockley (*2.4*) gave the first theory, in simplified form, of the field-effect transistor. This theory was completed and improved upon by Dacey and Ross (*2.5, 2.6*), who presented the first field-effect transistor structures and performances in 1953 to 1955, then by Teszner (*2.7, 2.8*), who demonstrated in 1958 to 1961 the structures and performances of the first field-effect transistor (called Tecnetron) working in the vhf band, and then by Teszner and Gicquel (*2.9*), with a presentation in 1964 of the first high-frequency power field-effect transistor (called Gridistor) structures, with pentode, triode, or mixed pentode–triode characteristics.

Many specific theoretical studies then appeared on channel pinching and saturation characteristic problems in pentode mode operation and equivalent circuit parameters, as well as power–frequency performance determination. Regarding the saturation mechanism, one should cite in particular Grosvalet et al. (2.10), Hauser (2.11), Grebene and Ghandhi (2.12), Kim and Yang (2.13), David (2.14), Kennedy and O'Brien (2.15), Chiu and Ghosh (2.16), Hower and Bechtel (2.17), and Lehovec and Seeley (2.18). An exhaustive review on this topic, due to Yang, has been published in a recent volume of this series (2.19), to which the reader will find complementary references.

Regarding the equivalent circuit parameter and power–frequency performance determination, one should cite Martin and Le Mee (2.20), Rose (2.21), Van Der Ziel and Ero (2.22), Johnson (2.23), Hauser (2.24), Teszner (2.25), Reiser (2.26), and Das and Schmidt (2.27).

Field-effect transistor operation can be controlled through pn or MS or MIS gate junctions. For the microwave power region, only the pn and MS types are to be considered at the present time. We shall now briefly consider the basic theory of these devices.

Figure 15 shows the elementary structures of the field-effect transistor in a planar (a) and cylindrical (b) configuration. The channel is cut off by the space-charge developed by gate–source bias voltage; it is said to be pinched off through the space-charge developed by drain–source bias voltage. Following Shockley's approximation (gradual case), the system would be one dimensional, with E_x assumed negligible compared with E_y, and $dE_x/dx = 0$. Then the Poisson equation of this system is reduced to:

$$dV^2/dy^2 \cong -\rho(y)/K \qquad (1)$$

where $\rho(y)$ is the charge density and K the dielectric constant.

Consequently, the cut-off and pinch-off voltage in the planar configuration is

$$V_{co} = \rho_0 a^2/2K \qquad (2)$$

while, in the cylindrical configuration, it is

$$V_{co} = \rho_0 r^2/4K \qquad (3)$$

where ρ_0 is the density of the space charge, in principle constant. For n-type material, commonly used for microwave transistors, $\rho_0 = q(N_D - N_A)$; in field-effect transistors $N_D \gg N_A$, so $\rho_0 \simeq qN_D$, q being the carrier charge.

In fact, these expressions are only available for gate–source bias, without

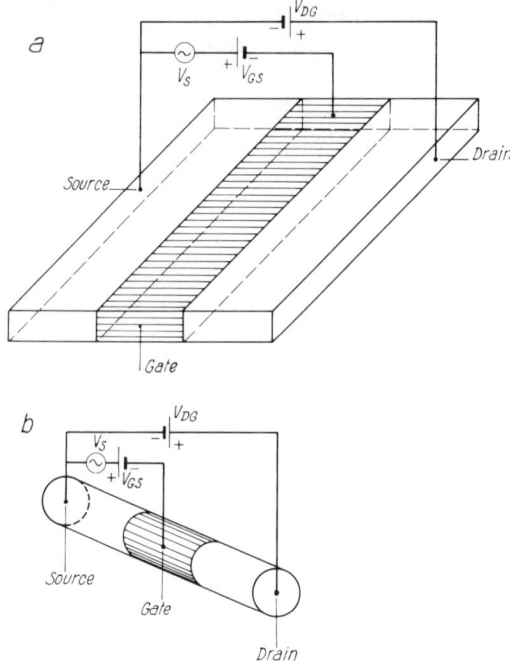

Fig. 15. Diagram of structure and circuit of field-effect transistor, N-type channel: (a) plane configuration, (b) cylindrical configuration.

drain–source biasing. Thus, they give only the cut-off voltage value and do not correspond to pinch-off voltage for the following two reasons:

a. Whatever the channel structure, it will not be pinched off by the drain–source voltage. It seems to be Teszner and Gicquel who first proposed (2.9), for the minimum channel width, that of the transition zone between a zone in equilibrium and a space-charge zone. The zone in equilibrium is then reduced to a plane in the plane configuration (laminar striction) and to an axis in the cylindrical configuration (centripetal striction). The width of this transition zone is equal to $2L_D$ and $2\sqrt{2}\,L_D$, respectively, where L_D is the so-called Debye length.*

b. In the specific case of microwave devices, the channel is generally short, its length-to-width ratio being relatively small. The gradual channel approximation is then no longer acceptable and the system should be considered two dimensional. Moreover, in the channel pinched condition, E_x at the narrow neck is comparable to E_y, whatever the channel length-to-width

---

* It will be recalled that $L_D = (KkT/q^2 N)^{1/2}$, where k is Boltzmann's constant, T the absolute temperature, and K, q, and N are as defined above.

ratio (*2.19*). This increase of E_x in the channel drain portion is a fundamental point of device theory. It results from nonlinear mobility of the carriers in the FET theory, explained first by Dacey and Ross (*2.6*), limited, however, to the tepid electron range. Then Teszner (*2.8*) proposed a demonstration of the necessary extension of this concept up to the hot electron range, at least for the narrow neck portion of the channel. This demonstration was based upon the cumulative process resulting from reciprocal action of the voltage drop increasing and the carrier mobility decreasing. Therefore, at least in this portion, the carrier-limiting drift velocity is attained and the greater part of the applied voltage is concentrated. This assumption has been qualitatively confirmed by the experimental results obtained by Tango and Nishizawa (*2.28*).

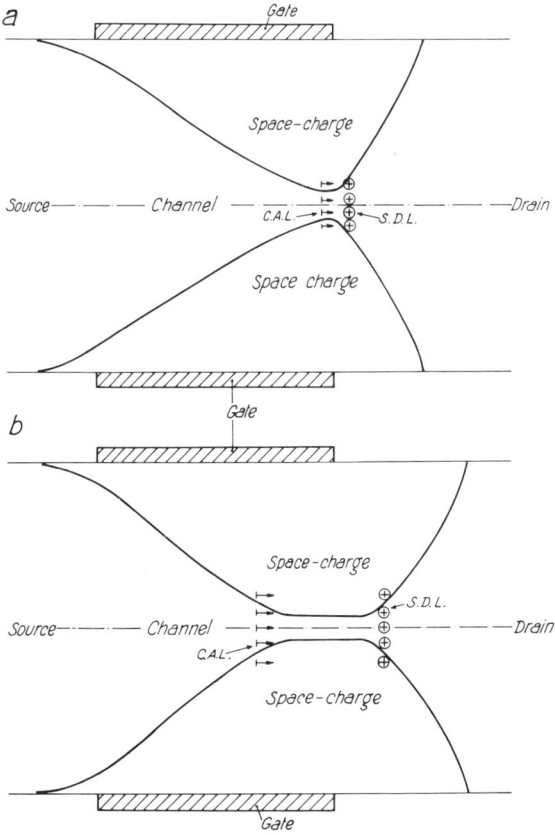

FIG. 16. Channel and space-charge configurations in field-effect transistor: (a) upon reaching the saturation point, (b) far beyond this point. C.A.L., carrier accumulation layer; S.D.L., space-charge developed layer.

Concerning the drain current saturation process in the pentode mode operation, drift carrier velocity saturation seems to be one of the main reasons for it (*2.10–2.12, 2.14–2.16, 2.18*), although this is not its only fundamental cause (*2.13, 2.17, 2.19*). In effect, a second necessary condition will be inferred from the following explanation of this process mechanism.

Figure 16 shows a channel and space-charge configuration upon reaching the saturation point (a) and far beyond the saturation attained (b). Figure 17 gives a sketch of the reciprocal variation within the narrow neck of the carrier concentration and the space-charge density.

We note that saturated carrier velocity is attained at least in the most pinched channel region. Then, if the channel current were confined within the conductive portion of the narrow neck with the decreasing carrier density as shown in Fig. 17, the current continuity would be broken. Thus it seems logical to note likewise, with Grebene and Ghandi (*2.12*) and Kennedy and O'Brien (*2.15*), that a carrier accumulation forms above the narrow neck with the opposite charge layer formation down this neck, as shown in Fig. 16a. Consequently, a space-charge limited current is produced and the current continuity is restored. It is noteworthy from our personal experience, that this current complement is absolutely necessary to explain

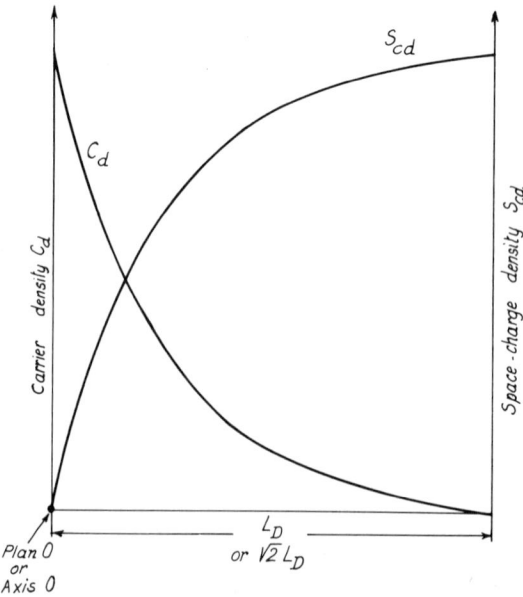

FIG. 17. Reciprocal variation of the carrier and space-charge densities within the most pinched channel region and beyond the saturation point.

the relatively high current saturation value obtained, primarily in the case of short-channel field-effect transistors.

When the drain–source voltage increases, the narrow neck length extends; our opinion is that this extension occurs in both directions, toward the source and toward the drain, as shown, for example, in Fig. 16b. The neck length and the carrier accumulation charge's simultaneous modulations are adjusted simply to maintain the current continuity and almost its saturation value, on the condition that the channel profile would be adequate. This condition may not be fulfilled, namely in the case of a channel with a small length-to-width ratio, i.e., $L/a < 2$, especially with a high-doped drain layer, and a divergent profile as in circular gate junctions (Fig. 18).

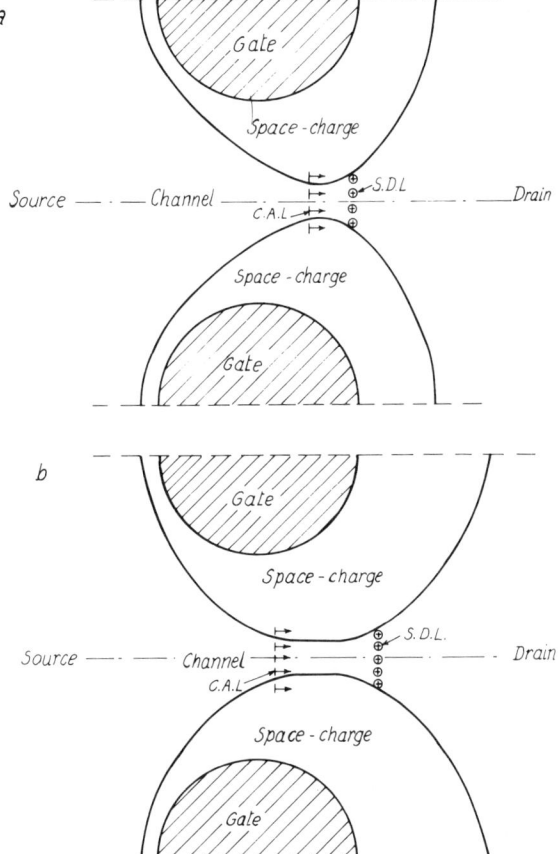

FIG. 18. Channel and space-charge configurations as in Fig. 16, but corresponding to the particular case of a circular gate junction.

However, this profile divergence effect may be corrected by an appropriate gradation of the channel doping concentration. In turn, it is obviously possible to derive a benefit from this effect through enhancing this divergence with a view to obtaining triode characteristics.* For the moment, however, such devices are not operational in the microwave power region. Thus, only the theory, structure, and performances of transistors operating in the pentode mode will be examined in this review.

First we shall consider the intrinsic frequency and frequency–power device limitations, and then the practical limitations, taking note of the parasitic elements of the transistor structure. As for bipolar transistors, we shall now determine (a) the cut-off frequency f_t, and (b) the maximum oscillation frequency f_{max}.

The cut-off frequency is essentially dependent on three particular times: (1) τ_{gc}, the gate–channel junction charging time, (2) τ_d, the dielectric relaxation time for the channel charge redistribution, and (3) τ_c, the carrier transit time through the channel. For this discussion we shall refer to Fig. 16.

For a signal applied in V_s, charging channel depletion layer, the time constant is

$$\tau_{gc} = r_c \times C_{gc}$$

where r_c is the equivalent channel resistance and C_{gc} the equivalent gate–channel capacitance. We shall now determine the expressions for these parameters.

The channel–gate equivalent circuit must be presented as a distributed nonuniform transmission line (2.8, 2.22, 2.25) (Fig. 19a). This presentation may be simplified (Fig. 19b), because the channel resistance concentration in the narrow-channel portion at the drain vicinity, especially for large-signal operation beyond the saturation point (Figs. 16b and 18b). τ_{gc} may then be given (2.8) by the approximate expression

$$\tau_{gc} = 0.05 r_c C_{GS} + 0.55 r_c C_{GD} \qquad (4)$$

where C_{GS} and C_{GD} are, respectively, the gate–source and gate–drain capacitances.

---

* Some investigators who produced such field-effect transistors called them "analog transistors" (2.29, 2.30). However, such an appellation was incorrect since the fundamental characteristic of analog transistor proposed for the first time by Shockley (2.31, 2.32)—as an analog to vacuum tubes—is that the semiconductor wafer has a balanced chemical impurity density in the region where the carriers flow, i.e., $N_D \simeq N_A$, corresponding to the intrinsic semiconductor definition. The semiconductor material forming the devices investigated by the authors cited above was absolutely extrinsic, with $N_D > 10^4 N_A$ or $N_A > 10^4 N_D$, very far from intrinsic level. Thus it is undeniable that these devices belong to the field-effect transistor category already described (2.9).

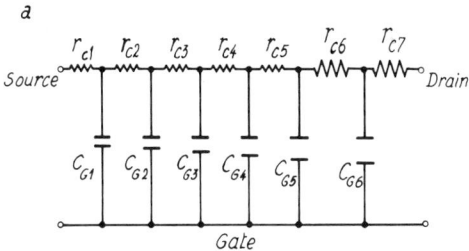

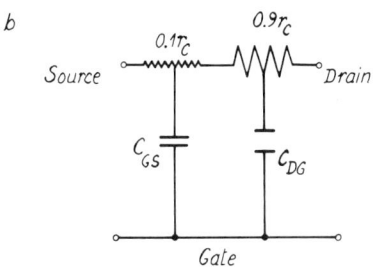

FIG. 19. Channel–gate equivalent circuit. (a) Distributed nonuniform transmission line. (b) With semilumped elements: $r_{c1}, r_{c2}, r_{c3}, r_{c4}, r_{c5}, r_{c6}, r_{c7}$, channel distributed resistances; $C_{G1}, C_{G2}, C_{G3}, C_{G4}, C_{G5}, C_{G6}$, gate-channel distributed capacitances; r_c, channel total resistance; C_{DG}, C_{GS}, drain-gate and gate-source capacitances, respectively.

Now assuming that the charging voltage is applied to the depletion layer, one must consider the second stage of the transistor operation. This stage will consist simultaneously of modulation of the depletion layers and of the carrier and opposite space-charge accumulated layers, and of adjustment of channel carrier concentration. This evolution will be performed through two simultaneous processes, i.e., the charge redistribution by dielectric relaxation and the carrier transit through the channel. We must then consider two characteristic times: The dielectric relaxation time of the semiconductor used, given by

$$\tau_d = K\varepsilon_0/qN|\mu| \tag{5}$$

where ε_0 is the permittivity of free space, and the transit time, for which only the transit time of the narrow neck, receiving by far the greater part ($\sim 90\%$), will be reckoned with. At the large-signal operating point (Figs. 16b and 18b) it may be assumed that the narrow neck length is $\leq 0.5L$, where L is the original channel length. Then

$$\tau_t = 0.5L/v_{sl} \tag{6}$$

Since both effects occur at the same time, only the effect of longest duration should be retained for the overall gate–drain delay-time determination. Now, for microwave power transistors, N_D can reach 10^{17} cm^{-3}, in the case of GaAs and Si, the only materials to be considered at the present time. For this N_D value, $\tau_d \geq 10^{-13}$ sec. On the other hand, L_{min} being ~ 0.5 μm, $\tau_t \simeq 5 \times 10^{-13}$ sec. Then, only the latter will be retained for the τ_{gd} calculation. Thus

$$\tau_{gd} \simeq \tau_{gc} + \tau_t \tag{7}$$

However, in much the same way as was proposed for bipolar transistors (cf. p. 143), at least, the main parasitic elements should be introduced in the formula giving the cut-off frequency. For this purpose we shall consider the FET equivalent circuit for a common gate (Fig. 20a) or for a common source (Fig. 20b) configuration. τ_{gc} will then be replaced by τ'_{gc}, expressed approximately as

$$\tau'_{gc} \simeq (r_{pG} + r_{pS})C_{pGS} + 0.05 r_C C_{GS} + 0.55 r_C C_{GD} + (r_{pG} + r_{pD})C_{pGD} \tag{8}$$

for the common-gate configuration given as an example,* where r_{pG}, r_{pS}, and r_{pD} are, respectively, the parasitic resistances of gate, source, and drain regions, C_{pGS} and C_{pGD} being the parasitic capacitances of the gate–source and the gate–drain.

Finally,

$$\tau'_{gd} = \tau'_{gc} + \tau_t \tag{9}$$

and

$$f'_t = \frac{1}{2\pi \tau'_{gd}} \tag{10}$$

Nevertheless, the field-effect transistor operation appears much simpler than that of bipolar transistor, and consequently τ'_{gd} should be shorter than τ'_{gd}, as defined by formula (11) in Section I. Then the maximum operating frequency would be correlatively higher. It would now be instructive to specify its possible value for large-signal operation with a corresponding operation domain in class A (Fig. 21), similar in such a case to that of the bipolar transistor.

Under these conditions $\tau'_{gc\,min}$ would be $\sim 10^{-12}$ sec for Si and $\sim 3 \times 10^{-13}$ sec for GaAs. Then, with τ_t at $\simeq 5 \times 10^{-13}$ sec as indicated

---

* The corresponding expression for common-source configuration differs only in some nonessential details.

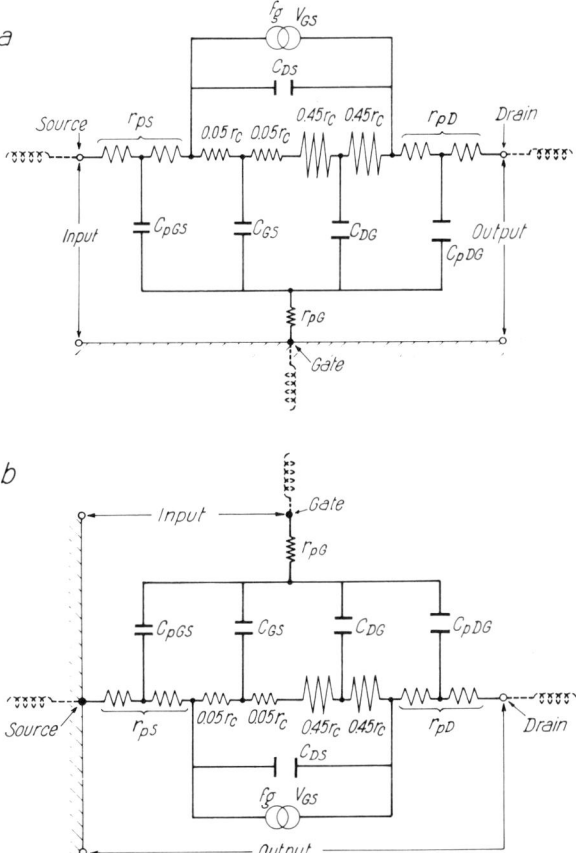

FIG. 20. FET equivalent circuit: (a) for common-gate configuration, (b) for common-source configuration. r_c, channel total resistance; r_{pS}, r_{pD}, r_{pG}, source, drain, and gate parasitic resistances, respectively; fg, forward transadmittance; C_{DS}, C_{DG}, C_{GS}, drain–source, drain–gate, and gate–source capacitances, respectively; pC_{DG}, pC_{GS}, parasitic drain–gate and gate–source capacitances, respectively.

above, one obtains $\tau_{gd\,min} \simeq 1.5 \times 10^{-12}$ sec for Si and $\tau_{gd\,min} \simeq 8 \times 10^{-13}$ sec for GaAs. Hence $f_{t\,max}$ would be respectively ~ 100 and ~ 200 GHz.

This significant difference between $f_{t\,max}$ values relating to these two materials results from substantially lower resistances in the GaAs devices through the much higher carrier mobility at low electric fields (as discussed in Section II,B,3).

For the f_{max} calculation, we shall refer to Fig. 20 and which will be

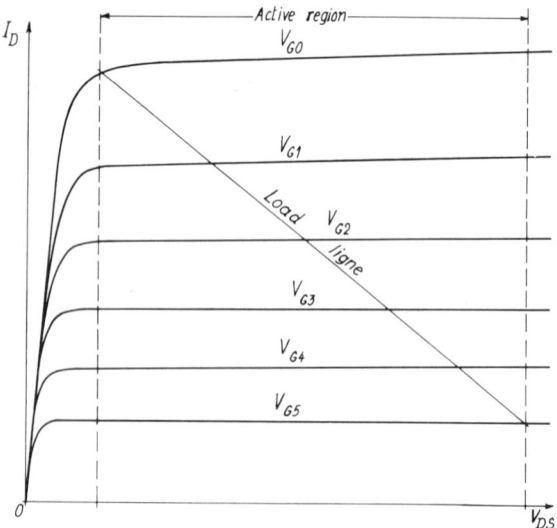

FIG. 21. FET dc characteristics for large-signal pentode operating mode in class A.

transformed into the equivalent π circuits presented in Fig. 22, the parameters of which give the unilateralized power gain U:

$$U = \frac{|y_{21} - y_{22}|^2}{4[\text{Re}(y_{11})\,\text{Re}(y_{22}) - \text{Re}(y_{21})\,\text{Re}(y_{12})]} \tag{11}$$

where $|y_{21} - y_{12}|$ is the transadmittance, the real part of which is the transconductance g_m.

The gain is higher in the case of a common-source configuration. On the contrary, the device is more stable in the common-gate configuration, which seems to be the only possible connection near f_{max}. The best method for f_{max} determination is the numerical calculation of U, f_{max} corresponding as usual to $U = 1$. This numerical calculation with optimized circuit parameters provides f_{max} about the same order as f'_t given above. These values obviously depend on the device structure and especially on its geometry. Two groups of problems must be solved:

1. The active element efficiency should be maximized, then the transconductance (per effective surface unit) should be as high as possible for a channel length as small as possible. In short, the intrinsic figure of merit given by the ratio $F_m = g_m/(C_{GS} + C_{GD})$ must be maximized.

2. The parasitic elements should be reduced as far as possible in order to minimize r_{pS}, r_{pD}, r_{pG}, C_{pGS}, and C_{pGD}.

In the following structure geometries discussion, we shall examine the solutions implemented with a view to solving both these types of problems.

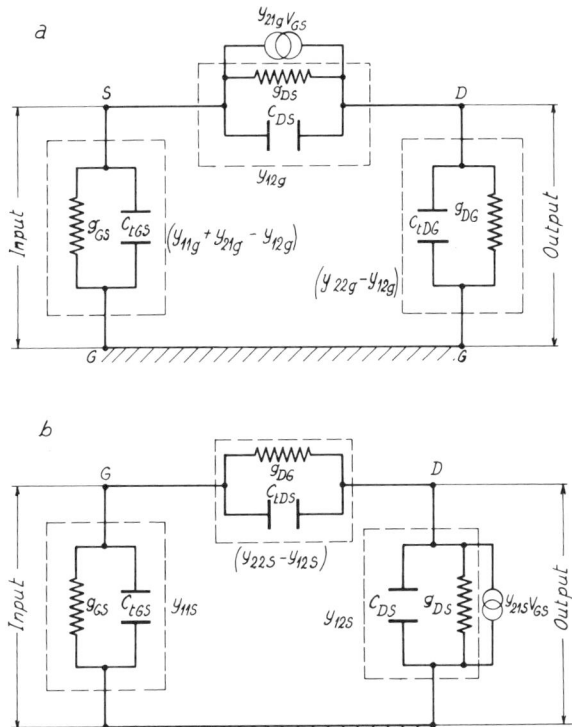

FIG. 22. FET equivalent π circuit for y parameter calculation of (a) common-gate and (b) common-source configurations. S, source; D, drain; G, gate; g_{DS}, g_{DG}, g_{GS} and C_{DS}, C_{tDG}, C_{tGS}, respectively, drain–source, drain–gate, and gate–source total conductances and capacitances.

2. Structure-Type Geometries

Two types of FET geometry have been developed for the microwave power region: (a) horizontal channel geometry and (b) vertical channel geometry. Two general problems are to be solved: (1) maximizing the active structure efficiency and (2) minimizing the parasitic elements. From this viewpoint, we shall examine both types of geometry successively.

a. Horizontal channel structure geometry. For the frequencies up to 1 GHz, the commonly used planar *pn* junction structure seemed not to be available for the microwave power region. The main reasons for this limitation were the excessive parasitic capacitance and parasitic resistance values. Thus, the only microwave power structures actually developed are the mesa or planar structures with MS gate-junction (called MESFET or Schottky-barrier FET) and the mesa structure with *pn* junction.*

---

* As indicated in Section II,1 only the *pn* or MS gate-junction FETs will be considered.

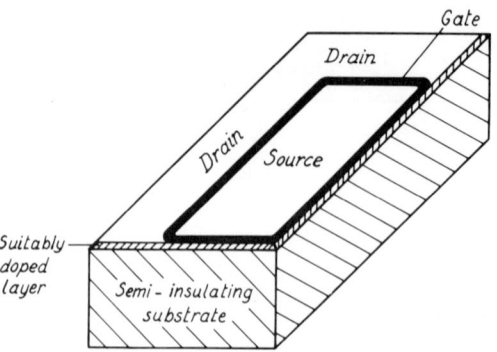

Fig. 23. Isometric view showing plan and cross section of a small-signal microwave MESFET structure.

The microwave MESFET was first presented in 1968 (2.33),* but it was only in 1973 that the microwave power MESFET appeared (2.34, 2.35, 2.37). In fact, the power structure is much more complicated than for small-signal operation, the latter consisting of a single metallic gate (Fig. 23) deposited on a GaAs layer grown on a semi-insulating substrate. Now for large-signal operation in the microwave region, a much more sophisticated gate

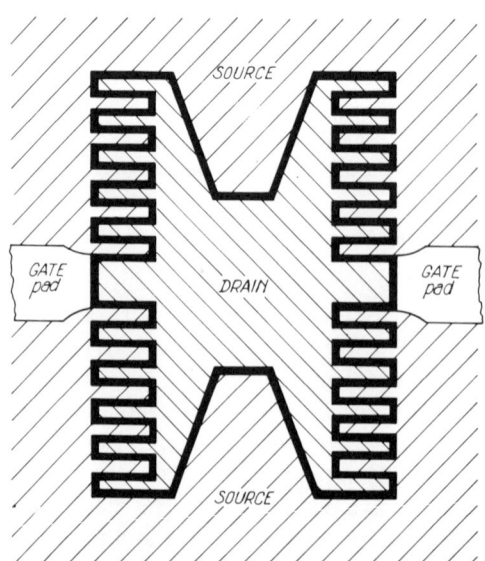

Fig. 24. Basic example of interdigitated microwave power MESFET structure geometry. Note that gate pads are overlaid on the source–drain layer and are insulated from this layer.

* However, the first MESFET was Tecnetron (2.7) developed for the VHF band in 1958.

geometry must be conceived. The fundamental problem is the considerable extension of the source periphery, as well as of the overall gate width. To attain this objective without excessive increase of the gate resistance, it is necessary to subdivide the gate into several pads connected in parallel. For this purpose, geometries similar to those for bipolar transistor have been developed, namely an interdigitated geometry (2.34–2.36) (Fig. 24), a mesh geometry (2.37) (Fig. 25), and the diamond-shaped drain area with closed cell structure (2.36) (Fig. 26). On the other hand, a new geometry, particular to this type of device, has been developed and called the multigate structure (2.38, 2.39) (Fig. 27).

The essential quality of this horizontal type of device is that the source–gate and drain–gate intrinsic capacitances are minimized. Moreover, the abrupt gate-junction provides the highest possible rate of depletion layer width variation versus gate–source voltage. Obviously then, g_m will be optimized and consequently also the figure of merit F_m, as defined above.

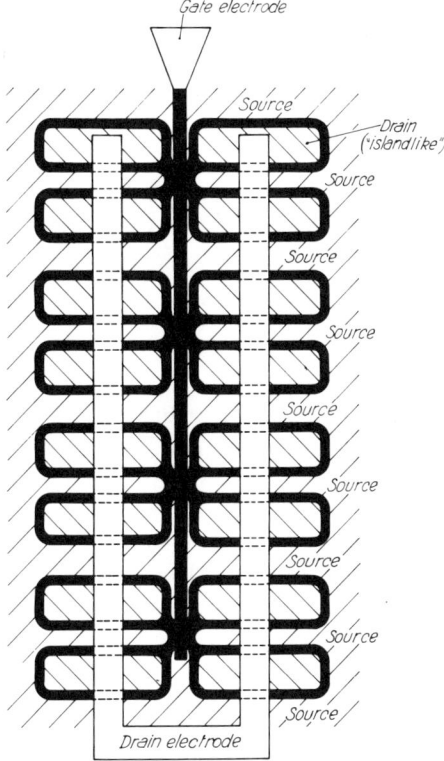

FIG. 25. Basic diagram of Mesh-source type MESFET structure geometry (gate and drain electrodes are overlaid).

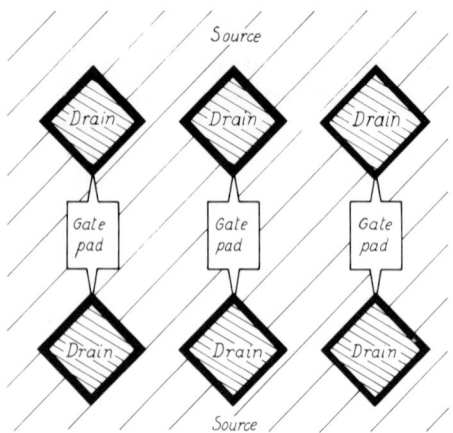

Fig. 26. MESFET structure geometry with diamond-shaped drain area (gate pads are overlaid).

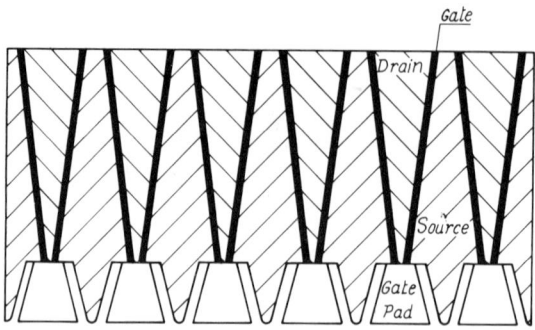

Fig. 27. Multigate MESFET structure geometry (source and drain electrodes are overlaying).

In turn, however, the g_m value is reduced, since the gate action is limited to one channel side. On the other hand, whatever the geometry used for transposing from a small- to large-signal structure, parasitic capacitances, resistances, and inductances result. The parasitic capacitances may exceed, even substantially, the intrinsic capacitance values. Finally, heat evacuation through the semi-insulating layer is deficient, especially in the GaAs device case. It is an important factor of power limitation.

More recently a different structure has been proposed (2.34) with MS gate-junction, and one has also been developed (2.40) with a *pn* gate-junction (Fig. 28). This structure is strictly derived from the structure called MEDC, which was presented earlier by Chappey *et al.* (2.41, 2.42). It consists of a monolithic gate forming the structure base layer, metallic or semiconductor highly doped. On this gate, a suitably doped N-type

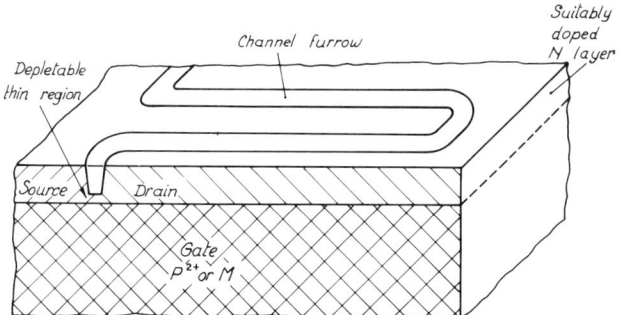

FIG. 28. Microwave power FET interdigitated geometry with channel furrow made on monolithic underlying gate.

semiconductor layer is deposited on which the source and drain areas are delimited through the channel furrow. The thermal conduction is thus greatly improved. On the other hand, this interdigitated structure pattern, obtained by free surface etching, is simplified and parasitic resistances are substantially reduced.

In contradistinction, the source–gate and drain–gate capacitances are greatly enlarged. Furthermore, the leakage risk on the free surface, despite the surface passivation process, imposes a stringent channel potential gradient limitation, which is an undeniable drawback for a power microwave FET. In effect, for a very high operating frequency and consequently short channel length, the drain admissible voltage should be very low and thus unsuitable for large-signal operation. Finally, concerning the active structure efficiency, the gate is still confined, as in the structures previously discussed, to one channel side, g_m being consequently limited.

b. Vertical channel structure geometries. The basic characteristic of this structure geometry is the plurality of vertical channels within an integrated grid. Vertical channel plurality was proposed in 1958 (*2.7*) and the integrated grid structure of this type of device (named Gridistor) was presented in 1964 (*2.9*).

Two types of channel cross-sectional shape have been developed: round and rectangular. It is noteworthy that the round configuration intrinsically ensures some remarkable particularities, namely, that for the same transverse dimension, the cut-off voltage is half that in the rectangular configuration. Consequently the transconductance is twice as high as the figure of merit; likewise the saturation of the pentode characteristics is better (*2.7*). On the other hand, however, the grid bulk surface is larger there than in the rectangular configuration, at least for a moderately active surface ratio (total channel cross section to total grid area). Thus, it proves a fundamental drawback for a microwave power structure, through an increase of the

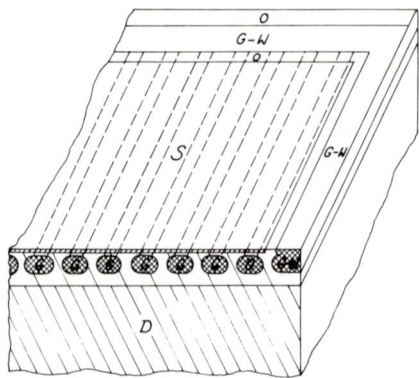

FIG. 29. Isometric view showing plan and cross section of a vertical channel FET structure obtained with conventional technology. S, source region; D, drain region; G, gate (grid); GW, gate connection wall; O, oxide layer.

parasitic gate–source and gate–drain capacitances (*2.43*). Therefore, we shall only consider the vertical structure with rectangular channel cross section. For this type of microwave vertical FET, Fig. 29 gives an isometric view showing plan and cross section of the structure obtained with conventional technology: diffusion and epitaxial deposition (see Section II,B).

In order to minimize the parasitic gate resistance, a distributed geometry equivalent to the interdigitated underwent early development (*2.43*). Figure

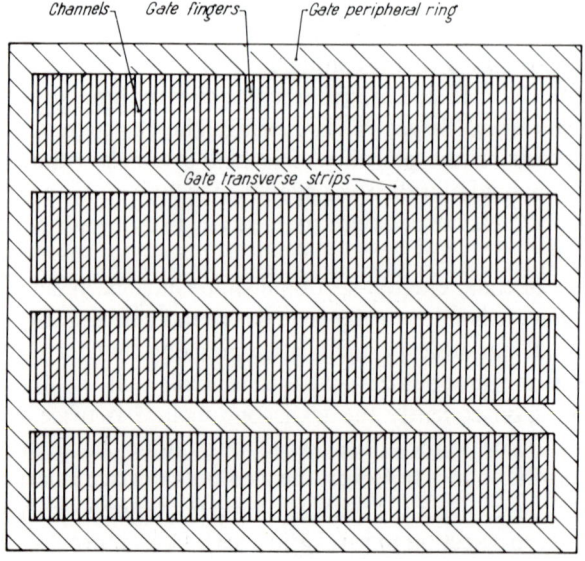

FIG. 30. Plan cross section of distributed Gridistor geometry.

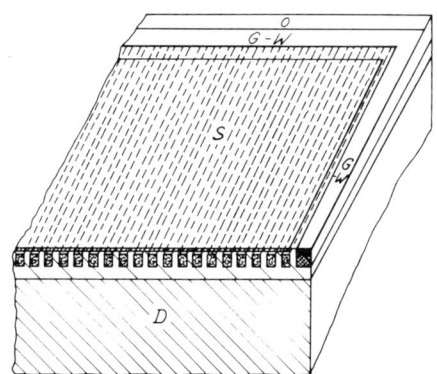

FIG. 31. Isometric view showing the improvement in the structure of Fig. 29 obtainable, in principle, through the ion implantation technology.

30 shows an example of such a geometry. On the other hand, the channel profile, junction abruptness, and active surface ratio may be improved through ion implantation technology (2.44) (Fig. 31). (This will be discussed in Section II,B.)

The essential advantage of the vertical channel geometry seems to be that its "volume structure" allows a better pellet exhaustion than the "surface structure." Consequently, for the same pellet surface a much higher frequency–power product, made possible through internal parallel connection of many channels, can be obtained. Obviously it will be dependent on the active surface ratio defined above, which results from the grid geometry resolution. Now, since its sophisticated structure is buried, it is easier there than on a passivated surface to realize and maintain a reliable high-resolution geometry. In fact, a grid resolution giving an active surface ratio of up to $\simeq 70\%$ seems to be a future possibility.

Other advantages of this structure are that the gate action stretches out over all the channel periphery, and the parasitic source and drain resistances are very low. On the other hand, the parallel connection of a plurality of channels being made through the structure itself, this device is practically exempt from external connections.

Its principal drawback, however, is that for a given active surface ratio, the parasitic gate–source and gate–drain capacitances are higher than for an equivalent surface structure.

The second disadvantage is the difficulty, at present, of applying GaAs (or InP) in this type of structure. For the moment only silicon is effectively used, and the series resistance of all low-field structure regions is at least four times higher than it would be with a gallium arsenide application.

Table II.1 presents a summary of the advantages and drawbacks of the structure geometries discussed above.

TABLE II.1

Structure geometry	Advantages	Disadvantages
Horizontal channel structure geometry MESFET, deposited on semi-insulating layer	Abrupt gate–junction, optimizing gate action efficiency per unit junction length Source–gate and drain–gate intrinsic capacitances minimized; minimizing of intrinsic resistances through use of semiconductor material with high carrier mobility (e.g., GaAs)	Junction–gate and consequently gate action limited to one channel side Notable parasitic resistances, inductances, and capacitances associated with transposal from a low- to a high-power structure Power dissipation limited through poor thermal conductivity, for the time being, problems in providing suitable reliability
MS or *pn* junction–gate being the basis of the drain–source layer	Thermal conductivity greatly improved; power structure geometry simplified and parasitic resistances reduced Minimizing of intrinsic resistances through using of semiconductor material with high carrier mobility (e.g., GaAs)	Intrinsic capacitances greatly increased Drain–source voltage strongly limited For the time being, problems in providing suitable reliability
Vertical channel structure geometry	Sophisticated geometry structure being embedded, relative ease in obtaining high-resolution geometry for this structure with suitable device reliability; ease of attaining high power without multiplying external connections Gate action stretching over all the channel periphery; low parasitic source and drain resistances	Relatively high gate–source and gate–drain parasitic capacitances For the time being, inability to realize the structure with GaAs or InP

B. Device Design and Fabrication

The related problems have already been discussed, and the FET structures performed have been described in many recent articles (cf. *2.34–2.40, 2.43–2.51, 2.54*). We shall examine successively the design and fabrication of horizontal (with both types of gate-junction situation) and vertical channel structures. For the present time, only the usual technology will be considered; however, the technologies currently in the experimental stage will be also discussed.

Apart from the semiconductor material presently used (GaAs for the horizontal and Si for the vertical channel structures), other choices will be considered, the application of which is in the experimental stage at present.

1. Horizontal Channel Structures.

a. Overlaid gate–junction structure. The geometries of this, the most commonly used, structure have already been shown in Figs. 24–27. However, for the fabrication process discussion, we shall also refer to the Figs. 32 and 33, which give the vertical cross sections available for all types of gate geometry.

In all cases the substrate is semi-insulating GaAs, generally Cr doped. Now, if the structure channel layer is deposited directly upon this substrate, the current leakage at the interface is not negligible and appears troublesome. To avoid this drawback a buffer layer interposition, of high resistivity but not a semi-insulating GaAs, has been proposed. On the other hand, many problems have appeared concerning the electrode alignment, especially for the gate electrode within the gate furrow, resulting in short-circuit risks. The self-aligned concept presented in 1971 (*2.46*) seems able to prevent this difficulty from arising.

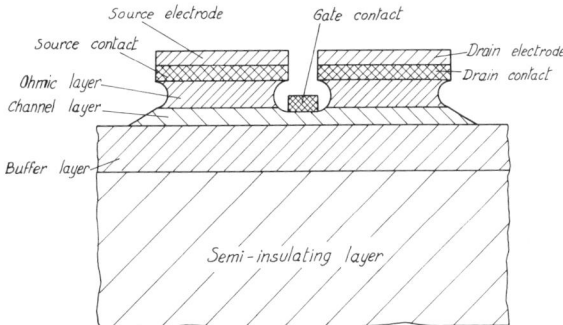

FIG. 32. Some construction details of MESFET horizontal mesa structure obtained through overlaid junction-gate self-alignment.

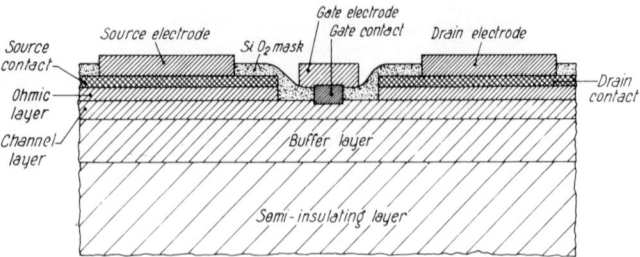

Fig. 33. Some construction details of MESFET horizontal mesa structure obtained through ion beam milling of overlaid junction-gate.

Figure 32 is related to a mesa structure with the most sophisticated structure and fabrication process to date, starting with triple epitaxial layer deposition. The first grown is the buffer layer. The second, the most essential, is the channel layer, necessarily N type, doped with sulfur or selenium with a concentration lying between 2×10^{16} and 1×10^{17} cm^{-3}, the choice being a function of structure geometry and of the channel thickness. The third layer is the ohmic layer for source and drain contacts: it is heavily N^{2+} doped, with a concentration of 5×10^{18} to 1×10^{19} cm^{-3}. The device's active area will be delineated afterward by removing these epitaxial layers from this area by mesa etching.

This active surface is metallized by evaporation of ~ 0.05 μm thick Au–Ge alloy film, followed by an ~ 0.5 μm thick Au layer. The FET pattern is then formed using a standard photolithographic technique, following, for instance, one of the geometries presented in Figs. 24–27. The metallic layer is then etched, forming a patterned layer, and is used as an etchant mask of the semiconductor layers in order to prepare an appropriate site for gate-electrode evaporation (e.g., also Au–Ge plus Au). It is noteworthy that the undercut of the metal as shown in Fig. 32 serves to automatically align the gate electrode in an accurate location between drain and source (2.46). Currently the gate length is 1–2 μm, for the structures operating up to the X band, the separation between source and gate being ~ 1–2 μm and that between drain and gate ~ 2–3 μm.

Finally, the respective electrodes (of gate, source, or drain) of the device subunits are interconnected. This process sequence is particularly delicate due to the concomitant risk of substantial degradation of the structure figure of merit. With a view to offsetting such a risk, this interconnection is performed for some structures partly on the ohmic side and partly on the semi-insulating side of the device; for other structures the connections are separated from the chip surface through an insulating layer (e.g., SiO$_2$). The latter technique may be extended by making the structure planar in order to minimize the leakage currents.

Figure 33 gives an example of such a structure, where the gate site may be delineated through ion-beam milling into the metal and subsequently into the semiconductor layers. On the other hand, the chemical etching of the epitaxial layers outside the active device area may be replaced by proton (ion H^+ or H^{2+}) implantation in order to form an amorphous, high-resistivity layer.

As indicated already, the aim of a buffer layer insertion is to minimize the leakage current on the bottom face of the channel layer. In fact, it seems to be one of the most delicate problems in this type of construction. However, it is not sure that this layer insert is sufficient to guarantee suitable reliability, the problem of the entire open surface effective passivation remaining to be solved in any case. On the other hand, some promising results have been already obtained without that layer. In that case obviously, double rather than triple epitaxy technology is employed.

Alternatively, it could prove interesting to aim at the complete elimination of epitaxial layer deposition, through ion implantation (*2.50, 2.51*). In fact, such technology would allow the channel to be obtained with a more exact thickness and carrier concentration than is currently obtained. The present investigation concerns, in particular, channel layer formation through implantation in the semi-insulating GaAs substrate of sulfur ions, e.g., with a dose of $\sim 5 \times 10^{12}$ cm^{-2} at 30 keV. After implantation, the samples are submitted to annealing treatment at 820 to 880°C providing, for treatment at the higher temperature, an electrical activation of up to 80 to 100% of the implanted ions. This technology is not yet perfected, but the performances obtained with the experimental samples are already quite similar to those of devices developed with common technology (with the exception of the noise figure, which is higher).

We shall briefly discuss the fabrication process of a new type of horizontal channel overlaid gate-junction MESFET, with two gates working successively and named dual-gate MESFET. This device is for the moment in the early experimental stage, at least as far as microwave power applications are concerned (e.g., *2.47, 2.48*). However, some appreciable benefits, especially improvement of the stability factor, may be expected from this experimental work.

Figure 34 gives a symbolic description and a conceptual vertical cross section corresponding to a fabrication process recently done for this device (*2.48*). The tetrode circuit (Fig. 34a) ensures a substantial feedback reduction, whence a potential possibility of power gain increase; moreover, two independent control gates provide increased functional capabilities. The design and fabrication process providing the structure shown in Fig. 34b use the gate self-alignment technique for both gate electrodes. This technique makes it possible to obtain submicrometer gates. The structure is formed as

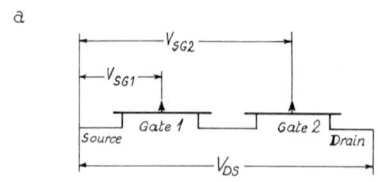

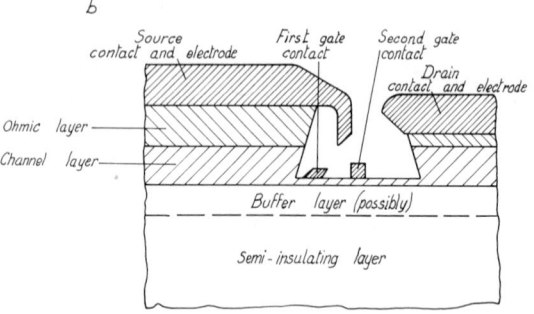

FIG. 34. Dual-gate MESFET. (a) Symbolic schema. (b) Conceptual vertical cross section corresponding to a structure created by the gate self-alignment technique.

usual upon a semi-insulating GaAs substrate, through double or triple epitaxial deposition. Then the deposited layers are etched up to delineate first the source and drain sites, which will be metallized, and second the dual-gate site. The wafer is later covered with a photoresistant film that is able to withstand ion-beam milling and is patterned to leave this film over the channel as well as over the source and drain contact regions. Then the wafer is properly etched up to the appropriate channel thickness as required, and both the gate sites are overhung through the metallic electrodes.

Subsequently, the first gate electrode is evaporated (with Ti–Au alloy) at a 45° angle to the normal to deposit this electrode under the edge of the source contact, and then the second gate electrode is evaporated (also with Ti–Au alloy) at a nearly normal angle of incidence. On the other hand, both gate pads are defined.

At present, only small-signal dual-gate MESFETs have been fabricated for the microwave region, and only as experimental samples. However, it also seems quite possible to fabricate the microwave power dual-gate devices, although with an even more complicated procedure.

b. Underlaid gate–junction structure. The only fabrication process described until now for this type of device (Figs. 28 and 35) is that of a structure formed from a double epitaxial deposition upon a P^{2+} GaAs substrate of a GaAs thin (~ 1 μm) "channel" layer doped N with 1 to 2×10^{16} cm^{-3} concentration and a very thin GaAs N^{2+} doped "ohmic"

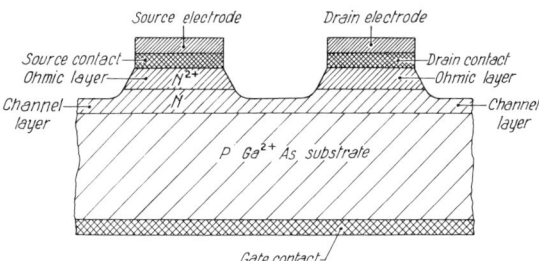

FIG. 35. Some construction details of horizontal channel underlaid gate-junction structure.

layer (2.40) (Fig. 35). The N^{2+} layer is then metallized (with Au–Ge and Au) and this metallization as well as both deposited layers are etched by ion beam, following an interdigitated pattern (Fig. 28) with adjustment of the channel thickness. The device is then delineated by chemically etching away those parts of the N^{2+} and N layers outside the active area. An alternate technique to this etching process may be selective proton bombardment of these layers in order to convert their inactive parts to semi-insulating material. Finally, the open surface of the N layer is passivated by SiO_2 sputtering, and at the rear the P^{2+} is metallized by gold evaporation.

The most delicate steps of this process are the channel delineating through ion-beam etching and its oversurface passivation. Concerning the channel delineating, it is important to avoid undercutting in order to reduce the parasitic resistance; this is the reason for the choice of ion-beam etching instead of chemical etching. Concerning the channel surface passivation, the problem seems to be even more delicate, because it is important to prevent any parasitic surface conductivity. Moreover, it seems to be more difficult to minimize the channel length than in the overlaid gate-junction structure. In contradistinction, the possibility of putting the active region in almost intimate contact with a high thermal conductivity heat sink is an assurance of better heat removal capability.

2. Vertical Channel Structures

a. *Structure performed in conventional technology.* Regarding the structure design, the internal (gate) geometry has been given in Figs. 29 and 30. The external geometry on the upper surface is shown in Fig. 36a. There is a dramatic difference between the resolution rates of these geometries, the high-resolution geometry being entirely embedded.

Concerning the fabrication process, we shall refer first to Fig. 36b. This process starts with double epitaxial deposition on the Si N^{2+} substrate (doped to $\sim 10^{19}$ cm^{-3}) and on the drain Si N layer, with a doping concen-

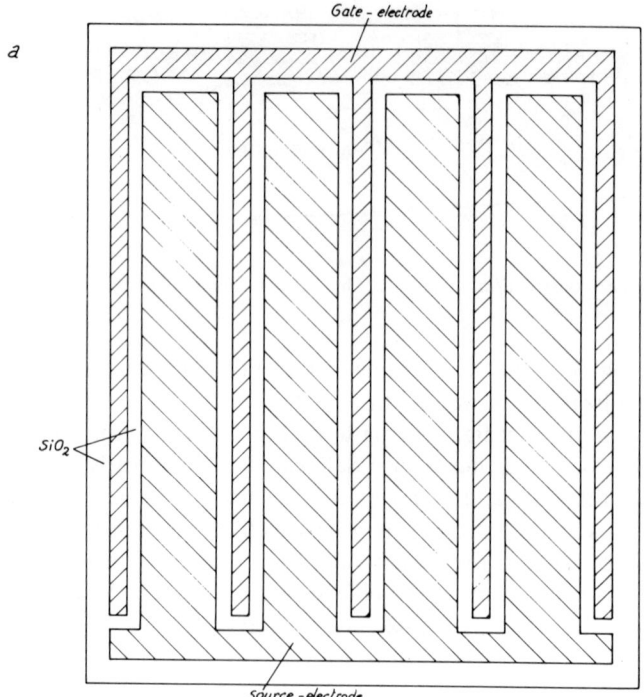
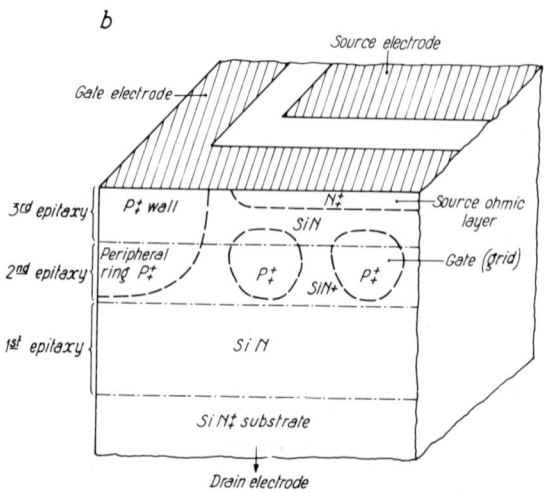

Fig. 36. Some construction details of vertical channel structure. (a) External geometry on the upper surface; (b) isometric view with vertical cross section (structure performed through conventional technology).

tration of some 10^{15} cm^{-3}, and on the "channel" Si N layer, with a doping concentration of 2 to 6 × 10^{16} cm^{-3}. The thickness of these layers will be ~2 and ~1 μm, respectively.

The gate pattern is then formed on the channel layer open surface using standard photolithographic techniques, and the gate of the kind shown in Fig. 30 is shaped by boron diffusion with a doping concentration of ~ 10^{19} cm^{-3}. Then, after surface cleaning, a third epitaxial deposition follows, likewise of Si N type, forming the source layer, the thickness of which is < 1 μm and the doping concentration intermediate between that of the drain and that of the channel. The aim of this doping gradation is to minimize the parasitic gate–drain and gate–source capacitances and at the same time to optimize the channel profile evolution with the gate biasing (2.43) as much as possible.

Subsequently the structure is completed by diffusion into the source layer, through an oxide mask, of boron in order to constitute the P^{2+} walls connecting to the gate peripheral ring and to transverse strips (see Fig. 30), and secondly through another mask, of phosphorus, to form the N^{2+} source contact area. Finally, the gate and the source areas on the upper surface are metallized, forming in this planar structure the gate and source electrodes, respectively, as shown in Fig. 36a, while at the rear the whole surface is metallized and forms the drain electrode.

This procedure is somewhat more complicated than those found in horizontal channel structures. Notwithstanding, the planar device performed in this way seems to be more reliable through full embedding of the high-resolution geometry structure. However, the resolution rate realizable with common technology is limited because the gate lateral diffusion rate is of the same order as that of vertical diffusion, as shown in Fig. 29. In fact, the grid body is thus spread out and the active surface ratio could not then practically exceed 25%. This drawback seems to be avoidable by using ion implantation technology.

b. *Structure performed through ion implantation.* Two alternate procedures should be considered. In one (Fig. 37a) the grid is formed by relatively deep (~ 1.2–1.5 μm) P^{2+} implantation, with a relatively high dose of boron ions implanted into Si N epitaxial layer of thickness ~ 3–4 μm, with some 10^{16} cm^{-3} doping concentration, deposited upon a Si N^{2+} substrate. The upper part of this implantation will then be compensated by N^{2+} implantation of arsenic, oxygen, nitrogen, or phosphorus ions with a suitable dose. Thus the grid embedding would be realized without source layer epitaxial deposition.

In the other one (Fig. 37b), the implantation is made in a Si N layer deposited upon a substrate Si N^{2+} as above, but with this layer being of lower thickness, i.e., ≤ 3 μm and with the implantation of boron ions depth

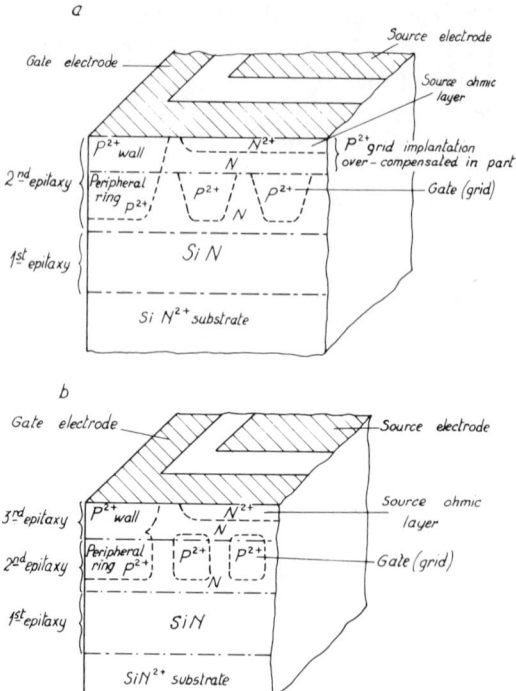

Fig. 37. Some construction details of vertical channel structure presently attainable through ion implantation technology: (a) grid P^{2+} deep implantation followed by N^{2+} shallow implantation; (b) grid P^{2+} shallow implantation followed by thin source layer epitaxial deposition.

being only ~ 0.5 μm. Later, the source Si N layer (with $5 \times 10^{15}-10^{16}$ cm^{-3} doping concentration) and thickness < 1 μm is formed by epitaxial deposition on this "gate-channel" layer. Thus the structure is realized by a combination of both techniques: ion implantation and epitaxial deposition.

We shall now discuss this procedural alternative. In the first case [described in (2.44)], the preliminary operations are diffusion into the Si N layer of P^{2+} walls for the grid connection and of N^{2+} source contact areas. The grid P^{2+} ion implantation is then performed through a chromium gold mask of thickness $\sim 0.6-0.8$ μm, with a dose of some 10^{14} cm^{-2} and with the relatively high energy $\sim 300-500$ keV N^{2+}. Ion implantation is later made through the same mask, with a sufficient dose for P^{2+} compensation and for source layer concentration within an upper limit of ~ 0.7 μm. Finally, the mask is removed and the operation of annealing the lattice disorder of this implanted wafer is carried out; generally a duration of 20 to 30 min and a temperature of $\sim 900°C$ for this operation are sufficient for attaining an electrical activation of $\sim 90\%$ implanted ions. This fabrication procedure is obviously completed by the source, gate, and drain electrode formation through metallizing (with a gold alloy).

Such a procedure appears attractive because it is only moderately complicated. However, two problems arise. First, in any case there is the current difficulty of obtaining a suitable and reproducible source layer N concentration. Second, in the present state of masking technology it seems to be impossible to produce a mask with the required thickness and abrupt edge. Now, tapered mask edges result in the formation of "horns" of high P doping concentration at the upper surface, and these P horns can not be properly compensated through N implantation. On the other hand, with the tapered mask edge the channel profile is degraded and the gate-junction becomes less abrupt.

Thus, for the moment the second procedure of the alternative seems to be preferable. In fact in that case, the source layer doping concentration will be quite reproducible. On the other hand, the mask being much thinner, i.e., < 0.5 μm, it is currently possible to make it with abrupt edges (2.52). Consequently, the implanted grid will likewise present an abrupt edge and an abrupt gate-junction. The annealing process of that implantation will be similar to that of the first procedure. The source layer is then deposited by epitaxy as indicated above, this operation being followed by the P^{2+} grid walls and N^{2+} source contact area diffusion. Finally, the source, gate, and drain contact areas are metallized.

In short, the first procedure would probably be the best technical solution in the future, supposing a radical improvement of technology of the patterned metal mask with abrupt edge and with suitable thickness. For the moment, however, the second procedure seems to be the most convenient and proper to provide substantial progress, with respect to the usual technology, in device performance. (This will be discussed later.)

3. *Semiconductor Material Choice Problems*

Concerning power–frequency performances, the material choice will depend essentially on the band gap E_g, on the mobility μ, as well as on its evolution with the electric field, and on the saturated drift velocity v_{s1}. As far as reliability is concerned, the most important problem is the capability of surface passivation. For the time being, three materials should be con-

TABLE II.2

		Si	GaAs	InP
Energy band E_g (eV)		1.12	1.52	1.29
Carrier mobility μ at 300°K (low doping concentration and low electric field) (cm^2/volt/sec)	electrons	1500	8500	4600
	holes	600	400	150
Saturated drift velocity (electrons) v_{ns1} (cm/sec)[a]		10^7	8×10^6	7×10^6

[a] Due to their higher mobility, only electrons are to be considered as majority carriers in microwave FETs.

sidered: Si, GaAs, and InP.* Table II.2 gives a comparison of E_g, μ, and v_{s1} for these three materials.

On the other hand, Fig. 38 shows three curves for electron drift velocity variation versus electric field for these materials. GaAs is obviously a more suitable material, due to its higher E_g (whence higher breakdown voltage for a given carriers concentration) and also higher μ; it is followed by InP, Si

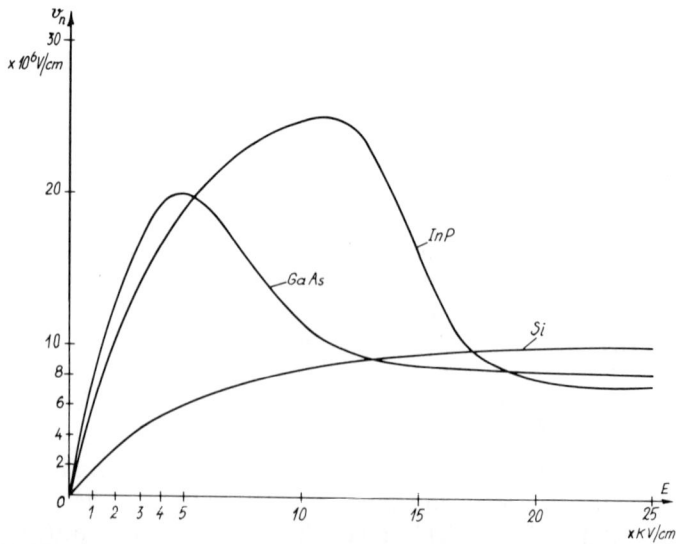

FIG. 38. Electron drift velocity v_n vs electric field E, variation for Si, GaAs, and InP.

* For the moment, only some preliminary theoretical and experimental investigations concerning the use of InP for microwave FET (at present small signal only) have been performed (2.54–2.56).

being the least convenient from this viewpoint. However, this advantage of GaAs and InP is somewhat weakened by their lower v_{ns1}.

In fact, as shown in Fig. 38, the drift velocity of GaAs reaches a much higher value prior to saturation. However, it happens for an electrical field much below operating values in microwave power FETs. The problem is similar for InP, although in that case the peak of drift velocity is even higher and corresponds to a higher operating field; but it still remains unsuitable for the microwave power FET operation.

The apparent contradiction between theory and experimental results with horizontal channel microwave power FETs, showing the definite superiority of GaAs versus Si, has been already pointed out (2.53). An explanation was sought, unsuccessfully however, in the transient peaks of drift velocity much higher for GaAs than for Si. Our opinion is that the transit time is not the performance-limiting parameter for the operating frequencies at the present time (see Section I,1). The effective performance limitation is due essentially to the charging time constants RC, which are determined mostly by the parasitic resistances and capacitances. These time constants are then ~ 5 times lower in GaAs and ~ 3 times lower in InP than in Si.

Evidently such advantages decrease as the parasitic resistances are reduced. Consequently, it would be probable that with device pattern improvement for higher and higher operating frequencies, where the parasitic resistances will decrease while transit time parameter importance will increase, the advantage of GaAs, as well as of InP, versus Si will decrease.

On the contrary, the GaAs and to a somewhat lesser degree, InP superiority in the E_g value would be better exploited by an improvement in technology. Simultaneously, some residual lack of reliability could be prevented through progress in technology.

C. Electrical Characteristics and Performances

As for bipolar transistors, we shall consider in turn transistor electrical characteristics and frequency–power performance in amplifier, oscillator, and subsidiary, switching operating modes; transistor thermal properties and reliability problems; transistor distortion and noise characteristics. Despite the relatively recent introduction of microwave power field-effect transistors, all the problems enumerated above have already been treated in many articles (cf. notably *2.34–2.40, 2.43–2.45, 2.49, 2.57–2.64, 2.66–2.82*).

1. Electrical Characteristics and Frequency–Power Performances

The overall preferential operating class for the field-effect power transistor is generally the class A, because output power in the cw operating mode is highest then, as opposed to the bipolar transistor case. This is easily comprehensible, since the advantage of the peak power's small increase under class B and especially class C operating conditions versus that under class A conditions is amply overcompensated by the average power's substantial reduction. In fact, the experimental data presently known concern, in particular, this operating case. Consequently, class A operating characteristics and performances will be given and discussed. In addition, however, some interesting results concerning the class B and AB operations will be mentioned. In all cases a large operating frequency bandwidth (of about one octave at 3 dB) is obtainable.

On the contrary, the packaging condition influence on the performance of field-effect transistors is similar to that of bipolar transistors. Package lead inductances and capacitances degrade device performances and counteract external input and output matching. Moreover, common gate (or common source) lead inductance is particularly inconvenient, notably through the troublesome negative feedback increase. Special efforts have been made to reduce its value as much as possible (up to < 0.04 nH in the 1 W transistor case). Thus, similar to the bipolar transistor case, the best solution would be package suppressing, by the FET incorporating a microwave integrated circuit.

At the moment, however, with microwave power FET development being quite recent, their incorporation in a integrated circuit is limited to some experimental samples. On the other hand, packaged devices with internally matched input and output are only in the laboratory stage. In fact, for the time being, commercially available microwave power FETs are only packaged devices (without internal matching). Evidently this contributes to widening the difference between the laboratory and commercial device frequency–power performances. In Tables II.3–II.5, we distinguish the laboratory devices' characteristics and performances from those of devices commercially available. On the other hand we shall discuss separately those devices differing in basic structure geometry with horizontal or vertical channels, and/or the semiconductor material applied: GaAs or Si.

Table II.3 gives for comparison the characteristics of the horizontal channel structure devices, applying GaAs as the basic material, and those of the vertical channel structure devices with Si as the basic material. In both cases the devices are, or have been, commercially available. Table II.4 gives the frequency–power performances of the horizontal channel structure GaAs devices, for both commercially available and laboratory samples. Table II.5 presents the frequency–power performances of the vertical channel structure Si devices. All the performances that will be given concern the amplifier operating mode under class A conditions. However as indicated above, the amplifier operating under class B or AB conditions will be also

TABLE II.3

Electrical Characteristics of Microwave Power Field-Effect Transistors (Developmental Types That Are or Have Been Commercially Available) for Class A and B Amplifiers and Oscillator Operations

	GaAs horizontal FETs	Si vertical FETs
Frequency range (GHz)	1–8	1–2.7
Drain–source breakdown voltage, BV_{DSO} (V)	> 15	> 40
Gate–source breakdown voltage, BV_{SGO} (V)	> 5	> 5
Drain–gate breakdown voltage, BV_{DGO} (V)	> 17	> 40
Gate–source cut-off voltage, V_{GCO} (V)	−4 typically	−3 typically
Saturation drain–current, I_{DSS} (A)	0.25–1	0.2
Drain–source operating voltage, V_{DS} (V)	8–12	28

TABLE II.4

Frequency–Power Performances of Microwave Power Field-Effect Transistors (Developmental Types That Are or Have Been Commercially Available or Laboratory Samples) for GaAs Horizontal FETs (High-Resolution Geometry), Class A Amplifier Operation

	Frequency range (GHz)					
	4[a]	6[a]	8[a]	9[b]	15[b]	18[b]	22[b]
Amplifier output power in cw operation, P_{out} max (W)	0.8–3	0.7–2.4	0.6–1.9	1	0.45	0.225	0.145
Amplifier power efficiency, η (%)	54–47	46–34	38–24	16.3	12.5	5.4	
Power gain (dB)	7–6	6.5–5	6–4	4.3	5.2	4.5	5.6

[a] Commercially available.

[b] Laboratory samples (2.57a, 2.57b), with three cells at 9 GHz, and two cells at 15, 18, and 22 GHz. In both cases the output powers are obtained at 1 dB gain compression.

TABLE II.5

FREQUENCY–POWER PERFORMANCES OF MICROWAVE POWER FIELD-EFFECT TRANSISTORS (DEVELOPMENTAL TYPES THAT ARE OR HAVE BEEN COMMERCIALLY AVAILABLE, OR LABORATORY SAMPLES) FOR Si VERTICAL FETs (MEDIUM-RESOLUTION GEOMETRY), CLASS A AMPLIFIER OPERATION

	Frequency range (GHz)					
	1		2		2.7	
Amplifier output power in cw operation, $P_{out\,max}$ (W)	1^a	4^b	0.5^a	1.5^b	0.2^a	1^b
Amplifier power efficiency, η (%)	31^a	28^b	23^a	26^b	16^a	14^b
Power gain (dB)	10^a	5^b	6^a	5^b	5^a	5^b

[a] Commercial (one cell).
[b] Laboratory (two cells) (2.43, 2.44, 2.65, 2.66). In both cases the output powers are obtained at 1 dB gain compression.

briefly discussed. On the other hand, the oscillator and switching operations will only be examined briefly.

Concerning the amplifier operation under class AB or B conditions, the peak output power may be then increased $\sim 20–30\%$ through amplifier power efficiency improvement [up to $\sim 70\%$ (2.57a)]. However, as indicated above, the average output power is finally reduced ($\sim 20–40\%$). Consequently, the operation under class A conditions is chosen, in general. This is a significant difference in comparison with the bipolar transistor amplifier operation.

With oscillator operation, on the contrary, the problems are quite similar to those already discussed for the bipolar transistor case. In fact, the field-effect transistor oscillator's operating effective maximum frequency is $\sim 15–25\%$ higher than that proved in the amplifier case. On the other hand the output power is 20–30% lower.

Similarly, the switching operation is a subsidiary problem as for bipolar microwave power transistors; however, intrinsically the unipolar transistor is especially capable of operating with particularly low turn-on and turn-off switching times. This directly results from eliminating the minority carrier storage and evacuation processes; consequently switching times are substantially reduced, namely below 100 psec. It is noteworthy that as early as 1971 switching times of ~ 300 psec had been obtained with a 4 GHz GaAs FET (2.69). Moreover, all the problems occurring in the bipolar transistors as a result of minority carrier injection during the turn-off switching operation are suppressed in unipolar transistors. Thus, a relatively high recovery voltage can be, in principle, sustained. But regarding the controlled power value, a unipolar transistor requires, in theory, for identical power, a larger device size than that of the bipolar transistor.

We shall now discuss the amplifier performance dependance on transistor structure as well as its geometry resolution, and the basic semiconductor material choice.

Concerning the f_0 and f_{max} values, it seems that the horizontal MESFET structure is particularly capable of attaining a very high operating frequency. As already discussed, this is due to the minimizing of source–gate and drain–gate capacitances (respectively, C_{GS}, C_{pGS} and C_{DG}, C_{pDG}, Fig. 20) and to the optimizing of structure geometry resolution. On the other hand, the metal–semiconductor gate-junction optimizes the gate action efficiency per unit junction length; however, with the gate action limited to one channel side, it is finally less efficient than a PN gate surrounding the entire channel periphery.

That is the case with vertical channel structure, which however, presents substantially higher source–gate and drain–gate capacitances than the horizontal channel structure. Furthermore, its structure geometry, refining up to and *a fortiori* below 1 µm resolution, requires a more sophisticated technology than that of the horizontal channel structure. This is one of reasons for the discrepancy in the performances (especially for laboratory samples) between Tables II.4 and II.5 in favor of horizontal channel devices.

The other essential reason for discrepancy is the use in horizontal channel structure of a semiconductor material with carrier mobility five times higher (GaAs instead of Si) still on a technological basis, for vertical channel structures. Thus, all the intrinsic resistances [r_{ps} and $0.1r_e$ as well as r_{pD} (see Fig. 20) and then Re(y_{11}) and Re(y_{22}) of Fig. 22 for given y_{21}] are minimized in the horizontal channel case with a consequent increase in f_0 and f_{max}.

The problem, however, seems to be quite different for output power optimizing. In fact, as explained in Section II,A,2, it seems in principle much easier to attain relatively high power output values with vertical rather than with horizontal channel structure. Yet, this structure superiority is counterbalanced by higher resistance parameter values, the consequence of which is some lessening of amplifier power output efficiency.

Concerning extrinsic elements (leads), which introduce parasitic parameters (resistances, inductances, and capacitances) that degrade FET performances, the problem seems to be similar to that in the bipolar transistor case. However, the microwave power FETs are very recent and few detailed data are known about these device integration in the amplifier circuit, with or without packaging. It seems proper nonetheless, to point out that this problem is even more important and more delicate to solve for this case than for bipolar transistors, because a much higher frequency level is already attained with the power FETs. Input and especially the output matching become particularly difficult to realize. The problem of eliminating or compensating extrinsic inductances and capacitances [as discussed already in Section I,C,1 (*1.57*)] becomes fundamental here for obtaining satisfactory performance.

Two other important problems, those of thermal behavior and reliability will now be examined.

2. *Thermal Properties and Reliability Problems*

It is commonly known that the unipolar transistor shows remarkable thermal behavior, due to minority carrier elimination. Through the drain current negative temperature coefficient, there is evidently no risk of thermal runaway and of the secondary breakdown phenomenon. On the other hand, drain current distribution inside a chip as well as between chips working in parallel is relatively self-regulated without any resistor introduction; likewise, the risk of hot spots developing is attenuated.

Thus, operating temperature is only limited by long-term reliability technological considerations and by drain-current and transconductance decreases, and consequently power output and power gain reduction with temperature rises. For the time being, it seems too early to specify a definitive limit for operating temperature values of microwave power FETs; presently, it seems to lie between 150 and 200°C. From this viewpoint, the relatively low thermal conductibility of GaAs (three times lower than that of Si) is a definite disadvantage for power devices.

For reliability problems, two cases should be distinguished: short-term reliability and long-term reliability. Short-term reliability corresponds to the degree of capability to withstand overvoltage and/or current surges, which may come from external origins or may result from circuit catastrophic degradation. It is noteworthy that the output shorting, particularly troublesome in the bipolar transistor case, is absolutely ineffectual for the FETs, through their intrinsic circuit configuration.

In fact, as explained above, the risks resulting from overcurrent are very limited. Thus, the most damaging are overvoltages, which may be conductive to a burnout in the regions where the highest electric field is concentrated. Thus, the damage is field-triggered at the edge of the active layer mesas, and the problem is to avoid, as much as possible, this high-field region formation. This seems to be more difficult in the microwave power FETs with horizontal channels, where the sophisticated geometry structure is on the open surface, than in the FETs with vertical channels, where such a structure is buried. In both cases suitable surface passivation is necessary, but it is so much easier to perform such passivation perfectly, as the structure geometry resolution is less.

Open surface passivation is likewise important for long-term reliability, as well for the actual service (*2.49*) and for the behavior in particularly degrading service conditions (for instance in space applications, where the FETs present an exceptionally high neutron radiation tolerance (*2.72*)).

Another long-term reliability problem is to avoid the failures resulting from characteristic gradual irreversible degradation or from electrical parameter instability. However, there are very few reliability data relating in particular to microwave power FETs; only some communications concerning microwave low-power MESFET reliability, with detailed experimental data, have been recently presented (e.g., *2.71*, *2.72*). We therefore confine ourselves to a brief general discussion on this problem.

Gradual characteristic degradation is most frequent concerning the drain–source current decrease, through the drain and source contact resistance progressive increase. Sometimes a nonohmic contact characteristic may likewise appear before the complete contact failure occurs. These MS contact degradations are often produced through metal migration, but other causes, e.g., improper choice of contact processing, may be possible. In particular, this could be the case of devices having widely developed MS contact peripheries.

Another failure that may gradually occur is in the MS diode gate. It could happen, e.g., through successive transient voltage spikes, gradually increasing the gate leakage current and finally causing gate burn-out.

Electrical parameter instability or reversible drift is characterized, notably, by the drain current variation with time, when the device is turned on into operating bias condition, this parameter recovering its original value some time after the device is turned off. It may be possible that such undesirable phenomena result from some anomalous density of deep-lying carrier trapping centers supposed to exist near the interface between the " channel layer" and the semi-insulating substrate (*2.71*). In that case, the introduction of a buffer layer could be beneficial.

Obviously, beyond disturbing the drain-current, these imperfections also disturb other device characteristics, such as amplifier power gain, and noise and distortion levels in transistor and oscillator operation.

3. *Noise and Distortion Characteristics*

Concerning noise characteristics (*2.73–2.80*), the following main sources of noise are to be considered: (1) thermal noise (already mentioned in the bipolar transistor case) in the channel, and source and drain resistances; (2)

induced noise in the gate circuit; (3) diffusion noise in the pinch-off space-charge region (*2.73, 2.78, 2.79*). An article (*2.79*) recently published in this series gives an overall review of these noise problems; subsequently, the reader will find therein references to this work. Shot-noise and flicker-noise, previously discussed, are eliminated; on the other hand, GR noise in the channel lines is only significant at the very low temperatures < 200°K (*2.74, 2.75*). On the contrary, a particular noise source should be taken into account in the GaAs (and InP) FETs; this noise results from so-called intervalley scattering, which removes electrons from regions of the conduction band where they can achieve relatively high velocities, and excites them to subbands where their velocity is much lower (*2.76, 2.80*). We shall now briefly review the noise sources particularly to the FET common-gate equivalent noise circuit in Fig. 39, which is derived from Fig. 20a.

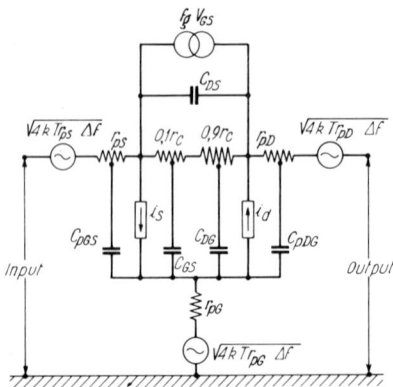

FIG. 39. FET equivalent noise circuit for common-gate configuration (derived from Fig. 20a).

Induced gate noise results from the noise voltage fluctuations along the channel, transmitted to the gate through the coupling capacitances C_{GS}, C_{pGS} and C_{DG}, C_{pDG} (cf. Fig. 39). As discussed above, the voltage distribution along the channel is absolutely nonuniform, the major portion being concentrated in the "pinch-off" region (*2.8, 2.9, 2.22, 2.25, 2.28*), where the noise temperature, dependent on applied field, is consequently increased (*2.78, 2.79*). The current i_s flowing from the source region to the gate is partially correlated with i_d, the drain output current; i_s may be given by a simplified expression (*2.73*):

$$\langle i^2 \rangle = 4kT_{\text{ngs}}(g_{\text{gs}}) \, \Delta f$$

where g_{gs} is the input conductance and T_{ngs} its equivalent noise temperature.

Diffusion noise is due to recombination centers in the space-charge region of an FET, especially that of the pinch-off region. It goes with the space-charge limited current between channel and gate, the latter superimposing on the induced current defined above.

The gate drain current may be expressed as

$$\langle i^2 \rangle = 4kT_{ngd}(g_{dg}) \, \Delta f$$

where g_{ds} is the output conductance and T_{ngd} its equivalent noise temperature.

The intervalley scattering noise in GaAs and InP FETs defined above results from the negative resistance that appears from the rise in carrier velocity. It is undeniable that this noise source exists, but its effect may be neglected by first approximation because the negative resistance region occupies in general a small space under the gate (*2.78*).

Thus, the expression of fundamental noise (except parasitic sources) becomes too complicated (*2.78, 2.79*) to be reproduced here, all the more so since it corresponds to only a minor part of the overall noise figure in the microwave power FET case. In fact, the greater part is due to the parasitic elements, notably r_{pS}, r_{pG}, and r_{pD} (*2.78*), the contribution of which is pointed out in Fig. 39. This is particularly accented in the power horizontal channel MESFETs, through the noise introduced by the chip parallel coupling external connexions.

Nevertheless, due in particular to the absence of shot-noise, the FETs present a notably lower noise level than bipolar transistors. Although the experimental results for the microwave power FETs are scarce for the time being, some recent data are very promising; this, one can expect a noise level between 5 and 10 dB, for 1 W output power in the X band.

Moreover, with the improvement of the structure geometry resolution, especially in the vertical channel FETs, and the structure geometry of coupling the chips in parallel, especially for the horizontal channel FETs, the parasitic noise sources could be substantially weakened. The overall noise figure would be then notably reduced.

The amplifier nonlinear distortion effects are lower overall in FETs than in the bipolar transistor case. However, the behavior difference is weaker than could be expected from the difference in the basic relationship between output current and gate voltage, and output current and base current, respectively, square versus exponential law. In fact, this transfer characteristic is quite different from $n = 2$; furthermore the relationship is nonlinear (*2.81, 2.82*). The reasons for this behavior are (1) the parasitic resistances, in particular r_{pS} and r_{pG} effects, (2) field-dependent mobility in the channel, (3) the carrier velocity saturation within a great part of the channel length, particularly in the short channel devices. Subsidiary, in the GaAs FETs, the appearance of negative resistance likewise generates distortion.

However, notably through the reduction of parasitic resistances, careful choice of the operating point parameters, and the appropriate limitation of the drain current excursion, suitable third-order intermodulation ratios for linear amplifiers have been obtained with quite acceptable power efficiency. Thus, for example, under class B conditions ($2.57a$), $I/C = -30$ dB with an added power efficiency of 35% at 4 GHz has been attained. Substantially lower intermodulation level, e.g., < -40 dB, may be reached with the power efficiency consequent reduction. On the other hand, further improvement of structure geometry, notably by minimizing the source–gate resistance, would provide a possibility of further lowering the I/C level and then maintaining relatively high power efficiency.

D. General Considerations and Conclusions

We shall try first, as for the other devices discussed previously, to classify the characteristics of microwave power field-effect transistors in terms of advantages and disadvantages. Their advantages may be summarized as follows:

1. Much easier application in the amplifier or switching circuits, due to their three-terminal configuration similar to that of bipolar transistors.

2. Operating frequency, the highest attained in the three terminal devices field, with relatively high output power.

3. Amplifier operation, in large bandwidth, in particular in class A.

4. Relatively high amplifier power gain.

5. The lowest amplifier signal distortion as well as noise level for all microwave power semiconductor devices.

6. The current temperature coefficient being negative, the risks of thermal runaway and secondary breakdown are practically eliminated. Also, operation is perfectly stable and, in the case of chips coupled in parallel, the power distribution is fair without any ballast resistances. On the other hand, immunity against amplifier or oscillator circuit output degradation effects is guaranteed.

7. Notably for space applications, satisfactory behavior to neutron or ionizing radiations.

On the other side, their disadvantages seem to be:

1. Their structure is most complex, compared to those of other microwave power semiconductor devices, for identical output power and frequency–power product.

2. The local current density is lower due to the characteristics of majority carrier devices;

3. By the same token, their operation in class C is relatively uninteresting.

4. Their long-term reliability is not yet positive because of their very recent development and complex structure.

We will then try to define the place occupied by the field-effect transistor in the microwave power semiconductor device family. The following discussion concerns, first the FET alone, for the microwave frequency range up to ~ 20 GHz, and second, the FET coupled with a varactor diode for the frequency range reaching ~ 50 GHz.

In the first frequency range, the comparison with microwave diodes, essentially IMPATT or TRAPATT diodes and TED, appears to be the advantage of FETs for similar output power. In fact, in oscillator and amplifier operation, FETs provide higher power efficiency, better thermal behavior, and a lower noise figure. Moreover, in amplifier operation the circuit is much simpler, the power efficiency is higher, and the signal distortion is lower. On the contrary, for the moment at least, output power values in oscillator operation are substantially higher with TRAPATT, IMPATT–TRAPATT, or IMPATT diodes. On the other hand, the power efficiency is likewise higher in TRAPATT diodes at the bottom end of the frequency range considered.

The comparison with bipolar transistors should be differentiated. On the one hand, bipolar transistors provide a higher output power in oscillator and amplifier operations at the bottom end of the frequency range, as well as higher power efficiency in amplifier operation. On the other hand, the FETs present a much higher operating frequency limit, fundamentally better thermal behavior, a lower noise figure, and a lower signal distortion. This statement can be seen to point to a distribution of application domains between these two types of device, to which we shall come back in the following section.

Regarding the FET–varactor coupling, it seems at present to be a possible solution for amplifier operation up to a frequency of the order of 50 GHz with an output power of ~ 100 mW. Obviously, in the oscillator case the use of IMPATT or eventually TED would be preferable if the problem of power efficiency were primordial.

The above considerations concern evidently the present situation. In the concluding section we shall give our views on the future development possibilities in the field of such devices.

REFERENCES FOR SECTION II

2.1. J. E. Lilienfeld, U.S. Patent 1,900,018 (appl. 1928).
2.2. O. Heil, U.K. Patent 439,457 (appl. in Germany, 1934).
2.3. W. Shockley, *Proc. IRE* **40**, 1365 (1952).

2.4. R. C. Prim and W. Shockley, *IRE Trans. Electron Devices* **4**, 1 (1953).
2.5. G. C. Dacey and I. M. Ross, *Proc. IRE* **41**, 970 (1953).
2.6. G. C. Dacey and I. M. Ross, *Bell Syst. Tech. J.* **34**, 1149 (1955).
2.7. S. Teszner, *Bull. Soc. Fr. Electr.* 13, Part 1, 682 (1958).
2.8. S. Teszner, *Onde Electr.* No. 409, p. 307 (1961).
2.9. S. Teszner and R. Gicquel, *Proc. IEEE* **52**, 1502 (1964).
2.10. J. Grosvalet, C. Motsch and R. Tribes, *Ann. Radioelectr.* **17**, 265 (1962).
2.11. J. R. Hauser, *Solid-State Electron.* **10**, 577 (1967).
2.12. A. B. Grebene and S. K. Ghandi, *Solid-State Electron.* **12**, 573 (1969).
2.13. C. K. Kim and E. S. Yang, *IEEE Trans. Electron Devices* **17**, 120 (1970).
2.14. P. David, *Onde Electr.* **50**, 38 (1970).
2.15. D. P. Kennedy and R. R. O'Brien, *IBM J. Res. Dev.* **14**, 95 (1970).
2.16. T. L. Chiu and H. N. Ghosh, *Solid-State Electron.* **14**, 1307 (1971).
2.17. P. L. Hower and N. G. Bechtel, *IEEE Trans. Electron Devices* **20**, 213 (1973).
2.18. K. Lehovec and W. G. Seeley, *Solid-State Electron.* **16**, 1047 (1973).
2.19. E. S. Yang, *Adv. Electron. Electron Phys.* **31**, 247 (1972).
2.20. A. V. J. Martin and J. Le Mee, *Int. Symp. Semicond. Devices, Paris, 1961*.
2.21. A. Rose, *RCA Rev.* **24**, 627 (1963).
2.22. A. van der Ziel and J. W. Ero, *IEEE Trans. Electron. Devices* **11**, 128 (1964).
2.23. E. O. Johnson, *RCA Rev.* **26**, 163 (1965).
2.24. J. R. Hauser, *IEEE Trans. Electron Devices* **12**, 605 (1965).
2.25. S. Teszner, *Ann. Telecommun.* **26**, 303 (1971).
2.26. M. Reiser, *IEEE Trans. Electron Devices* **20**, 35 (1973).
2.27. M. B. Das and P. Schmidt, *IEEE Trans. Electron Devices* **20**, 779 (1973).
2.28. H. Tango and J. Nishizawa, *Solid-State Electron.* **13**, 129 (1970).
2.29. R. Zuleeg, *Solid-State Electron.* **10**, 449 (1967).
2.30. J. Nishizawa, T. Terasaki, and J. Shibata, *IEEE Trans. Electron Devices* **22**, 185 (1975).
2.31. W. Shockley, *Proc. IRE* **40**, 1311 (1952).
2.32. Western Electric Company, U.S. Patent 756,339 (appl. 1952).
2.33. K. E. Drangeid, R. Jaggi, S. Middelhoek, T. Mohr, A. Moser, G. Sasso, R. Sommerhalder, and P. Wolf, *Electron. Lett.* **4**, 363 (1968).
2.34. M. Fukuta, H. Ishikawa, K. Suyama, and M. Maeda, *Int. Electron Devices Meet. Washington, D.C.* 13.1 (1974).
2.35. T. G. Blocker, H. M. Macksey, and R. L. Adams, *Int. Electron Devices Meet., Washington, D.C.* 13.2 (1974).
2.36. J. A. Angus, R. S. Butlin, D. Parker, R. H. Bennett, and J. A. Turner, *Proc. Eur. Microwave Conf., Hamburg* B 5.1 (1975).
2.37. M. Fukuta, T. Mimura, I. Tujimura, and A. Furumoto, *Int. Solid-State Circuits Conf., Philadelphia* THAN 7.6 (1973).
2.38. L. S. Napoli, R. E. Debrecht, J. J. Hughes, W. F. Reichert, A. Dreeben, and A. Triano, *Int. Solid-State Circuits Conf., Philadelphia* T.H.A.M. 75 (1973).
2.39. L. S. Napoli, J. J. Hughes, W. F. Reichert, and S. Jolly, *RCA Rev.* **34**, 608 (1973).
2.40. C. Vergnolle, R. Funck, and M. Laviron, *Int. Solid-State Circuits Conf., Philadelphia* W.P.M. 7.3 (1975).
2.41. J. Delmas and M. Chappey, *Vide* No. 109 (1963).
2.42. M. Chappey, A. M. Dauge, J. Delmas, and P. Durand, *Onde Electr.* **45**, 305 (1965).
2.43. S. Teszner, *IEEE Trans. Electron Devices* **19**, 355 (1972).
2.44. D. P. Lecrosnier and G. P. Pelous, *IEEE Trans. Electron Devices* **21**, 113 (1974).
2.45. I. Drukier, R. L. Camisa, S. T. Jolly, H. C. Huang, and S. Y. Narayan, *Electron. Lett.* **11**, 104 (1975).

2.46. M. C. Driver, H. B. Kim, and D. L. Barrett, *Proc. IEEE* **59**, 1244 (1971). (Letters.)
2.47. S. Asai, F. Murai, and H. Kodera, *IEEE Trans. Electron Devices* **22**, 897 (1975).
2.48. R. H. Dean and R. J. Matarese, *IEEE Trans. Electron Devices* **22**, 358 (1975). (Correspondence.)
2.49. D. R. Ch'en, H. F. Cooke, and J. N. Wholey, *Microwave J.* **18** (Nov.), 60 (1975).
2.50. R. G. Hunsperger and N. Hirsch, *Solid-State Electron.* **18**, 349 (1975).
2.51. W. Kellner, H. Kniepkamp, D. Ristow, and H. Boroffka, *Int. Electron Devices Meet., Washington, D.C.* 11.3, 238 (1975).
2.52. H. I. Smith, F. J. Bachner, and N. Efremow, *J. Electrochem. Soc.* **118**, 821 (1971).
2.53. J. G. Ruch, *IEEE Trans. Electron Devices* **19**, 652 (1972).
2.54. J. S. Barrera and R. J. Archer, *IEEE Trans. Electron Devices* **22**, 1023 (1975).
2.55. T. J. Maloney and J. Frey, *IEEE Trans. Electron Devices* **22**, 357 (1975). (Correspondence.)
2.56. T. J. Maloney and J. Frey, *Int. Electron Devices Meet., Washington, D.C.* 13.4, 296 (1974).
2.57a. H. C. Huang, I. Drukier, R. L. Camisa, S. T. Jolly, J. Goel, and S. Y. Narayan, *Int. Electron Devices Meet., Washington, D.C.* 11.2, 235 (1975).
2.57b. J. E. Adair, *Microwave J.* **19** (May), 38 (1976).
2.58. R. L. Camisa, I. Druckier, H. C. Huang, J. Goel, and S. Y. Narayan, *Int. Solid-State Circuits Conf., Philadelphia* WPM 7.5 (1975).
2.59. R. L. Camisa, J. Goel, and I. Drukier, *Electron. Lett.* **11**, 572 (1975).
2.60. P. Bura and D. Cowan, *Int. Solid-State Circuits Conf., Philadelphia* THPM 14.1 (1976).
2.61. M. Fukuta, K. Suyama, H. Suzuki, Y. Nakayama, and H. Ishikawa, *Int. Solid-State Circuits Conf., Philadelphia* THPM 14.5 (1976).
2.62. C. A. Liechti, *Int. Solid-State Circuits Conf., Philadelphia* WPM 7.2 (1975).
2.63. W. Jutzi, *Proc. IEEE* **57**, 1195 (1969). (Letters.)
2.64. P. Durand and J. Laplanche, *Electron. Lett.* **8**, 306 (1972).
2.65. P. Da Rocha and J. L. Le Bail, Personal communication (1974).
2.66. R. Funck, Personal communication (1973).
2.67. M. Maeda, K. Kimura, and H. Kodera, *IEEE Trans. Microwave Theory Tech.* **23**, 661 (1975).
2.68. A. Y. Takayama, A. Higashizaka, R. Yamamoto, and Takenchi, *Int. Solid-State Circuits Conf., Philadelphia* THPM 14.4 (1976).
2.69. L. S. Napoli, W. F. Reichert, R. E. Debrecht, and A. B. Dreeben, *RCA Rev.* **32**, 645 (1971).
2.70. D. A. Abbott and J. A. Turner, *Int. Electron Devices Meet., Washington, D.C.* 11.4, 243 (1975).
2.71. H. Kozu, I. Nagasako, M. Ogawa, and N. Kawamura, *Int. Electron Devices Meet., Washington, D.C.* 11.5, 247 (1975).
2.72. B. Buchanan, R. Dolan, S. Roosild, *Proc. IEEE* (Lett.), **55**, 2188 (1967).
2.73. A. van der Ziel, *Proc. IEEE* **58**, 1178 (1970).
2.74. C. F. Hiatt, A. van der Ziel, and K. M. van Vliet, *IEEE Trans. Electron Devices* **22**, 614 (1975). (Correspondence.)
2.75. K. M. van Vliet and C. F. Hiatt, *IEEE Trans. Electron Devices* **22**, 616 (1975). (Correspondence.)
2.76. W. Baechtold, *IEEE Trans. Electron Devices* **19**, 674 (1972).
2.77. G. Lecoy, D. Rigaud, and D. Sodini, *Solid-State Electron.* **17**, 11 (1974).
2.78. H. Statz, H. A. Haus, and R. A. Pucel, *IEEE Trans. Electron Devices* **21**, 549 (1974).
2.79. R. A. Pucel, H. A. Haus, and H. Statz, *Adv. Electron. Electron Phys.* **38**, (1975).
2.80. J. Frey and T. J. Maloney, *Int. Electron Devices Meet., Washington, D.C.* 11.6, 251 (1975).
2.81. J. S. Vogel, *Proc. IEEE* **55**, 2109 (1967).
2.82. R. B. Fair, *IEEE Trans. Electron Devices* **19**, 9 (1972).

Present Trends and Future Development Prospects for Microwave Power Semiconductor Devices

I. Present Trends

We shall point out successively the situation in the fields of two- and three-terminal devices. The most promising two-terminal devices are IMPATT diodes. They are particularly attractive in the oscillator-mode application; the frequency range presently extends from 6 to 110 GHz, with maximum output power (in cw operation) of ~ 10 W at the bottom end and ~ 0.3 W at the top end of this frequency range. Results at higher frequencies are presently limited to a few tens of milliwatts. The preferential material for their fabrication appears to be GaAs. However, the lower thermal conductivity of this material versus Si appears to be an intrinsic drawback. For frequencies less than 20 GHz this problem may be circumvented by a multimesa structure. For higher frequencies in the millimetric range, Si remains quite competitive especially in the double-drift technology, which may be more easily achieved in this material. However, regarding the major drawback of IMPATT diodes for oscillator applications, i.e., the high noise level, GaAs provides slightly better performances than Si. Consequently the present trend to improve the IMPATT diode characteristics as a power source goes along with the improvement of GaAs technology.

The lowest microwave frequency (1–5 GHz) would be preferably covered with TRAPATT (or IMPATT–TRAPATT) diodes, guaranteeing high output power (but particularly in the pulsed-wave operation) and high power efficiency, in any case where the high noise level peculiar to these devices can be accepted. In any other case, the problem can be effectively solved by bipolar transistors. We will come back to this problem later.

Regarding three other types of devices, which were discussed in Part I:

1. TEO is presently limited to applications in the centimetric region requiring low output power and admitting low efficiency, mainly for intruder alarm applications. In the millimetric range, power applications are currently limited to pumps for parametric amplifiers.

2. BARITT diodes are only attractive because of their noise level, even lower than that of TEO: in contradistinction, their power output and efficiency in the microwave range are definitely too small for power applications.

3. Finally, varactor diodes being complementary devices will be considered together with three-terminal devices, which need such a complement for the time being.

On the other hand, the diode application in the amplifier operation, in our opinion, seems to be because of circuit complication, narrow bandwidth, high noise figure (as long as IMPATT diodes are concerned), and signal distortion, a temporary solution. Meanwhile in the high power–frequency range, the use of three-terminal devices, with or without varactor coupling, is increasing.

The last few years have been marked by a remarkable development in the microwave power field of three-terminal devices, bipolar and particularly field-effect transistors, with enlarged possibilities in amplifier and oscillator applications. Bipolar power transistors seem to be limited in frequency, for the moment, to 10 GHz, with $P_{out} \sim 1$ W, and in power to ~ 50 W (in cw operation) at 1 GHz. Field-effect transistors have already gone beyond 20 GHz, with $P_{out} \sim 0.2$ W, and as for output power are limited to ~ 3 W at 6 GHz and ~ 4 W at 1 GHz (still in cw operation). The situation is moving so rapidly, particularly for FETs, that it is impossible to indicate performance limits unreservedly.

This is likewise the case for the device structure and technology (geometry type and resolution), as well as the material used (GaAs or Si). So the highest performances reached today are those of GaAs MESFETs with horizontal channel structure. Now the vertical channel field-effect transistors (VFET Gridistor) are evolving in a parallel way with a view to improving this structure, in principle, for high-power applications. This is also the case of bipolar transistors; however, their operating frequency limit seems to be staying definitely lower than that of field-effect transistors. The trends for both transistors are presently to refine the structure geometry as much as possible in order to increase the collector, or drain, current density as much as possible and, at the same time, to minimize the parasitic parameters. On the other hand, one is seeking technology adaptation with a view to using the most convenient semiconductor material for transistors.

Consequently, with such remarkable progress, the varactor operating preferential area should climb in a parallel direction. Then the frequency range would be between 20 and 100 GHz, if possible. For the moment, it is limited to ~ 60 GHz; we shall consider hereafter its future possibilities.

II. Future Development Prospects

We shall consider in order oscillator and amplifier applications. For oscillators, the most attractive among the devices known at the time is the IMPATT diode, on condition that its high noise level be accepted. This device still presents considerable potential for further progress. Thus, after extending the double drift-zone structure to the highest frequency range > 100 GHz, it could be still improved by alternation of doping levels: low–

high-low, as indicated in Fig. II.1f of Part I. Then the structure could be refined through thickness layer reduction (e.g., up to 0.3–0.1–0.3 μm) using, e.g., epitaxial deposition technique at relatively low temperatures (800–850°C) or by using ion implantation. Such a structure should require an advanced level of technology. Concerning material, it seems that for the highest frequency GaAs or Si could both be used, the choice depending only on technological facilities. As far as performance is concerned, it seems very probable for a medium run to go beyond 200 GHz with output power of ~ 1 W. For the lowest frequency, it is probable that an output power of 50 W would be attained at 5 GHz, by using GaAs material, with series-stacked multidiode units, or with diode parallel coupling, used with ballast resistors.

Now it remains possible that for the bottom end of the frequency range the TRAPATT diode will still be used, likewise on condition that its very high noise level be accepted. Its use could be interesting, particularly in pulsed-wave operation, in which a peak power of several kilowatts could be obtained.

When a low noise level is required, with relatively low output power the TEO may be suggested, as technological progress on this device will take place. It would be quite possible, notably through improving the quality of the active layer and of the contacts with the N^+ layer. Then an output power of > 1 W at the bottom end of the frequency range and 0.2–0.3 W at the top would be possible. On the other hand, the operating frequency limit could approach 100 GHz. Extensive industrial development may take place for TEO in the millimeter range if collision avoidance radars find their place in the automobile industry. Besides GaAs TEO, InP TEO may at last find a limited place in the power semiconductor market for low duty cycle pulsed applications. However, the TEO applications will stay limited, compared to those of FET and bipolar transistors.

In fact, the breakthrough recently achieved in the field of microwave power FETs is still operating. Concerning the GaAs horizontal channel (GaAs MESFET) structure, the possibility of geometry refining up to 0.5 μm resolution and even lower, and on the other hand, of improvement of the gate pads in parallel coupling, in order to minimize the parasitic parameters, could lead to a substantial performance increase. It seems possible to attain in the near future an operating frequency of 40 to 50 GHz with 1 W of output power.

Regarding the vertical channel structure (Gridistor or derived structures) the fundamental problem is the application of ion implantation technique with parallel geometry refinement, e.g., the grid thickness (channel vertical length) reducing (up to 0.5 μm) as well as the channel width (up to 0.2 μm), and channel profile improvement. Then, the intrinsic parameters would be optimized and the parasitic parameters, resistances and capacitances, would be substantially reduced. It is noteworthy that especially through the reduction of parasitic resistances, the advantage of GaAs versus Si is weakened* and it would then be possible, using Si material, to attain an operating frequency of ~ 30 GHz with output power of 1 to 2 W. If the use of GaAs could be demonstrated, this limit would become ~ 45 GHz. On the other hand, this structure would be capable of providing higher output power (up to ~ 20 to 30 W) at the bottom end of the frequency range. Then it seems possible to cover a large microwave frequency and power range with these two types of structure. Moreover, the desirable quality of very low noise figure—< 5 dB—even at the top of this frequency range would be obtained, this being possible through minimizing the parasitic resistance.

Bipolar transistors likewise present an appreciable potential for further progress. Thus, the emitter aspect ratio could be increased up to ~ 20/1, with simultaneous improvement of the geometry resolution up to 0.3 to 0.4 μm; the base width could be then reduced to < 0.1 μm, maintaining the lateral resistance value. The collector depletion layer thickness could be reduced up to 0.2 to 0.3 μm. Then the operating frequency limit would be increased up to 15 to 18 GHz with Si as the material used. If the use of GaAs were possible, especially in a heterojunction type configuration GaAlAs–GaAs, this limit could become 20–25 GHz and even higher. In both cases, the power output would be ~ 1 W. It is noteworthy that considerable development of the output power ability can be foreseen at the bottom end of the frequency range. Thus, at 1 GHz, a power output of ~ 200 W in cw operation, and ~ 1 kW in pulse-wave operation could be expected, for a medium run.

The varactor diodes could keep almost their present place, but with increased frequency values, i.e., in the range ~ 30–~ 100 GHz. The problem then will be to increase simultaneously the varactor capabilities, which seems possible, particularly with snap varactor diodes.

Finally, amplifier applications could be exhausted almost exclusively by the field-effect or bipolar transistors with varactor as complement, if necessary. Only in the frequency range exceeding 100 GHz and on the other hand for output power higher than indicated above, the application of IMPATT diodes (or eventually TRAPATT diodes) could be justified. The frequency limit given for oscillator application must be reduced by 15 to 20% for amplifier application. In return, the output power value will be increased by 15 to 25%.

---

* As explained in Section II,B,3.

Electron-Bombarded Semiconductor Devices

DAVID J. BATES, RICHARD I. KNIGHT, AND
SALVATORE SPINELLA

*Watkins-Johnson Company,
Palo Alto, California*

AND

ARIS SILZARS

*Tektronix, Inc.,
Beaverton, Oregon*

I. Introduction 221
   A. Historical Perspective 222
   B. Principle of Operation 224
   C. General Characteristics 227
II. Analytic Foundation 232
   A. Basic Analysis of the EBS Target 232
   B. Risetime Optimization 234
   C. Output Current Optimization 235
   D. Diode Output Capability 236
III. Physical Limitations to EBS Capabilities 238
IV. Target Assembly 246
   A. The Semiconductor Diode 246
   B. Mounting Substrate 258
V. Some Device Configurations and Performance 260
   A. Planar-Gridded Amplifiers 260
   B. Deflected-Beam Amplifiers 266
   C. Signal Processors 271
VI. Life and Reliability 276
VII. Future Potential 277
   References 280

I. INTRODUCTION

This chapter describes electron-bombarded semiconductor, or EBS, power and signal processing devices. Our discussion is restricted to devices having an electrical input and an electrical output where the semiconductor is an active element. Since most of the commercially available devices are for power amplification and control, the emphasis will be on these devices. The principle of operation, a brief analysis, device configuration and performance, semiconductor processing, reliability considerations, and future expectations are discussed. Some historical information is also presented as well as a brief comparison with more familiar devices.

The first commercially available EBS devices have demonstrated that EBS has considerable promise as an important new electron device for power amplification and control. In addition, they show further promise for very versatile low-power multifunction signal-processing devices. Although work was initiated in the latter 1940s on an early form of the EBS device, development effort on the device languished for two decades. Over the past eight years, development has progressed at the Watkins-Johnson Company, Palo Alto, California, with support from the U.S. Department of Defense and the National Aeronautics and Space Administration. There has also been work at many other United States laboratories, but at present the only viable EBS programs are at Tektronix and Watkins-Johnson Company. At present, Tektronix has one EBS product on the market, a scan converter, and Watkins-Johnson offers a variety of power amplifier and power control devices for sale.

The EBS is a hybrid vacuum tube–semiconductor device, where both technologies are of essentially equal importance. Present designs are based on the excellent analytical foundation and initial experimental verification thereof, provided by Gibbons, Norris, Sigmon, and others, at Stanford University during the period 1966–1970 (*1–3*). As amplifying devices, semiconductors and vacuum tubes are rivals in the marketplace, and the designers of the individual devices are by necessity adversaries. Keeping in mind then that EBS devices have two natural rivals, we will present the case for EBS devices without deliberate overstatement. Those EBS's that are for optical imaging and display and mass memory devices are excluded.

A. Historical Perspective

The ability of radiation to alter the state of matter was discovered near the start of the 20th century, when energetic photons produced in one spark gap were observed by Roentgen to alter the conductivity of the air in another spark gap. The earliest recorded experiments of conductivity changes in a semiconductor due to electron irradiation appear in a paper by Kronig (*4*), published in 1924.

The first systematic investigation of conductivity changes due to electron bombardment of semiconductors was started in the late 1940s at Philips Laboratories, New York. Rittner reported in 1948 (*5*) on experiments on bombarding polycrystalline selenium with electrons up to 2 keV in energy. Rittner and his associates continued the work for approximately three more years. A number of patent applications were filed and a total of five patents (6–10) were issued to Rittner and his associates.

At about the same time, experiments were being conducted at the Bell Telephone Laboratories on the effect of energetic electrons interacting with insulators and semiconductors. No papers were written by the Bell group, but one patent resulted (11).

In 1950 work was initiated at the University of London under the direction of Ehrenberg. In 1951, results of bombarding both selenium and copper oxide at beam energies up to 90 keV were reported (12). Work continued for several years although no device work was pursued.

In the early 1950s, Moore of the RCA Princeton Laboratory worked on EBS-related phenomena. One paper was published (13), and two patents on devices were granted to the RCA workers (14, 15). The results of bombarding point contact diodes biased in both the forward and reverse directions in the low-field region were reported.

The first published work describing successful realization of EBS devices was by Brown of IBM in 1963 (16), who measured 300 V output across a 94 Ω load with a risetime of 4 nsec. In 1966 he described work on the development of a 64-position decoder–driver for use with magnetic core memories (17). Each output element was capable of 150 V output into a 50 Ω load with 3 nsec risetime and had an average power capability of 15 to 20 W. Limited-life tests were also conducted. The IBM effort constituted the first known successful EBS device demonstration.

In 1966 at Stanford University, Gibbons initiated work that resulted in a strong technical foundation being developed for EBS devices. Norris and Gibbons developed a simple but very useful theory, described the design of many interesting device configurations, and demonstrated the feasibility of a number of these configurations (18). This work provided a strong foundation for the renewed EBS interest by a number of organizations including the work started in 1968 by Watkins-Johnson (19). A number of other laboratories were actively investigating EBS devices, apparently all basing their work on what was being demonstrated at Stanford. One of these efforts to develop a high-power rf amplifier at the General Electric Laboratory, Syracuse, resulted in the first report of significant pulsed and cw rf output power from an EBS (20). The first prototype devices were introduced on the market by Watkins-Johnson Company in 1973 (21).

From the history recited, it can be seen that the development of EBS devices has not progressed in an orderly and rapid fashion from discovery through commercial exploitation. Because of the high technology, and even more significantly the dual technology requirements of the EBS, development has required funding beyond the capability of industry to support by itself. In addition, the majority of the initial device applications appeared to be primarily military. Prior to the Stanford work, the technical foundation was sufficiently primitive that only the IBM effort met with even limited success. In the period after the Stanford work, the United States was engaged in a costly war, which further deprived both industry and the U.S. Department of Defense of the needed research and development funds to assure rapid development of the new device, even though its capabilities were needed by developmental military systems. Continued economic difficulties and uncertain military support in the mid-1970s continue to impede the device's progress and restrict the work to only two laboratories. As history has shown, if the development effort is too low to provide continuing growth, industry will turn its attention elsewhere, and often a very promising device will die from lack of interest. EBS is now in the critical state where it must move ahead or it will inevitably decline.

It should be noted that the hybrid character of the device has been the cause of many of its problems, since the various groups handling EBS have been faced with the difficulty of classifying it either as a "tube" or a "semiconductor." It has, in essence, been treated as the illegitimate offspring of both technologies and often as an outcast. Depending upon the viewpoint, the EBS has been alternately judged to have combined either the best or the worst of tubes and semiconductors. In some laboratories its development was assigned to a tube group, in some to a semiconductor group, and in others even to a committee composed of people from both disciplines. Likewise in the various U.S. Department of Defense agencies, the device had to be assigned to a group whose responsibility was either to tubes or semiconductors, and often the EBS ended up as an orphan with no one wanting to assume responsibility for the device. The situation at present is, however, somewhat improved over that of the past.

B. *Principle of Operation*

The EBS principle can be utilized for broadband frequency or time domain amplifiers, pulse modulators, harmonic generators, real-time samplers, signal processors, and many other devices (22). Multifunction performance, in which the EBS performs several functions with a single device, is also possible.

The EBS is a two-port active device where the modulated electron beam acts as a control element by injecting carriers into a back-biased semiconductor diode. The basic principle of the EBS amplifier is illustrated in Fig. 1. The key elements are a means for generating an energetic electron beam, an rf input circuit for modulating the electron beam, and an rf output circuit consisting of a semiconductor target and rf matching circuitry. When biased as shown, the energetic electrons will strike the diode with an energy

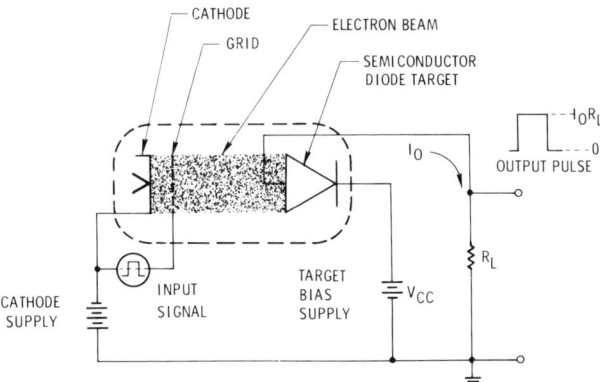

Fig. 1. Basic elements of an EBS. Electrons strike the semiconductor target at high energy, leading to amplification through charge multiplication. The current injected into the diode is controlled by the signal input to the grid. Output can be either polarity with respect to ground.

typically 10–15 keV. The diode is reverse biased well below its avalanche threshold. In the absence of an electron beam, there is only a negligible target leakage current. With no input signal, either a static beam current or no beam current is present. When a signal is introduced, the semiconductor is illuminated by varying electron beam currents, and since the target current is proportional to the beam current, the device acts as a linear amplifier.

The injection of carriers into a typical diode is accomplished by bombarding the top metal contact with the energetic electron beam, as shown in Fig. 2. The bombarding electrons lose some energy penetrating both the top

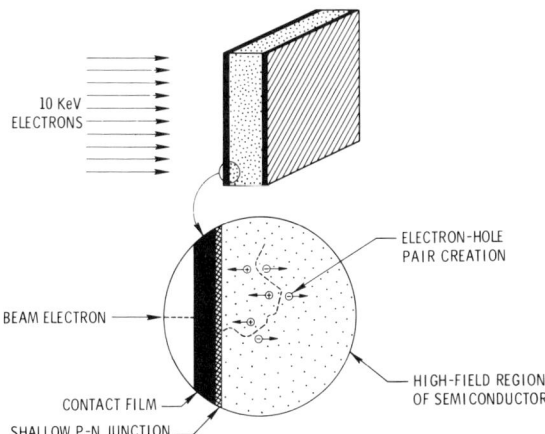

Fig. 2. Electron beam injection and carrier pair multiplication occurs near the shallow junction. The incident electrons typically lose 3–5 keV energy penetrating through the top contact. Electron–hole pair creation energy is 3.6 eV.

metal contact and the highly doped junction region, typically 3–5 keV. Since the pair-creation energy in silicon is approximately 3.6 eV, each 12 keV incident beam electron can produce thousands of carrier pairs in the diode. This results in current amplification in the semiconductor of 2000 or more. Electrons move through the depletion region of the semiconductor, causing current to flow in the load while the holes are swept back to the bombarded contact, thereby providing current continuity. Unlike conventional semiconductor devices, the EBS input and output circuits are isolated and therefore the EBS can be analyzed in terms of the separate characteristics of the input and output circuits.

The basic output response characteristics of a lumped-element EBS device are determined primarily by the semiconductor diode and output circuit elements. The frequency response is limited by the transit time for charge carriers through the diode, and by the frequency dependence of the load impedance in conjunction with the diode capacitance. The peak pulsed output power is limited by breakdown voltage, drift velocity, and space charge within the diode depletion region, as described for time domain operation by Norris (23).

The input circuit also plays a role in the frequency response depending upon the type of modulation structure. The two basic modulation

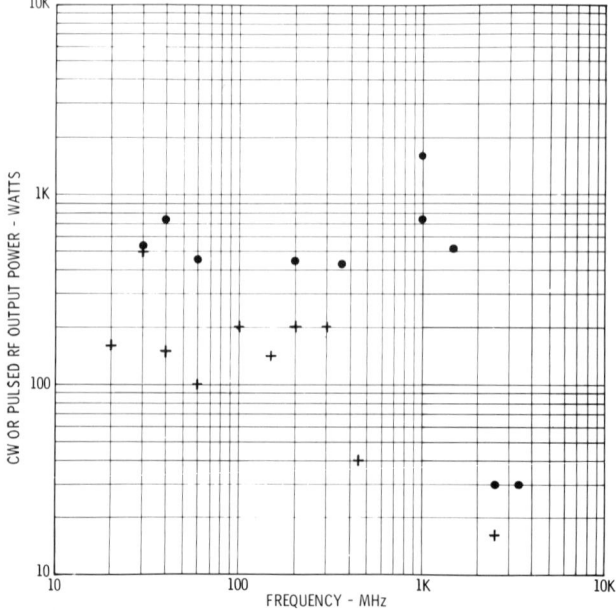

FIG. 3. Highest measured cw (+) and pulsed rf (●) output power vs frequency from EBS amplifiers as of October 1976. Both density- and deflection-modulated devices are represented.

techniques used in an EBS are density modulation and deflection modulation. In the former, the total beam current is varied by a modulating signal, and in the latter the position of the beam is varied relative to the semiconductor target. The type of modulation determines the device configuration, and therefore the design and performance of the device (24).

C. General Characteristics

The EBS is a hybrid electron device, thus inviting comparison with both tube and semiconductor devices. It would perhaps be desirable to present several graphs comparing demonstrated EBS performance with that obtained from power grid tubes, klystrons, transistor amplifiers, etc. However, since EBS's are only now merging from the laboratory such a comparison is not meaningful. The highest reported cw and peak rf power outputs versus frequency for EBS amplifiers are shown in Fig. 3. At the present time EBS performance is rapidly being improved, and thus it is expected that these results will be surpassed in the future. Figure 4 shows the theoretical peak rf power capability as a function of low-pass bandwidth of an EBS.

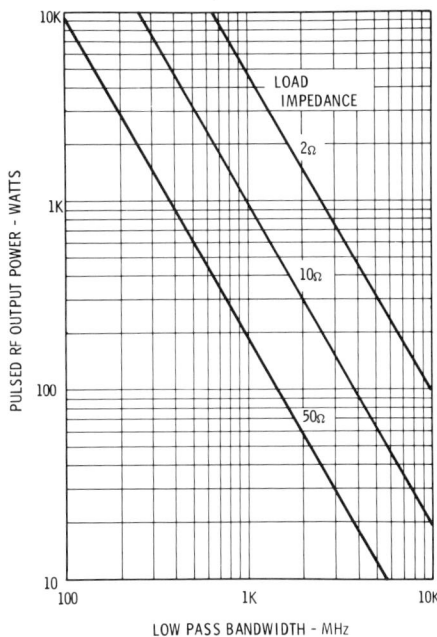

FIG. 4. Analytically predicted peak-pulsed rf output power capability of a low-pass EBS amplifier as a function of bandwidth for various load impedances presented to the diode. Typically $P(\Delta f)^2 Z$ is approximately 10^{22} W/Hz^2 Ω.

The EBS can be characterized as a linear two-port amplifier with exceedingly high isolation between output and input. Since ideally there are no low-field regions in an EBS diode, low-field parasitics are minimized. In addition the carriers travel at the high-field saturation velocity in the EBS diode. The gain bandwidth of the EBS is relatively high since there is high gain due to current multiplication in the semiconductor.

The EBS also has limitations that may make it unattractive in some applications. It requires a power supply of 10 to 15 kV, albeit at low current levels. In existing devices, the electron beam is generated by a thermionic emitter and therefore requires a heater supply floating at high voltage. The heater power will not substantially degrade the overall efficiency of cw power devices but the heater power and warm-up delay make the device less attractive for low-power, low-duty devices.

Since the EBS is basically a hybrid device, one can separate the vacuum tube elements and their design from the semiconductor element and its design to a considerable degree. Figure 5 shows some of the devices that

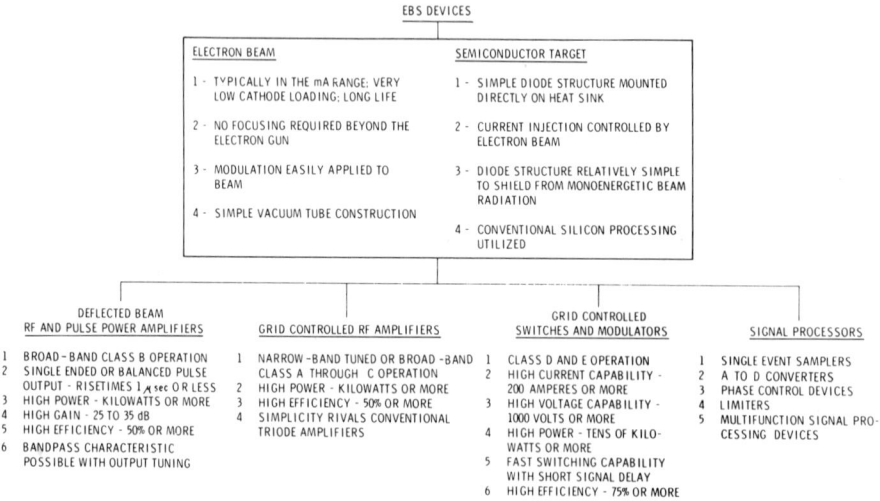

FIG. 5. Summary list of advanced lumped-element EBS devices resulting from the combination of electron beam and semiconductor technologies.

result from the various ways of combining of the electron beam and semiconductor elements comprising an EBS, illustrating both density and deflection modulation devices.

The basic elements for generating an energetic density-modulated beam is the planar or convergent-flow triode. Because of the extremely high current gain in the semiconductor target, some of the severe design constraints encountered in conventional triode are relatively unimportant. For example, the output circuit portion of the device is the semiconductor diode and not the grid-to-anode space; thus the grid–anode transit angle has negligible effect on overall device performance. The density-modulated EBS is a very simple, compact low-cost device. The greatest disadvantage is the necessity to couple the input signal to cathode potential.

One form of a deflected beam EBS device is shown in Fig. 6, where an

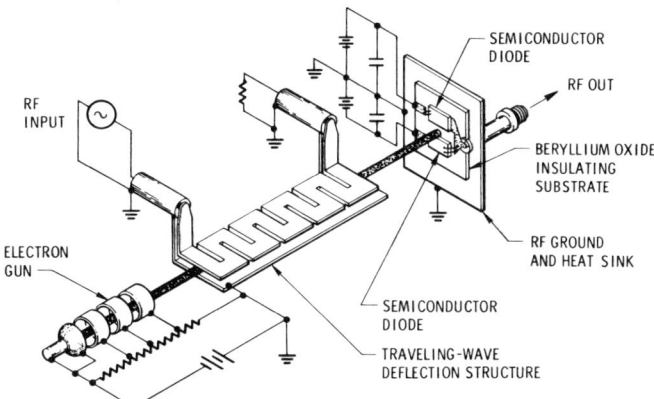

FIG. 6. Deflection-modulated EBS amplifier using a cylindrical beam electron gun. An input signal to the distributed deflection structure is used to control the amount of beam current intercepted by the class B connected-diode target.

electrostatically focused circular electron beam is projected through a traveling-wave deflection structure before bombarding a class-B-connected, dual-diode semiconductor target. Unlike a density-modulated EBS, the electron gun and deflection structures can be spatially and electrically isolated so that all the beam modulation can be applied at ground potential. As the beam passes through the deflection structure, it receives a small transverse velocity; the resulting angular deflection and the drift distance determine the linear deflection at the semiconductor target. A multiplicity of deflection structures in the two orthogonal transverse directions can be used to impart very complex deflections to the electron beam. The deflected beam EBS is suitable for very broad-band, linear, high-power amplifiers and is the most versatile configuration, but it is larger and more complex than a density-modulated EBS.

Uniform current density across the electron beam is important for both types of EBS modulation in order to produce uniform current density in the semiconductor. However, for a deflection-modulated beam it is doubly important to have a sharply defined planar beam edge so that in the absence of an input signal there will be no beam current intercepting the diode. Also, to have a linear transfer characteristic, the fraction of the beam current intercepted by the semiconductor iode must be directly proportional to the area intercepted, requiring a uniform current density. Both planar triode and convergent-flow triode guns have previously been designed for use in high-voltage switching tubes as well as for forming electron beams for linear-beam microwave tubes and will not be described (25, 26).

For typical deflected beam EBS power amplifier applications, the electron beam should be capable of being focused at a distance of 10 to 20 cm from the gun exit plane with a focused beam diameter in the order of 0.2 to 1 mm, or in other words, focused at a distance from the gun exit of 100 to 1000 beam diameters. In addition, for power devices the target current density must be on the order of 1 to 5 mA/mm^2, for a 10 kV beam. These requirements are not satisfied with conventional multielement electrostatic focused cross-over guns, such as used in cathode-ray tubes. Cross-over guns generally produce a gaussian current density profile and are deficient in the amount of current density that can be generated at beam currents above 1 mA. Nor are conventional Pierce-type guns suitable because of the lack of control of the beam current uniformity over a wide range of beam currents and the lack of focusing and accelerating electrodes beyond the anode. Thus, deflection-modulated EBS's required the development of a new type of electrostatically focused gun that could provide either circular or rectangular electron beams. A new type of electron gun, called the laminar flow gun, was developed for EBS's in two different versions, for producing either circular or rectangular electron beams (27–29). Figure 7 shows the salient difference in the beam envelope of the cross-over and laminar flow guns.

One of the most interesting and unique properties of an EBS device is the extreme variety of configurations possible for configuring and/or combining the individual diodes that comprise the semiconductor portion of the EBS target. Single-diode targets have found practical application for both pulse and rf amplification, for example, in both a 1000 V modulator EBS, used for controlling the beam current in a gridded TWT, as well as a cw microwave amplifier operating at 2.5 GHz.

Most EBS devices use two or more diodes. Figure 8 shows a six-diode array, used in a high-power wide-band rf amplifier, where a rectangular electron beam is used to illuminate this target. A large target area can be provided either by a single large diode or by interconnecting a number of small diodes, in series or parallel, or even in combination. Adding diodes in series is equivalent to adding capacitances and voltages in series, while adding diodes in parallel is equivalent to adding capacitances and currents in parallel. Thus, for high output currents and low-load impedance one

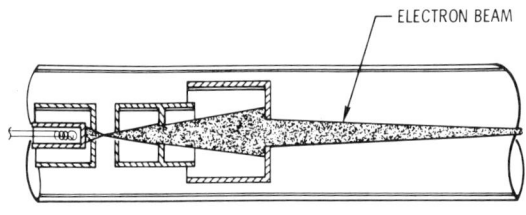

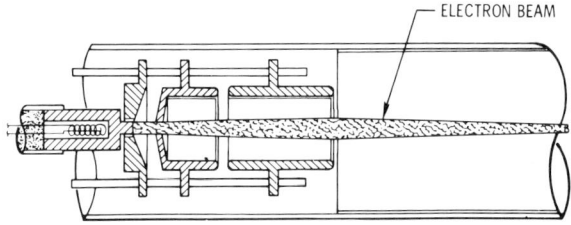

FIG. 7. Comparison of the beam envelope of a typical CRT cross-over gun compared to the laminar flow gun used for deflected-beam EBS amplifiers.

FIG. 8. A 2 × 3 diode array used in a high-power wide-band deflected-beam EBS rf amplifier. A thin rectangular beam is deflected between opposite diode triplets, which can either be connected in parallel or in series as shown.

would connect diodes in parallel, while for high voltages and large load impedances, series connecting is preferable.

The number of individual semiconductor diodes used in the EBS output can be made very large for a variety of purposes requiring a single input but multiple output elements, such as signal processing or sampling of very fast nonrepetitive signals. Multiple-target devices can have very fast access times because the electron beam can be scanned over very large numbers of diodes and large target areas in extremely short times. In fact, the time to deflect the beam from one diode to another can be made considerably shorter than the time it takes for a signal, traveling at the free-space velocity of light, to travel from one diode to the other. This is no violation of relativistic principles since the transverse " velocity " of the beam is equivalent to a phase velocity, which can be arbitrarily large. In addition, electron beam deflection structures can have bandwidths of dc to over 5 GHz.

As can be seen from the device listing in Fig. 5, the combination of an electron beam and semiconductor target provides a very versatile capability, which can be applied to many new practical devices. The signal processor devices have received the least attention during the past few years but it is our belief that a few years from now more EBS will be signal processors than power amplifiers and power control devices. One can visualize practical signal processor devices that could perform many of the functions of semiconductor integrated circuits with speeds an order of magnitude better than those of the existing devices.

II. Analytic Foundation

A. Basic Analysis of the EBS Target

In general, the peak power and bandwidth of all devices is limited by the dielectric breakdown field and by saturated drift velocity of charge carriers. The familiar expressions $P_0 f^2 Z = K$ (where P_0 is peak power, f the frequency where the response drops by a specified amount, Z the load impedance, and K the power bandwidth figure of merit) as well as $V_m f = \text{const}$ (where V_m is the peak voltage) are based on these parameters (30). The capability limits of EBS devices are similarly derived from these fundamental constraints. An additional limit is imposed on average power P_0 by the heat dissipation capability of a given device.

The basic EBS analysis, treating the time domain operation of a *pn* junction diode reverse biased into the high-electric-field region was conceived by Gibbons and Norris. The low-pass frequency domain, or rf, operation can easily be deduced from this analysis, since the product of minimum risetime and low-pass rf bandwidth is nearly equal to 0.35. It is assumed above a certain field value E_s that the carrier velocity can be considered constant, that the junction depth can be neglected, and that pair creation occurs close to the bombarded contact. These assumptions allow the analysis to be presented in general terms without consideration of specific diode excitation or geometry. When the semiconductor diode is reverse biased, the depletion region of the *pn* junction extends throughout the semiconductor, establishing a high-field drift region. In the absence of electron bombardment, the voltage–current characteristics of the target resemble those of a typical diode, with large conduction for small forward voltages and small conduction for large reverse voltages below the avalanche breakdown voltage.

The injection of carriers into the diode is as shown in Fig. 2. Most of the beam electrons penetrate the metal contact and enter the semiconductor with considerable remaining energy, while a few percent are lost as secondary electrons. The energy of the penetrating electrons is primarily dissipated by the creation of hole–electron pairs through momentum transfer to valence band electrons. The mean pair creation energy is 3.6 eV for silicon, and for a 10 keV electron carrier pair creation occurs to a depth of approximately 1.5 μm. Since the carrier pairs are created in a region of high electric field, causing rapid separation of electrons and holes, the probability of recombination is low; consequently one electronic charge traverses the load circuit for each carrier pair created. The semiconductor top contact metallization and junction region are structured to minimize beam absorption. Typically 3–5 keV beam energy is lost in this region. Therefore, with a 10 keV incident electron, 5–7 keV is a useful pair creation energy, and current gains typically on the order of 1500 to 2000 can be expected, with correspondingly higher gains for higher beam energies.

The carrier current *i* produced by the motion of a charge *Q* with velocity *v* will be

$$i = Qv/w \tag{1}$$

where *w* is the separation of the metal contacts. Assuming that the carrier drift velocity remains constant across the drift region, the carrier current is proportional to the total amount of moving charge in the drift region.

In the basic analysis, the finite penetration depth of the electron beam is ignored, and accordingly it is assumed that all the carrier pair creation occurs at the edge of the drift region. In this case, the device current is produced by the carriers that traverse the drift region, and the carriers of opposite polarity have no effect but to provide continuity of charge.

B. Risetime Optimization

In general, the risetime of the target will be determined by a combination of the transit time and capacitance effects. Since these two effects vary inversely with respect to each other as a function of w, it is possible to minimize the overall risetime by properly choosing the drift region width.

If transit time were the only limitation, the carrier current would respond with a risetime of

$$t_{r\,min} = 0.8w/v_0 \qquad (2)$$

where $t_{r\,min}$, the minimum risetime, is defined here as the time it takes the current to rise from 10 to 90% of its final value and v_0 is the constant carrier drift velocity.

The analysis only considers the response for transit angles less than 180° and is not applicable to such cases as non-transit-time-limited microwave devices. The current produced in an external load impedance is affected by the charge stored in the diode capacitance, and an equivalent circuit that shows the relation between carrier and load currents is shown in Fig. 9. In

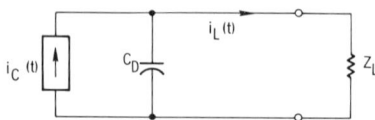

FIG. 9. Schematic representation of a simple target output circuit. The diode itself appears to be a current source shunted by a capacitance. Current is delivered to an external load impedance Z_L.

the case of a resistive load impedance, the carrier current appears to be exciting an RC network such that

$$t_{r\,min} = 2.2 Z_L \varepsilon A/w \qquad (3)$$

where Z_L is the load resistance, ε the dielectric constant of the semiconductor, and A the effective area of the diode. For purposes of simplicity we will use the active area as the effective area even though this leads to predictions that are generally optimistic. Typically for radiation-resistant planar pn junction diodes, the effective area will be 20–100% greater than the active area, depending upon diode size. The factor 2.2 in Eq. (3) is the result of our specifying 10–90% risetime for an RC-coupled low-pass circuit.

The inverse relationship between transit time and diode capacitance with drift region width of the diode can be used to optimize a diode to yield a minimum risetime response into an external load impedance. Norris has presented two very useful expressions for silicon at normal room temperature. The optimum drift region width is found to be

$$w_{\text{opt r}} = 38.9 \left(\frac{v_0}{10^7 \text{ cm/sec}} \frac{\varepsilon_r}{11.5} \frac{A}{1 \text{ mm}^2} \frac{Z_L}{50 \text{ }\Omega} \right)^{1/2} \quad (4)$$

and the minimum risetime, obtained when $w = w_{\text{opt r}}$, is given by

$$t_{r\text{min}} = 425 \left(\frac{10^7 \text{ cm/sec}}{v_0} \frac{\varepsilon_r}{11.5} \frac{A}{1 \text{ mm}^2} \frac{Z_L}{50 \text{ }\Omega} \right)^{1/2} \text{ psec} \quad (5)$$

where these expressions have been normalized so that the radicals have a value of unity for the case of a 1 mm^2 silicon diode biased to velocity saturation driving a 50 Ω load.

The response of an optimized diode can be quite fast; in fact, the risetime or bandwidth capabilities can exceed those of common lumped-element beam-modulating structures, making it necessary to employ traveling-wave modulation structures or special techniques to obtain an EBS device whose primary response limitation is due to the diode.

C. Output Current Optimization

The output current of the diode will be limited by the drift velocity and maximum field, while the output voltage will be limited by the avalanche breakdown at zero current and minimum voltage drop across the diode at maximum current. The average power will be limited by the rate at which heat can be removed from the target.

Energy saturation occurs when the electric field is reduced to a low value at some point in the drift region, increasing transit time and causing charge storage effects. The result is introduction of nonlinearities common to all energy-saturating devices. One of the interesting features of the EBS diode is that it can be designed for high-level operation with impressive gain, risetime, and bandwidth values with large output voltages and currents, yet with relatively low distortion.

In order to maximize the peak output current capability of a given diode geometry, the carrier drift velocity should be as high as possible to minimize the space charge in the depletion region. In addition, the carrier velocity should be independent of variations in electric field. Electrons drift in silicon with relatively high velocity and exhibit the desired strong velocity saturation effects, thereby meeting our requirements.

Assuming the bias voltage is limited to 60% of the theoretical bulk breakdown voltage $V_B(w)$, the target voltage will limit the peak output current i_l to approximately

$$i_l \simeq (1/2Z_L)V_B(w) \tag{6}$$

Since the breakdown voltage is proportional to the drift width, from voltage considerations, the peak output current is proportional to drift width.

The output current limitation arises from the fact that high values of carrier space charge in the diode reduce the field to a low value at the injecting edge of the drift region. Again, assuming the diode is biased at 60% of its breakdown voltage, the space charge limits the current to approximately

$$i_l \simeq 2v_0 \varepsilon A V_B(w)/w^2 \tag{7}$$

The two effects of space charge and voltage reduction combine to determine the output limitation of the diode. However, since these two effects vary inversely with w, it is possible to maximize the overall output capability by properly choosing the drift region width. Combining Eqs. (6) and (7) gives the expression for the drift width to yield peak output current, or

$$w_{\text{opt }i} = (2v_0 \varepsilon A Z_L)^{1/2} \tag{8}$$

In terms of normalized parameters, Eq. (8) becomes

$$w_{\text{opt }i} = 31.9 \left(\frac{v_0}{10^7 \text{ cm/sec}} \frac{\varepsilon_r}{11.5} \frac{A}{1 \text{ mm}^2} \frac{Z_L}{50 \, \Omega} \right)^{1/2} \, \mu\text{m} \tag{9}$$

Finally, the semiconductor doping density must be chosen properly since it determines the density of fixed space charge in the depleted drift region. For a diode with a given area and Z_L, there is an optimum doping for each value of w.

D. Diode Output Capability

When w is given in Eq. (9), the optimum doping produces the electric-field profile shown in Fig. 10 when the diode is biased properly. It is interesting to observe that the electric-field profile in an optimized device pivots about its value at $x = w$ as the device current is increased. When the limiting current for linear operation i_l is reached the field is constant across the drift region. The peak linear output voltage is typically 0.87–0.92 of the diode bias voltage for this condition.

The semiconductor doping necessary to provide the electric-field profile in Fig. 10 can be computed from Gauss' law; the differential from Gauss' law is

$$\Delta E/\Delta x = (E_m - E_s)/w = \rho/\varepsilon \tag{10}$$

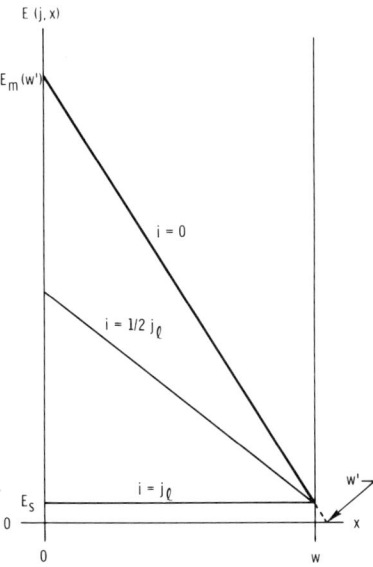

FIG. 10. Electric field profiles for various values of diode current. The output is optimized when the load impedance is adjusted so that the pivotal point is located at $x = w$.

where ρ is charge density and ε the silicon permittivity, equal to 1.02×10^{-10} F/m.

Since the output current capability of a fixed-area diode has been maximized for a given load impedance, the output voltage and power of the device have also been maximized.

The linear output capabilities of a silicon diode are given approximately by Eq. (5):

$$i_{l\,\mathrm{max}} \cong 61.5 \left(\frac{v_0}{10^7\ \mathrm{cm/sec}} \frac{A}{1\ \mathrm{mm}^2}\right)^{3/7} \left(\frac{\gamma^2}{0.5}\right)^{4/7} \left(\frac{1\ \Omega}{Z_L}\right)^{4/7} \tag{11}$$

$$v_{l\,\mathrm{max}} \cong 61.5 \left(\frac{v_0}{10^7\ \mathrm{cm/sec}} \frac{A}{1\ \mathrm{mm}^2} \frac{Z_L}{1\ \Omega}\right)^{6/14} \left(\frac{\gamma^2}{0.5}\right)^{4/7}\ \mathrm{V} \tag{12}$$

$$p_{l\,\mathrm{max}} \cong 3780 \left(\frac{v_0}{10^7\ \mathrm{cm/sec}} \frac{A}{1\ \mathrm{mm}^2}\right)^{6/7} \left(\frac{\gamma^2}{0.5}\right)^{8/7} \left(\frac{1\ \Omega}{Z_L}\right)^{1/7}\ \mathrm{W} \tag{13}$$

where γ is a constant determined by the permissible avalanche multiplication of the particular diode, which requires that $V_{CC} \leq \gamma^2 V_B(w)$, where V_{CC} is the diode bias voltage.

If γ^2 is taken as 0.5, then the optimized peak pulse power of 1 mm² silicon diode into a 1 Ω load is 3780 W; from Eq. (5), the risetime would be approximately 60 psec, which is equivalent to a peak rf power of 475 W with a low-pass bandwidth of 5.8 GHz. It is evident that even a small silicon diode has high output capability. However, the pulsed output capability of the diode is considerably above the cw output capability, which is determined by thermal limitations.

Equations (5) and (11)–(13) of the linear analysis can be conveniently combined into a series of diode capability curves showing the peak current, voltage, and power capabilities of a semiconductor target as a function of risetime, device area, and load impedance. These curves, presented in Figs. 11–14, are based on the simplifying assumptions of linear operation and are very useful for the initial evaluation of EBS device capability. In terms of more familiar figures of merit, Eqs. (11)–(13) predict values of Pf^2Z of approximately 10^{22} W-Hz²-Ω, and voltage and current risetimes of 600 V/nsec and 600 A/nsec, respectively, into optimized load impedances.

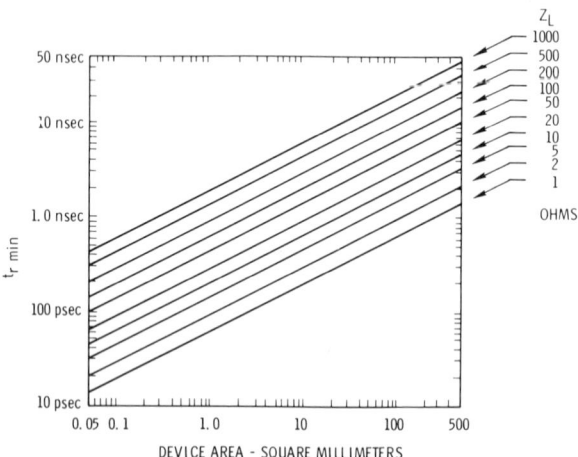

Fig. 11. Optimized lumped-target risetime shown as a function of effective diode area for various load impedances. This figure shows that subnanosecond risetimes are readily obtained. The effective area referred to in this figure is the one that determines the capacitance of the target.

III. Physical Limitations to EBS Capabilities

The EBS diode can be represented by a current source shunted by a capacitance C_D; for operation at high frequencies the capacitive reactance of

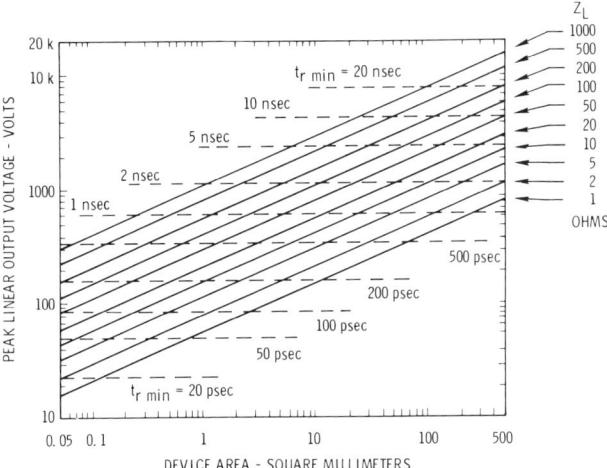

FIG. 12. Linear output voltage capability of an optimized silicon lumped target as a function of active device area for various load impedances. The constant-risetime contours give the minimum risetime, corresponding to the A and Z_L, and are drawn with the assumption that the effective area is equal to the active area.

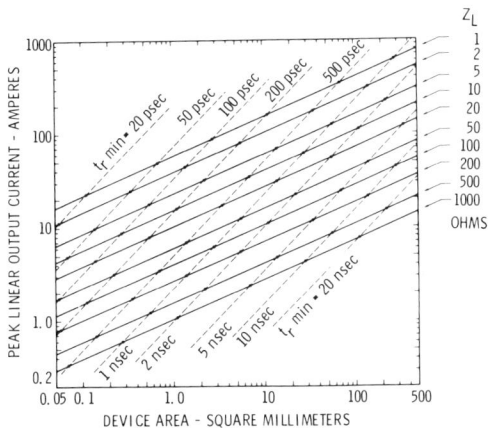

FIG. 13. Linear output current capability of an optimized silicon lumped target as a function of device area for various load impedances. Since the target area and load impedance determine the minimum target risetime, constant-risetime contours are drawn in this figure on the assumption that the effective and active areas of the target are equal. In the event of significant difference between these two areas, the active area is to be used with this and the preceding two figures and the risetime contours are to be disregarded.

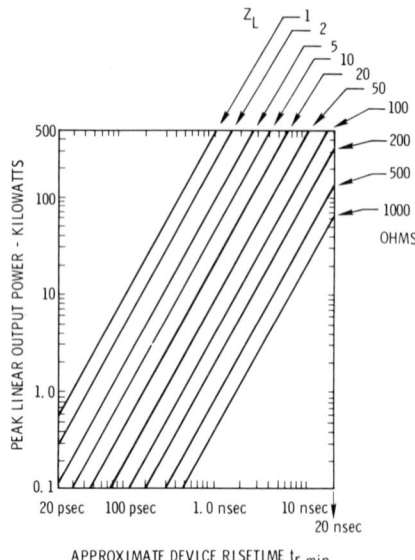

Fig. 14. Linear output power capability of an optimized silicon lumped target as a function of optimized risetime for various load impedances. This power is the product of the output voltages and currents given in the preceding two figures and represents the ultimate peak pulsed power that can be obtained without reference to thermal limitations. The effective diode area is assumed to be the active area, that is, there is no excess diode capacitance.

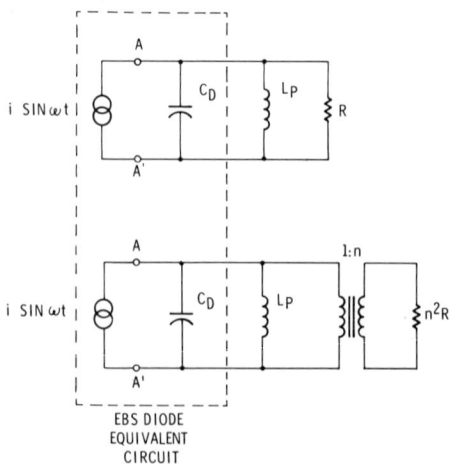

Fig. 15. Schematic representation of two bandpass output circuits where the diode capacitance is resonated with an inductance in the circuit.

the diode can be matched, for example, by resonance with a parallel inductance as shown in Fig. 15. The upper frequency limit is then set by the transit time of the carriers through the diode depletion region. Figures of merit for the power–frequency K and power–bandwidth K' capabilities of the diode can be concisely described by Pf^2Z and $P(\Delta f)^2Z$, respectively, where f is the maximum frequency of operation, Δf the bandwidth, P the maximum output power, and Z the real part of the impedance presented to the current source across the terminals $A-A'$ (in the figures, $Z = R$). K is a characteristic of the diode and is not changed by circuit impedance-matching techniques such as the transformation shown in the lower figure, since the terminal impedance presented across $A-A'$ is unchanged (unless explicitly stated otherwise when K is used, it refers to both figures of merit). However, transformations of this type are useful since the terminal impedance (R) across $A-A'$ can be made lower than the external amplifier load impedance (n^2R), which is most commonly chosen to be 50 ohms for coaxial transmission lines. Thus, while K is a function of both fundamental materials limits and fabrication technology, the actual amplifier output power depends on the use of circuit-matching techniques such as impedance transformation or power combining. A cw figure of merit cannot be simply deduced, since the cw power output is thermally limited and is dependent on other factors in addition to the fundamental materials limits. The linear theory presented earlier projects a maximum materials-limited value for K of approximately 10^{22} W Hz^2 Ω for silicon EBS targets. For comparison with actual devices, this theoretical capability translates into 3300 W peak output power at 1.0 GHz with a 3 Ω terminal impedance. An actual amplifier designed for those performance levels achieved 1500 W peak output power, which corresponds to K of 4.5×10^{21} W Hz^2 Ω. For reference, a cw amplifier produced 16 W at 2.5 GHz with a 15 Ω load for a cw K value of 1.5×10^{21} W Hz^2 Ω. These performance levels show that (1) EBS amplifiers have already demonstrated significant power capabilities, and (2) considerable growth remains before the limits set by the theoretical materials capabilities are reached.

Improvements in both diode design and fabrication technology are required in order to approach the theoretical limits of the linear model. Also, a number of performance-limiting factors are not included in the linear model; for example, total diode capacitance is greater than that of the active area only, carrier velocity is reduced under large signal or elevated temperature conditions, and resistive losses become significant at microwave frequencies.

One of the primary limitations arises from the constraints imposed on the design of the perimeter of the diode. If the energetic electrons strike the passivating insulator at the diode perimeter a net positive surface charge will result, as shown in Fig. 16. If this positive surface charge is formed in a high electric field region of the p^+–n junction, both diode leakage and breakdown voltage will be degraded. Specially designed perimeter configurations in-

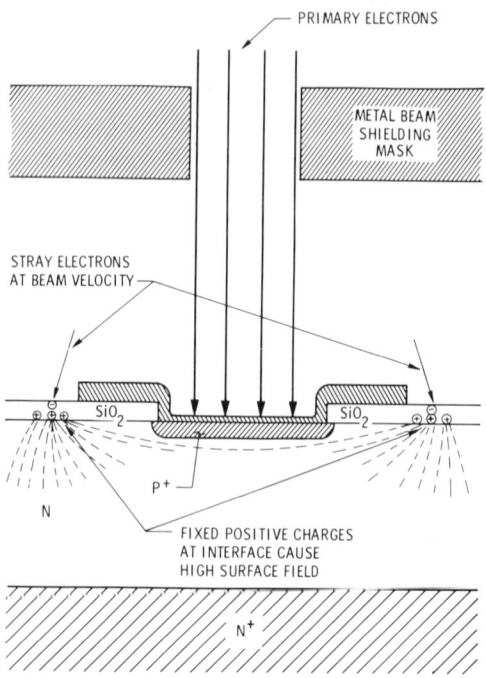

FIG. 16. Effects of stray secondary electron and X-ray bombardment of the passivating dielectric, which leads to positive trapped charge concentration at the dielectric.

cluding guard rings, and thick oxide and metallization layers are used to minimize or eliminate radiation-induced degradation. However, this results in an increase in the capacitance of the EBS diode.

The total diode capacitance C_t is larger than the capacitance of the junction area alone, C_j, by the excess capacitance ratio m, where $m = C_t/C_j$, typically m has ranged from 1.8 to 2.5. Excess capacitance decreases the reactance of the diode, resulting in a decrease in the diode Q (Q is the diode reactance/diode series resistance). For narrow-band amplifiers the circuit losses are proportional to Q^2, causing the amplifier efficiency to be reduced by a term proportional to m^{-2}. Thus, both the efficiency and the output power of narrow-band EBS amplifiers are degraded as a result of excess capacitance.

Excess capacitance also strongly affects broad-band amplifiers, since K' is proportional to m^{-2}. The maximum K' demonstrated by a broad-band

EBS amplifier is 3×10^{20} W Hz^2 Ω, well below the projected theoretical capability of 10^{22} W Hz^2 Ω, partially as a result of excess capacitance. Recently it has become possible to reduce the excess capacitance by a factor of 1.5, which will result in a significant increase in broad-band amplifier capabilities.

Series resistance in the diode or the matching circuitry also lowers the Q and becomes increasingly important as the frequency increases. Diode resistance can occur in the back contact n^+ layer and metallization, but the major portion occurs in the top contact p^+ and metallization layers. Typical top contact resistance is in the order of 0.1 to 0.6 ohms per square, which translates to 0.1 to 0.8 Ω series resistance. The effect of series loss and excess capacitance is such that the efficiency of existing devices at 2.5 GHz is the order of 35 to 40%, while those at 0.2 GHz are in the order of 50 to 60%.

For a given diode thickness and doping level, the diode perimeter design affects the breakdown voltage, which determines the maximum voltage swing, and limits the peak power output. Guard rings, field plates, or a combination of both are used to maximize the breakdown voltage.

The linear theory assumes that the pair creation depth is small compared to the diode thickness. This assumption is no longer valid for diodes thinner than about 20 μm, corresponding to devices operating at 500 to 1000 MHz or above, since the average pair creation depth is typically 2–3 μm. A more accurate model using a large-signal computer program has been developed that allows simulation of the diodes under both static and dynamic conditions. Deep injection results in some increase in power–frequency capabilities when compared with the shallow linear theory since the mean carrier transit time is reduced. However, the dominant effect of deep carrier pair creation is that the holes, as well as the electrons, affect the shape of the internal electric field profiles. In order to maximize the output current and efficiency of thin diodes in the presence of deep injection, it is necessary to move the metallurgical junction well within the diode. In practice this requires the use of silicon wafers with both n- and p-type epitaxial layers and requires a more complex diode peripheral geometry.

During high average power operation, the temperature of the EBS diode increases, with the amount determined by the thermal impedance of the complete target, which consists of the diodes, mounting substrate, and heat sink. As the temperature of the diodes increases, the carrier drift velocity decreases, strongly affecting the amplifier output since the semiconductor figures of merit both decrease with the square of the drift velocity. As shown in Fig. 17, the electron velocity at 2 V/μm drops by 33% for an increase from 300 to 430°K, decreasing the output capability by 55%. Thus, it is extremely important to minimize the thermal impedance in order to maintain the lowest possible diode temperature. Special techniques have already been

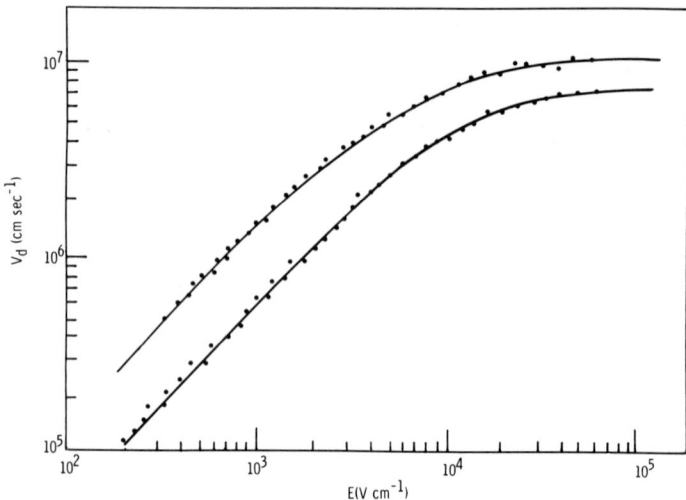

FIG. 17. Electron and hole drift velocity in silicon as a function of electric field at three different temperatures. The points are the experimental data and the continuous line is the best fitting curve.

developed to achieve void-free low-thermal-impedance die bonding. In the future other techniques such as plated heat sinks and very thin diodes are expected to be used in order to maximize the average power capabilities.

The linear theory assumes that the semiconductor carrier velocity remained constant during operation, which is not true for operation either at elevated temperatures or at low electric fields. Carrier drift velocity variation at low fields degrades amplifier linearity; however, the resulting increase in distortion is often accepted in exchange for increased amplifier efficiency. For example, at room temperature the electron drift velocity saturates at 10^7 cm/sec at a field of 2 V/μm. If the minimum field is reduced from 2 to 1 V/μm, the carrier velocity drops by 20% while the diode dissipation is reduced by about a factor of two.

The linear theory does not apply with EBS amplifiers used as high voltage pulse modulators. When operated into a highly capacitive load, large output currents flow during the turn-on transient with the steady-state currents much lower. The EBS diode operates in the high-field region during much of the turn-on transient. However, as the output voltage rises, the electric field in the diode is reduced to such a low level that the electrons and holes are no longer swept out of the junction region. A substantial diffusion current develops, and recombination becomes significant. The deviation from linear operation is so great that the EBS is best described by a composite nonlinear model utilizing a piecewise-linear function during turn-on and steady-state operation and by a charge control model for turn-off.

The linear theory can be used to compare the potential of alternate semiconductors to silicon. Ideally, a semiconductor for use in EBS diodes should have the following characteristics: high breakdown voltage, high saturation velocity, high low-field mobility, high thermal conductivity, and low pair creation energy. Some of the compound semiconductors have certain properties that surpass silicon by a substantial margin; however, the overall capability is dependent upon both electrical and thermal properties. In addition, some of the desired characteristics are mutually contradictory such as high breakdown voltage and narrow bandgap. An analysis of the cw power capabilities of various semiconductors is summarized in Fig. 18. As

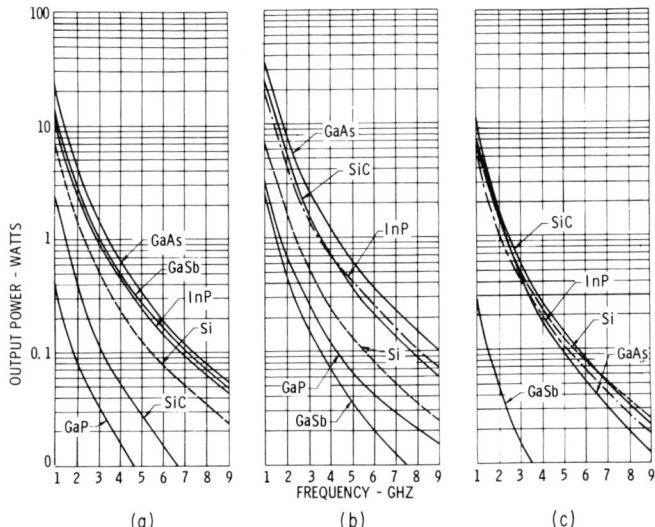

FIG. 18. Comparison of cw power capability of various compound semiconductors to silicon for various approximations. (a) Power dissipation of 20 W/mm^2 and the same operating temperature for all materials; (b) the higher heat dissipation capability of the wider-band-gap semiconductors; (c) both the gap dependence and thermal conductivity of the semiconducting materials.

shown, SiC, GaAs, InP, and Si all have comparable capabilities for cw operation, while the compound semiconductors appear to have advantages for pulsed operation. Experimentally, both GaAs and GaAsP diodes have been tested in EBS devices. No comparative analysis of relative capabilities could be made due to the relatively primitive experiments that were performed; however, the diodes did demonstrate useful performance capabilities and were reported to be electrically stable.

At this time the optimum material for EBS diodes is unquestionably silicon because of the extremely well-developed fabrication technology that makes it possible to fabricate sophisticated designs reproducibly and at relatively low cost. Silicon will continue to be the dominant material for EBS diodes so long as performance is limited primarily by fabrication technology. However, in the future, when devices approach the basic materials limits, alternative semiconductors having greater materials-limited capabilities will become progressively more attractive, assuming that the technology is already developed for other devices. By themselves EBS devices could not provide economic justification for developing compound semiconductor technology.

IV. Target Assembly

The target assembly, which is that portion of the output circuit within the vacuum, usually consists of one or more semiconductor diodes, a mounting substrate that acts both as a heat sink and part of the electrical circuit, and various elements of the circuit itself.

A. The Semiconductor Diode

The semiconductor diode, being the active element that provides both the current amplification and the voltage stand-off, is the heart of an EBS device. During the early years these diodes created the weak link that severely limited the performance and stability of EBS devices, and as a consequence, hampered many development efforts. Unpassivated, bare-junction mesa diodes were formerly used. Modern EBS semiconductor diodes are passivated and radiation hardened and thus are stable and reliable. The performance of an EBS device is still mainly limited by the semiconductor diode and considerable effort is ongoing in order to reduce these limitations.

An EBS semiconductor diode differs greatly from most conventional semiconductor diodes because of its unique operating environment and unusual operating mode, especially when used for power applications. Since the diode is inside a vacuum tube, it must be compatible with vacuum tube materials and processing techniques. It must withstand high-temperature vacuum bakeout, high electric fields in a high-vacuum environment, and deposition of partially conductive surface films without any change in its electrical characteristics. This requires the use of passivated-junction diodes. When operating, the diode is irradiated with energetic electrons and X rays that generate hole/electron pairs within the oxide or glass used to passivate the surface. In the normal p^+–n diode, this requires that the oxide–silicon interface be protected so that energetic electrons and X rays cannot penetrate into the high-field region near the interface. In addition, the electron–hole pairs must be generated in the high-field region of the diode, which means that the incident electrons should suffer a minimum energy loss in penetrating through the top metal contact and the low-field junction layer. These requirements are inherent for diodes in all EBS devices. In addition, large surface areas, high breakdown voltages, and low thermal impedances are required from EBS diodes designed for power amplification and control. The impact of these requirements on the design and fabrication of both the semiconductor diode and the vacuum tube is discussed in the following paragraphs.

During the fabrication of a conventional power tube, the entire vacuum assembly is generally baked out at temperatures of 400 to 500°C for periods of 24 to 48 hours prior to sealing it off. This practice is incompatible with target assembly and diode materials and fabrication. In fabricating the target assembly, the semiconductor die is attached with Au–Ge eutectic preform and all other elements are attached with a Au–Sn eutectic preform. Since these alloys have melting temperatures below 400°C (346 and 280°C, respectively), the target assembly would be damaged if subjected to normal bakeout practices. In addition, under these conditions, metals used on the diode top contact would overalloy into the silicon, severely degrading the diode characteristics. When processing an EBS, the bakeout temperature is lowered to about 300°C (during fabrication the Au–Sn alloy is allowed to become Au rich, raising its melting temperature well above 300°C), with the addition of an extended period of target outgasing by the electron beam with an appendage ion pump attached. Although this substantially increases the vacuum processing time, it has proven to be adequate.

Early EBS devices used bare-junction mesa diodes with little success. Applying a high electric field across a bare junction in a vacuum causes the diode to degrade due to high-field surface migration effects, rendering the diode useless since it becomes extremely leaky and its reverse breakdown voltage continuously decreases. The use of passivated planar diodes removed this problem. In addition, impurities evaporate onto the diode surface due to outgasing of the various tube elements during bakeout and cathode activation. Some of these evaporants are alkali ions that are positively charged and readily drift through undoped silicon oxide. Thus passivation oxide must be capable of immobilizing alkali ions to prevent them from migrating to the silicon interface. Phosphorous-doped silicon oxide is effective in immobilizing alkali ions so that EBS diodes with this passivation show excellent stability through both tube processing and cathode activation as well as throughout operating life.

During normal operation an EBS diode is irradiated with a 10–20 keV electron beam. In the design of the tube a beam mask is included that prevents other than the active area from receiving direct beam illumination. However, secondary electrons and X rays are generated that can still strike the diode surface outside the active area, as illustrated in Fig. 16. When they penetrate into the passivating oxide layer, hole–electron pairs are created. Since electrons are mobile in the oxide, whereas holes are virtually immobile, the electrons will drift out of the oxide into the silicon, leaving behind an immobile net positive charge. This trapped positive charge accumulates in the oxide, increasing the high electric field along the junction periphery, which decreases the reverse breakdown voltage of a p^+–n diode, which is the preferred-polarity diode. A non-radiation-hardened p^+–n diode with a reverse breakdown voltage of 300 to 400 V will degrade to 150 to 200 V when exposed to a few hours of electron beam bombardment.

The most effective technique for minimizing or eliminating the effect is to prevent the hole–electron pairs from being generated in the immediate vicinity of the oxide–silicon interface by use of thick oxides and metals. One such diode consists of depositing 4 μm thick phosphorous doped silicon oxide over the surface outside the active area, depositing 800 Å of aluminum over the active area, and covering the periphery of the diode with an aluminum ring 4 μm thick and 75 μm wide (see Fig. 19a). The thick metal ring helps to shield the oxide over the junction edge from electron and X-ray bombardment, as well as providing a contact pad for bonding wires. The thick oxide not only helps to prevent the ionizing radiation from penetrating clear to the interface, it also reduces the excess capacitance associated with the metal ring. These diodes are both simple and inexpensive to fabricate and are used on most commercially available EBS devices. However, while it is quite effective, it does not provide absolute protection from ionizing radiation penetrating through the passivating oxide. After beam exposure, the reverse breakdown of a diode fabricated this way will stabilize at about 90% of its initial value. This reduction is of no concern to the EBS user but is a constraint on the designer.

Another method, which utilizes an integral gold beam shield, provides complete protection from any irradiation and is used in designs where no decrease of reverse breakdown voltage is possible in order to meet the required design value. This consists of the same thick oxide as before, a Ni–Si film 0.02 μm thick over the active area, a composite layer of Ti–Pt–Au, 0.1–0.2–1.0 μm thick, respectively, along the junction periphery, and an integral gold beam shield 150 μm wide and 10 μm thick suspended above the junction periphery by plated gold pillars 10 μm tall along the composite layer, as shown in Fig. 19b. This thick integral beam shield is not placed in contact with the oxide film, but suspended above it in order to

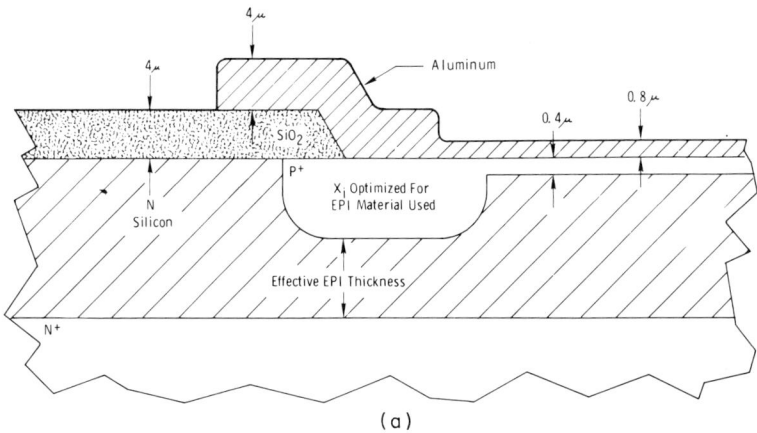

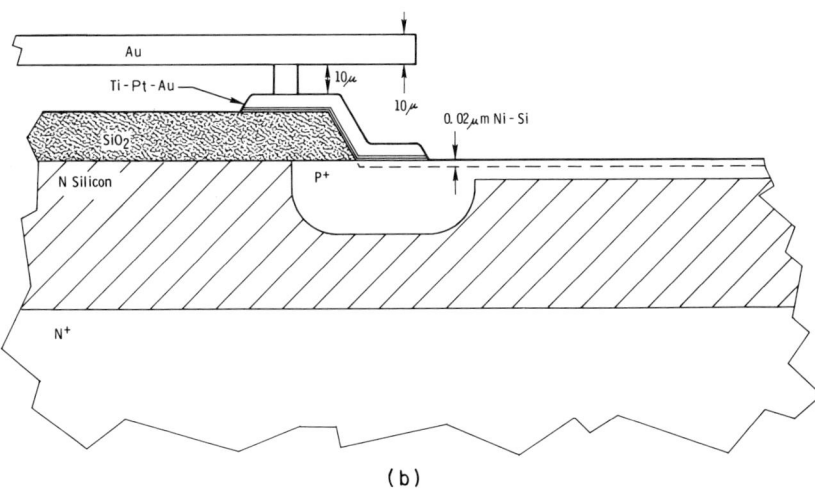

FIG. 19. (a) Cross section of the edge of an EBS diode utilizing aluminum metallization showing the thick aluminum ring along the periphery, thick phosphorous-doped silicon oxide layer in the field, and thin aluminum layer over the active area. (b) Cross section of the edge of an EBS diode utilizing an integral thick, gold beam shield showing the pillars used to support the beam shield, Ti–Pt–Au metallization along the periphery, thick phosphorous-doped silicon oxide layer in the field, and thin nickel-silicide layer over the active area.

minimize the excess capacitance associated with it. Nickel silicide is chosen for the top metallization because of its compatibility with the various etches used in fabrication of the integral gold beam shield. This method, however, is complex and expensive to fabricate and, as will be explained later, suffers in other regards due to the use of nickel silicide as the top metallization. Since the penetration of electrons into silicon is very limited (the range of a 15 keV electron being approximately 3 μm in silicon), it is necessary that the top contact metallization be very thin and that the rectifying junction comprising the active area be very shallow. Typical junction depths used for the active area of an EBS diode range from 0.2 to 0.3 μm. This shallow rectifying junction may be formed either as a Schottky or as a p–n junction.

A p–n junction can be formed by introducing a p dopant into n-type silicon (p^+–n junction) or by introducing an n dopant into p-type silicon (n^+–p junction). No special processing is required to radiation harden an n^+–p diode. In an n^+–p junction the polarity of the internal electric field is such that neither the leakage current nor the breakdown voltage is affected by electron bombardment, since the electric field induced by trapped positive charges in the oxide weakens the internal field at the silicon–silicon oxide interface. As advantageous as this may sound, p^+–n diodes are preferred and used almost exclusively. This is because p^+–n diodes operate on the transport of electrons that have a higher low-field mobility than holes (1350 vs 475 cm^2 V^{-1} sec^{-1}, respectively) and a higher carrier velocity, which saturates at high fields. This implies that both the Pf^2Z product and the linearity of a p^+–n diode are greater than for an n^+–p diode.

In analyzing the operation of an EBS diode, one finds that only those carriers generated in a depleted, high-field semiconductor region contribute to device operation. Carriers generated in an undepleted region serve to lower its effective resistivity but do not contribute to current amplification. The energy expended by an electron in such a region in the semiconductor is lost with regard to current amplification and is referred to as "dead loss."

Schottky diodes offer the possibility of reducing both the dead loss and the rf loss associated with the p^+ region of a p^+–n junction. Thus, for a given incident beam energy, the current gain of a Schottky diode can be greater than that of junction diodes. Schottky barrier EBS diodes should therefore be given strong consideration in designs where high current gain, low beam voltages, or low rf losses are essential. However, Schottky diodes have not yet proven to be a viable choice. Their application to large-area EBS diodes is still in question due to their inherent higher reverse-leakage currents. This current for a Schottky barrier includes a term resulting from the thermionic emission of majority carriers into the semiconductor that is typically several orders of magnitude larger than the diffusion–recombination current found in a p–n junction. This current increases rapidly with temperature and can drive a device into thermal runaway when the power dissipated in the device exceeds a critical value. Schottky diodes, however, are expected to receive more attention in the future, especially when compound semiconductors are applied to EBS's.

Even p–n diodes achieve current amplification factors of 2000 at incident beam energies of 10 to 12 keV, and at 20 keV the amplification is over 4000. This is sufficient for most applications and there has been little motivation to use a Schottky barrier merely for higher gain.

In order to reduce the rf losses associated with the p^+ region of a p^+–n junction, a metal film is deposited over the active area. This film, however, adds additional dead loss, thereby further lowering the current gain of the device. An engineering tradeoff must be made with respect to material parameters, ease of processing, control of film thickness, current gain, and rf loss consistent with the required device parameters. For example, where rf loss must be minimized, increased loss of beam energy must be tolerated.

A useful figure of merit (31) to evaluate different materials for the top contact metallization is defined as $M = \sigma/\rho$, where σ is the electrical conductivity of the material and ρ its mass density. Materials with a higher value of M have a more favorable combination of beam penetration and electrical conductivity. Table I compares values of M for various materials. It is clear from looking at Table I that highly doped p^+ silicon ($NA \approx 10^{20}$ atoms/cm^3) is a poor choice by itself and that a metal conductive film is necessary. The values of conductivity and density in this table are for bulk materials. Both of these parameters can be considerably different for thin films, depending on the deposition process; also, both backscattering and secondary emission loss of electrons increases with atomic number Z,

TABLE I

ELECTRICAL CONDUCTIVITY, DENSITY, AND FIGURE OF MERIT FOR VARIOUS METALS THAT MAY BE USED FOR TOP METALLIZATION

Material	Electrical conductivity ($\times 10^4$/ohm-cm)	Density (g/cm^3)	Figure of merit ($\times 10^3$ cm^2/g-ohm)
Aluminum	38	2.7	140
Silver	63	10.5	60
Molybdenum	17	10.2	17
Nickel	15	8.9	17
Chromium	7.7	7.8	10
Nickel silicide	~4.0	~5.0	~8
Platinum	9.4	21.4	4.4
Platinum silicide	~3.0	~12.0	~2.5
Silicon (p^+)	0.1	2.3	0.45

making the heavy elements even less favorable than indicated. Table I should therefore only be used as a qualitative measure.

Aluminum has the best figure of merit and is the most frequently used contact metallization for commercially available EBS devices. As discussed previously, however, this metallization/shielding scheme does not provide complete radiation protection, and sometimes integral gold beam shields with nickel silicide as the top metallization are used for additional protection. Nickel silicide has an appreciably lower figure of merit and these diodes suffer from both higher dead loss and higher surface resistance. In most designs, the higher surface resistance is a more serious handicap than the dead loss. Work is presently in progress to develop an integral beam shield process utilizing a more favorable material over the active area and both silver and molybdenum appear to be the next best candidates if an aluminum-compatible system is not feasible.

High reverse-breakdown voltage is required in an EBS diode to exploit its maximum power capability. Typical reverse-breakdown voltages range anywhere from 100 to well over 1000 V. Present EBS diodes are fabricated using planar, thermally diffused, junction technology, resulting in a junction with curved sides having a radius of curvature approximately equal to the junction depth. This curvature concentrates the electric field in that region such that all other parameters remaining the same a small radius of curvature (shallow junction) results in high local electric field, which leads to a low breakdown voltage (*32*). Thus, a large radius (deep junction) is required to achieve high breakdown voltage. Figure 20 shows the relationship between

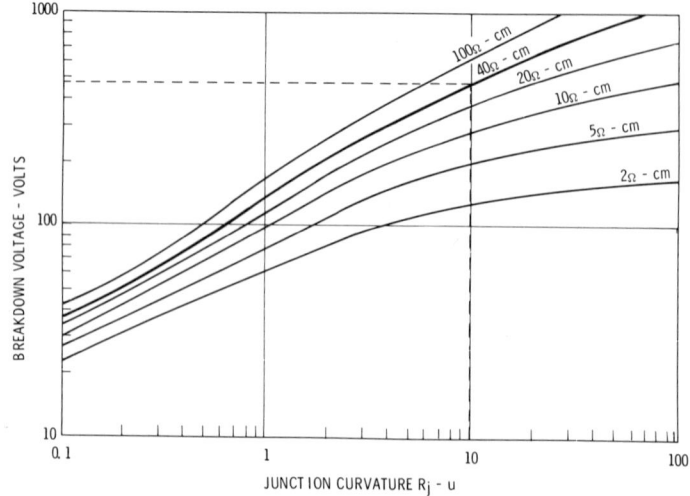

FIG. 20. Avalanche breakdown voltage vs junction curvature (depth) for cylindrical, abrupt *p–n* junctions in silicon.

breakdown voltage and junction depth for varying doping densities, and that an 0.3 μm junction depth as is required for an EBS diode will result in a breakdown voltage of only 65 V even in 40 Ω cm material. This is far too low, since an optimally designed diode will generally require at least 400 V when using 40 Ω cm material.

A guard ring, which is a separate deeper diffusion overlapping the shallow diffusion around the periphery of the active area, is used so that the diode periphery will have a larger radius of curvature. Referring again to Fig. 20, it is seen that if the guard ring is made 10 μm deep, a breakdown voltage of 480 V would result on 40 Ω cm material. While the guard ring decreases the electric field at the junction periphery, thereby increasing the breakdown voltage, it also increases the total junction area and reduces the depletion width region under the guard ring in epitaxial layers, both of which contribute to increasing the excess capacitance.

On thick bulk material it is possible to design a diode to a specific breakdown voltage within the constraints described. However, most EBS diodes are fabricated on epitaxial wafers having a finite depletion width. For example, the epitaxial layer might be 40 μm thick, 40 Ω cm, n-type, deposited on an n^+ ($\rho = 0.02$ Ω cm) substrate. On such a structure (p^+-n-n^+) the breakdown voltage is proportional to the width of the n region as long as the space charge layer reaches all the way through it (a properly designed and biased EBS diode meets this condition). The dimension affecting the breakdown voltage is not the initial epitaxial thickness but the remaining thickness underneath the guard ring, as shown in Fig. 19a. Thus, for a given initial epitaxial layer, the deeper the guard ring, the thinner the effective epitaxial layer, and vice versa.

These two parameters, effective epitaxial thickness and effective radius of curvature, can separately determine the breakdown voltage. However, there exists an optimum guard ring depth that will yield the maximum breakdown voltage. Figure 21 shows the 40 Ω cm curve from Fig. 20 with the addition of a plot of the breakdown voltage as a function of guard ring depth for a 40 Ω cm, 40 μm thick epitaxial layer. This clearly shows that the maximum breakdown voltage of 500 V can be achieved in a diode having a 12 μm guard ring. When the guard ring depth is less than the optimum value, breakdown is initiated by avalanche multiplication at the diode surface due to enhancement of the electric field there. When the guard ring depth is greater than optimum, breakdown will occur under the guard ring. At the optimum guard ring depth, breakdown will occur in the two regions simultaneously. For more critical designs, the total out-diffusion of the n^+ region into the epitaxial layer, as well as the amount of silicon consumed during oxidation, must be considered in arriving at the final effective epitaxial thickness.

If the maximum voltage achieved with an optimum guard ring depth is

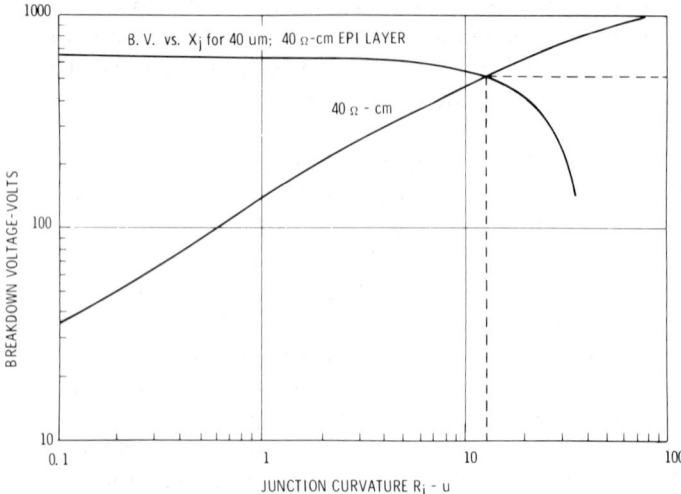

Fig. 21. Avalanche breakdown voltage for both a plane, abrupt junction on a 40 Ω cm, 40 μm thick epi layer and for a cylindrical, abrupt junction on 40 Ω cm bulk vs junction curvature (depth). The intersection marks the highest breakdown possible by using a guard ring of the appropriate depth.

still less than the desired voltage, it can be further increased by using a shallower guard ring and a specially designed surface topology with an overlay metallization. Extending the top contact metallization beyond the edge of the junction acts as a field plate that reduces the enhancement of the electric field at the surface. In effect, the radius of curvature is made to appear larger than the guard ring depth. This translates the curves on Fig. 20 to the left by an amount that is a complex function of surface topology, overlay, width, and junction depth, so that it is usually determined empirically. Some general guidelines, however, are found to exist:

1. The metal overlay should be close to the silicon–silicon oxide interface in the area immediately above the junction edge.

2. It should extend beyond the junction periphery about the same distance that the space charge layer would at breakdown in the absence of the field plate.

3. The surface topology should change gradually from thin to thick.

4. The edge of the overlay should be far away from the silicon–silicon oxide interface.

5. Its effectiveness is inversely proportional to junction depth.

(See Fig. 22.) Field plates generally yield a higher breakdown voltage than guard rings, but they also result in considerably higher peripheral capaci-

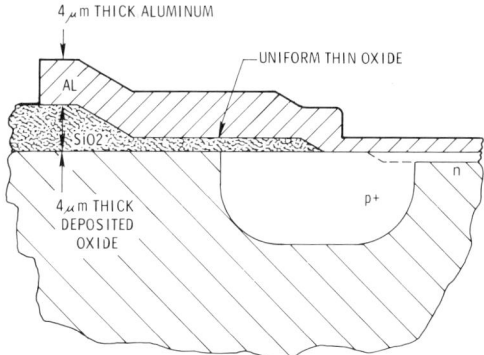

FIG. 22. Cross section of the edge of an EBS diode using aluminum metallization showing a possible surface topology for use when a field plate is necessary to achieve higher breakdown voltages.

tance. As a result, guard rings are used in broad-band designs, where bandwidth is the primary consideration, while field plates are more attractive for high peak power, narrow-band devices.

In designing an EBS diode, it is necessary both to minimize excess capacitance in order to optimize the power–frequency and power–bandwidth capabilities, and to achieve the proper value of breakdown voltage consistent with the required peak output power. Some combination of guard ring, field plate, and beam shield is required to provide the necessary breakdown voltage that is stable under radiation. However, all three decrease the power–frequency capability of the design by contributing to the unwanted excess capacitance. For calibration, it is found that the field plate capacitance can be as much as 40% of the total capacitance and that the guard ring plus the beam shield capacitance can be as high as 15% of the total capacitance. Clearly, a field plate should be used only when necessary in achieving the required breakdown voltage, and the diode should not be designed for higher breakdown voltage than necessary.

Until recently typical excess capacitance ratios of 2 to 2.5 were common. However, diodes have been recently built with excess capacitance ratio of 1.5 or less. Newer, improved designs, using tighter tolerances and narrower guard ring widths, are being developed that will lower the ratio even more.

EBS diodes used for power amplification and control require large active areas with typical values ranging from 1 to over 20 mm^2. In addition, the junction is required to be extremely shallow throughout these large areas. This places severe requirements on the bulk and surface quality of the substrate used, fabrication environment, and processing techniques in order to fabricate EBS diodes with acceptable performance and yield.

As discussed previously, there are several semiconductors having material properties that make them suitable for EBS device fabrication (GaAs, $GaAs_{1-x}P_x$; GaSb, InP, SiC, and Si). However, when one considers the material quality required from the semiconductor, none of the compound semiconductors are competitive with silicon at the present time, since both its material and processing technology are far more advanced than any of the others. Silicon crystals can be routinely manufactured that have defect densities several orders of magnitude lower than other materials; silicon epitaxial layers are easier to grow and hence more reproducible; silicon processing is much better documented and therefore easier to design and reproduce; and low-defect silicon wafers are larger and hence more economical to manufacture. So from a manufacturing point of view, silicon is the obvious material choice, especially when one considers that present EBS devices are far from being materials limited. Thus, at present the motivation to use materials other than silicon is strictly academic. In fact, nearly all the analysis and experiments with EBS devices to date have used silicon and all EBS devices available commercially use silicon exclusively.

Although excellent quality bulk silicon wafers with zero dislocations are available, local resistivity variations of up to a factor of 3 : 1 are common. These variations produce high local electric fields, which cause local breakdown and the formation of hot spots. Large-area, high-breakdown-voltage diodes are especially vulnerable to this problem. Recently, neutron-doped bulk starting material with excellent resistivity uniformity has become commercially available and is under evaluation.

Commonly, epitaxial starting material also presents a problem for large-area diodes, in that it usually has up to two order of magnitudes greater dislocation density than that found on the bulk starting material. An epitaxy manufacturer (33) has recently made commercially available dislocation/slip-free epitaxial material, and results with this material are very encouraging.

Depending on the frequency and breakdown voltage requirements, the silicon depletion region width ranges from 5 to 250 μm. For depletion widths from 5 to 50 μm, an epitaxial layer of the appropriate resistivity is grown over a heavily doped, low-resistivity substrate. For widths over 100 μm, bulk material is ground down to the required thickness after the top surface processing is completed. The range from 50 to 100 μm is fabricated by either method, both of which suffer from low yields since it is extremely difficult to produce a high-quality, thick epitaxial layer, or alternatively to handle thin bulk wafers.

EBS diodes have been fabricated from both (111) or (100) oriented silicon material. There are slight differences in oxide growth rates, diffusion rates, and positive surface-state densities between the two materials, but no

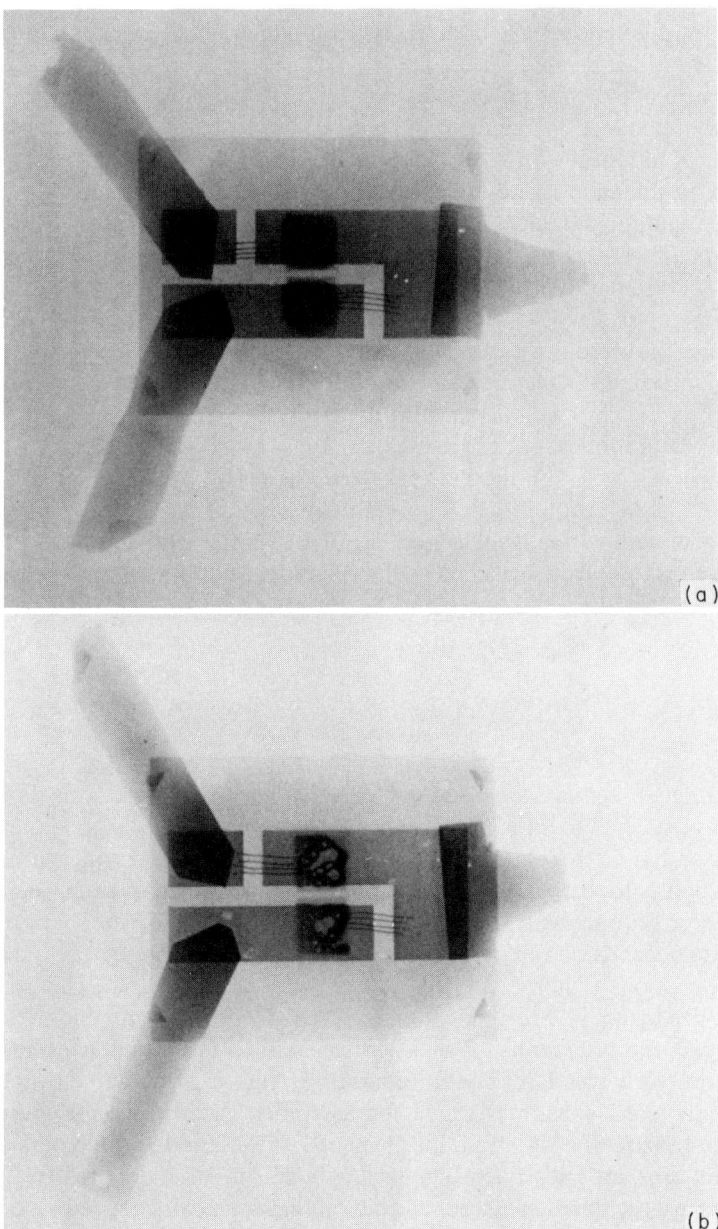

Fig. 23. X-ray radiograph of diodes bonded to metallized pads on a BeO substrate. Top picture is of a void-free die attachment and bottom picture shows a number of large voids in the die bond.

significant differences have been observed in the quality of diodes fabricated on the two different materials.

Because of the large power densities, it is necessary to minimize the diode thermal impedance by minimizing its thickness since it has the highest impedance of any of the series elements of the target configuration. Beside the obvious problems in handling thin chips, a lower limit is imposed by the need to maintain sufficient strength to withstand the large shear forces due to thermal mismatch between the die and the mounting substrates. Depending on diode area, typical die thicknesses vary from 4 to 6 mils. It is also necessary to achieve void-free attachments of the die to the mounting substrate or a localized hot region will appear above the void since it has a high thermal impedance. One successful method that has been used is to metallize the back of the chip with 0.1 and 1.0 μm thick Ti–Pt, respectively, and braze it to the metallurgical substrate with an Au–Ge eutectic preform. After the die has been attached to the BeO substrate the result is checked for voids by taking an X-ray radiograph. The resultant picture will clearly show the presence of void with a resolution of less than 1 mil. Figure 23 shows both a void-free and poor die attachment. Development continues, however, for a higher-temperature, void-free die attach process.

B. Mounting Substrate

In order to achieve the electrical isolation usually required, the diode is normally attached to an insulating substrate, which must be a good thermal conductor, have a thermal expansion close to that of silicon, be mechanically strong, be easily metallized, and be compatible with high vacuum. Beryllia meets these requirements quite well and is commonly used. The shape of the substrate depends on the number and shape of the semiconductor diode, the interconnections required, and the other circuit elements used. The beryllia is metallized with Ti–Mo–Au–Cu films 0.1, 0.2, 6.0, and 6.0 μm thick, respectively, and the interconnecting pattern as well as the die attach pads are delineated with standard photolithography. Figure 24 shows a variety of commonly used substrates made for assembly. After the semiconductor diode is brazed, the other circuit elements, as well as the interconnecting platinum tabs, are subsequently brazed with an Au–Sn solder preform. Connections are then made with either gold or aluminum wire depending on the diode metallization. Figure 25 shows a completed target assembly utilizing eight semiconductor diodes and eighteen MOS bypass capacitors. The entire assembly is then mounted in the vacuum tube and the various connections are made.

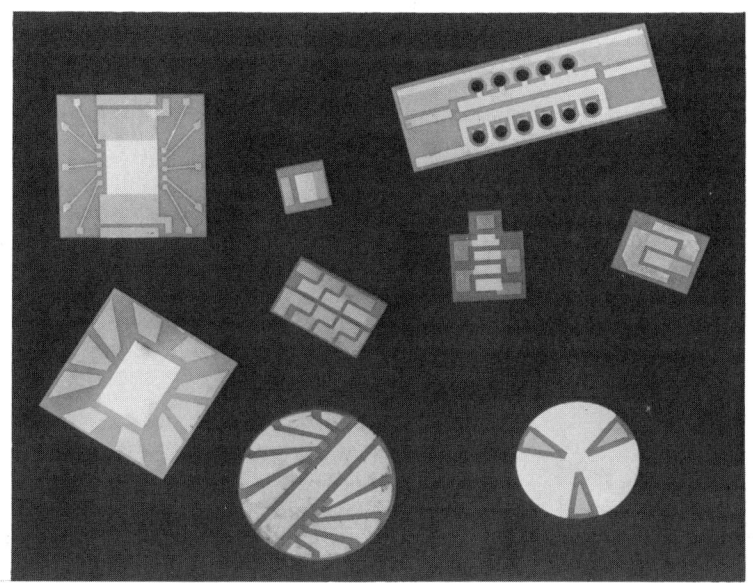

FIG. 24. A variety of metallized beryllia substrates for targets used in different EBS tubes.

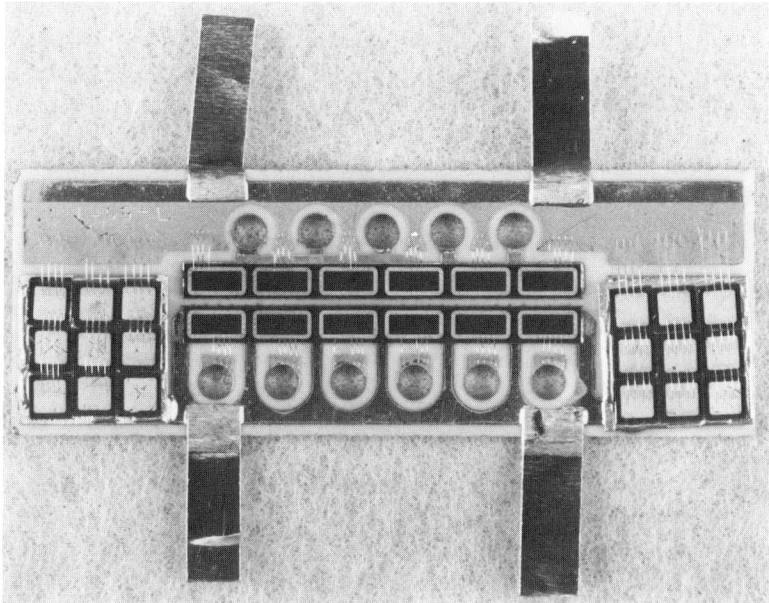

FIG. 25. Completed target assembly utilizing 8 semiconductor diodes and 18 MOS bypass capacitors; this configuration utilizes a class B circuit that has produced over 500 W of cw output power.

V. Some Device Configurations and Performance

The choice of device configuration is determined by a number of considerations. Performance characteristics such as operating frequency, bandwidth, gain, and linearity are primary factors to be evaluated. Specific applications also affect the choice through limits on device size, interface requirements, simplicity, and cost. These trade-offs not only determine the basic configuration, but also the choice of the many design options available, which are listed in Fig. 5.

A. Planar-Gridded Amplifiers

Both the source of electrons and means of modulation are combined in the form of a triode electron gun in the planar-gridded EBS amplifier, shown in Fig. 1. Close-spaced planar grids can easily be designed to provide transconductance densities of over 10^{-4} mhos/mm^2 and overall transconductances of 10^{-2} mhos, using well-known design techniques (34). The overall transconductance density g_t for the EBS is $g_t = ag_e$, where a is the diode current gain, typically 1500–3000, and g_e the transconductance density of the triode gun. The overall transconductance density can thus be as large as 0.15 to 0.3 mhos/mm^2 with total transconductances as high as 10 mhos, or three orders of magnitude higher than a conventional power triode. Thus, the gain bandwidth of the density-modulated EBS can be approximately a thousand times greater than its conventional counterpart.

The overall amplifier power gain is then given by

$$\text{Gain} = (ag_e A)^2 R_{in} R_L$$

where R_L is the output load impedance, R_{in} the input impedance, and A the diode area. Generally the area of the electron gun and the diode are the same, but in some cases there may actually be beam convergence such that the beam from a large cathode area is focused onto a smaller-area diode.

The maximum value of R_{in} is determined by the fraction of the electron beam that is intercepted by the grid and is typically greater than 500 Ω. Frequently a shunt resistance is used to reduce R_{in} to a value substantially less than the maximum in order to increase the input bandwidth or to provide a convenient impedance match to standard input driver circuitry. The choice of load impedance is determined by the application and may range from a few ohms for rf amplifiers to 100 Ω for video instrumentation amplifiers, and it may be as large as several hundred ohms for high-voltage pulse generation. Typical devices provide power gains of 10^2 to 10^4 (20 to 40 dB) under the above conditions.

1. Video Amplifiers

The planar-gridded EBS has a number of characteristics that make it attractive for fast-pulse applications. As shown in Fig. 26, these devices are compact and rugged. In addition, they use relatively simple power supplies since they do not require multiple voltages for electrostatic beam focusing. The gridded configuration does have two significant limitations: (1) the grid

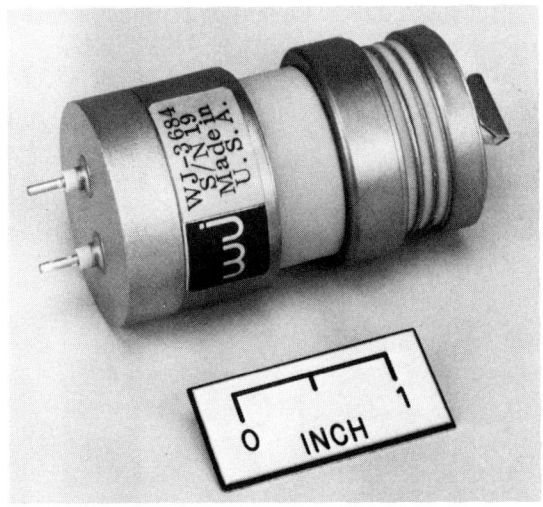

FIG. 26. Planar-gridded EBS used for high-voltage modulator. Target bias and output terminals are shown on the left.

introduces a parasitic capacitance that must be matched for operation at frequencies above about 200 MHz, resulting in a gain–bandwidth tradeoff for broad-band devices, and (2) the input signal applied to the grid must be dc isolated from the output signal by an amount equal to the bombarding voltage, which is typically 10 to 12 kV. Frequently the input signal is coupled by means of an isolation transformer, and for some pulse applications it is common to "float" the input pulse generator at the cathode voltage and to trigger the generator with signals transformer-coupled from ground.

Planar-gridded EBS amplifiers are used for both linear and saturated video amplifiers, the latter for voltage switching, and for rf amplifiers. The primary differences between the different devices are in the design of the semiconductor diode and the input and output matching circuitry. The simplest is a video amplifier, illustrated by the high-current pulse amplifier where the linear theory predicts that over 600 A of peak output current can be delivered into a 1 Ω load with a risetime of approximately 1 nsec or $di/dt = 6 \times 10^{11}$ A/sec. The actual device produces output pulses in excess of 100 A with a risetime of less than 2.2 nsec, corresponding to a di/dt of 5×10^{10} A/sec. The peak output power is over 10 kW and the power gain is approximately 30 dB. This device is suitable for fast modulation of injection lasers, where high-current pulses with extremely fast rise and fall times result in a substantial increase in the average optical output over that achieved with thyristors.

The linear theory also predicts that fast-risetime video amplifiers can achieve voltage slew rates of $dv/dt = 6 \times 10^{11}$ V/sec. A planar-gridded

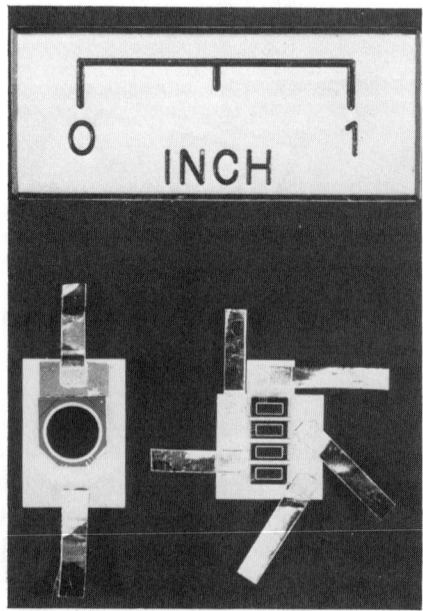

FIG. 27. Comparison of the four-diode series-connected target with the high-voltage single-diode target.

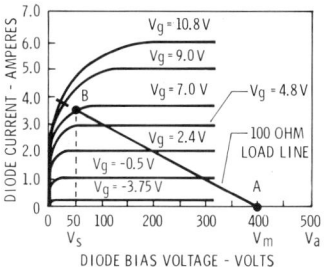

FIG. 28. The EBS characteristic curves show the diode output current as a function of diode bias voltage for varying levels of input signal.

amplifier produces 400 V at 4 A with a 1.5 nsec risetime or slew rates in excess of 2×10^{11} V/sec. A higher-voltage device has demonstrated 1000 V and 7 A peak output.

Over 1600 V has been obtained by operating several diodes in a series-connected array, as shown in Fig. 27. However, it is necessary to ensure equal voltage division across each diode to prevent avalanche breakdown of any of the diodes, and this is accomplished using an external resistive voltage divider. A disadvantage of series-connected arrays is the decrease in amplifier gain, since the full output current must be generated in each element of the array. As a result, single diodes are used where practical and series-connected arrays are used only when the required output voltage cannot be obtained with a single diode.

The overall characteristics of the planar-gridded EBS can be seen by referring to the characteristic curves shown in Fig. 28, where these particular curves were taken for an EBS rated at 400 V, 4 A peak output. The EBS behaves as a "stiff" current source over most of the dynamic range. Maximum output power is developed when the EBS is operated at maximum output current into a load impedance that develops the maximum possible output voltage. For example, when the EBS, with a semiconductor diode with a junction area of only 2 mm^2, is operating at point B on the 100 Ω load line, the peak output power is 1.6 kW and the peak diode dissipation is 200 W.

It is frequently desirable to operate the EBS in the "saturation" region to the left of point B, due to the low power dissipation that results. Recombination effects are significant in this region as shown by the fact that the output current drops even when the electron beam current is held constant (constant grid voltage). As a consequence, charge storage takes place within the diode when operating in the saturation region. It is possible to make rapid transition during turn-on from point A to an operating point within the saturation region; however, turn-off will be much slower, due to charge storage within the junction. For the device shown, operation in the active region produces rise and fall times of 1.5 to 2 nsec. When operating in the saturation region, the risetime is affected only slightly but the fall time can increase to several hundred nanoseconds. External circuit techniques, similar to the use of a collector clamp diode on a transistor, permit operation in the saturation region with minimum fall time degradation (35).

2. rf Amplifiers

An EBS video amplifier has an RC-coupled low-pass frequency response given approximately by $\Delta f = 0.35 t_r^{-1}$, where Δf is the frequency at which the output power has dropped 3 dB below the low frequency value and t_r the 10–90% output pulse risetime. The bandwidth over which an amplifier produces the maximum output power can be increased through the use of more complex matching circuits; even so, the maximum amplifier bandwidth is limited, with ideal matching, to $\Delta f_B \leq (4R_L C)^{-1}$, where R_L is the impedance appearing across the diode terminals (A–A' in Fig. 15) and C the diode capacitance (36). A typical matching network consists of a five-element Chebychev or minimum-ripple network. It is worth noting that the matching impedance used is not a conjugate match, which would dissipate half of the total power generated by the diode, but instead is designed to present a constant real impedance across the diode terminals. The load impedance is chosen to be much larger than the effective diode resistance so that high circuit efficiency is obtained.

The capacitance of the diode can be resonated at a higher frequency f_0, as shown in Fig. 15, to obtain the same bandwidth Δf centered at f_0. Figure 29 shows a schematic of a narrow-band, high-peak-power amplifier in which the diode is matched using a quarter-wave cavity. This amplifier can be operated with the electron beam illuminating the semiconductor diode for conduction angles (fraction of an rf cycle) ranging from 0 to 360°. Operation at a 360° conduction angle, referred to as class A operation, results in a maximum power conversion efficiency in the diode of only 50%. Operation at a 180° conduction angle ("linear" class C) results in a maximum diode efficiency of 78.5% at the same peak output power that can be obtained under class A conditions. Operation at transit angles less than 180° (class C) can result in even higher efficiency, although the peak output power, linearity, and dynamic range are degraded. For the above reasons, gridded rf amplifiers are typically operated at a transit angle of approximately 180°. A specific narrow-band amplifier has produced 1500 W of peak output power with a bandwidth of approximately 40 MHz at 1 GHz. The amplifier is fabricated with the EBS diode located on the center conductor of a coaxial transmission line. A simple coaxial cavity, mounted external to the

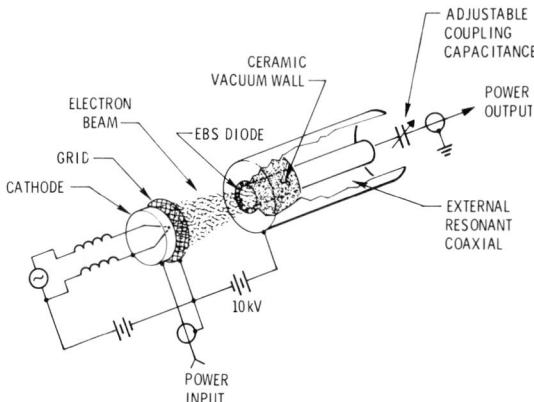

FIG. 29. The grid-controlled L-band amplifier, shown schematically, provides over 20 dB gain and peak output power of up to 1500 W. The grid-controlled EBS amplifier is considerably smaller than a deflected-beam amplifier of equal gain.

vacuum, can easily be replaced to change the band center frequency without changing the rf output power at any frequency up to approximately 1.3 GHz, where the output power begins to drop due to carrier transit time in the diode.

This amplifier illustrates the interrelationships between practical design considerations and the more fundamental diode K, or Pf^2Z, and K', or $P(\Delta f)^2 Z$, capabilities described previously. It was observed that present technology permits achieving a K value of 4.5×10^{21} for pulsed amplifiers. If the resistance presented across the diode were 50 Ω, the predicted output power at 1 GHz would be only 90 W, whereas in practice the impedance is 3 Ω. The quarter-wave cavity not only "resonates" the capacitance of the diode, but it transforms the low impedance presented to the diode to the 50 Ω level. When operating at this much lower impedance level, an output power of 1500 W is achieved. The potential bandwidth Δf of this amplifier having a diode capacitance of 90 pF and a 3 Ω terminal impedance, which could be achieved with a multielement matching network, is 900 MHz, resulting in a K' product of 3.6×10^{21}. However, the amplifier was only capable of 40 MHz bandwidth due to the limitations of the single-element quarter-wave transformer, resulting in an actual K' value of 7.2×10^{18} or 0.2% of the potential capability.

The above illustration has shown that the device designer is able, within a broad range, to perform tradeoffs that may sacrifice bandwidth for lower terminal impedance and thus higher peak power for matching network simplicity. Similarly, average power capabilities may be sacrificed for a simpler mechanical design. Each application imposes its own weighting factors on the tradeoffs, resulting in many specific configurations within a generic device type, as shown in Fig. 5, all constrained by the same basic limits but reflecting widely different emphasis on the optimization parameters.

B. Deflected-Beam Amplifiers

The deflected-beam EBS amplifier offers unique performance capabilities because the electron beam deflection mechanism is linear, is capable of extremely broad-band operation (dc to over 5 GHz instantaneous bandwidth), and can be driven from ground into a matched 50 Ω line. This type of EBS device does not have a conventional vacuum tube counterpart; without the current gain resulting from the semiconductor diode, the device is impractical. The transconductance density of the meander line deflection structure is determined by the voltage required to deflect the electron beam across the EBS diode: $g_e = I_e S/W$, where g_e is the meander line transconductance density in mhos/mm^2, I_e the electron beam current density in A/mm^2, S the meander line sensitivity in mm/V, and W the width of the diode in mm.

The maximum value of g_e for a beam with circular or square cross section is limited primarily by transverse space charge spreading of the electron beam (37). As is the case of the planar-gridded amplifier, the total transconductance density is the product of the input transconductance density and the diode current gain: $g_t = aI_e S/W$. Typically, transconductance is of the order of 0.1 to 1 mhos for circular beams and 1 to 10 mhos for wide rectangular beams. For present amplifiers the resulting input power requirements is between 0.5 and 2.0 W when the deflection structure is a 50 Ω meander line. Since a rectangular electron beam can easily be extended in width, higher-power amplifiers are typically made by extending the target length (corresponding to the beam width or longest dimension of the beam cross section), rather than width, resulting in no increase in drive power even though the output power, and hence the gain, is increased. Typical deflected beam amplifiers provide 25 to 30 dB gain for 500 W output.

The semiconductor target is subject to similar design tradeoffs and modes of operation described for the gridded amplifier. Deflection modulation simplifies some circuit configurations for extremely broad-band or high-frequency amplifiers. A single diode can be operated in class A or class C by choosing the electron beam deflection angle, which in turn controls the conduction phase angle. In class A, the electron beam covers half of the diode with no input signal; in class C, the beam edge is located statically at or outside the diode edge. Figure 30 shows two diodes in a class B amplifier that are located on either side of the electron beam and illuminated during alternate half-cycles. The class B circuit provides the same maximum

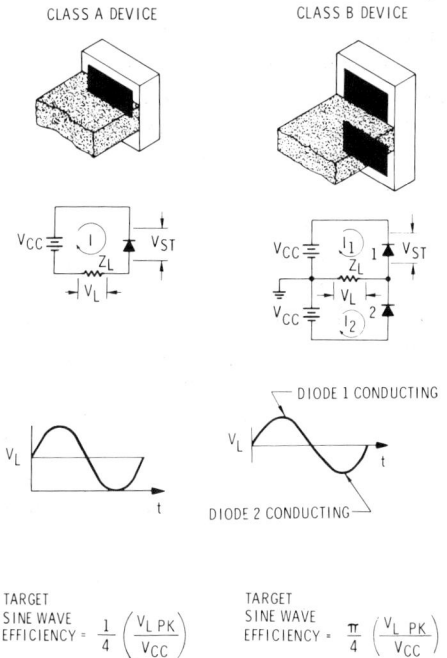

FIG. 30. Schematic target connections and target efficiency relationships for class A and class B targets, together with the zero-signal beam illumination for a deflected-beam amplifier.

efficiency as linear class C; however, class B operation provides a number of other advantages such as improved linearity and harmonic suppression. Alternately, class B operation can be achieved by connecting two separate devices in push–pull. However, locating both diodes within the same vacuum envelope permits operation at higher frequencies than would be possible with two separate devices.

Deflected beam amplifier capabilities include true dc response so that they can be used in fast-risetime, linear video amplifiers and very broadband and bandpass rf amplifiers. They are suited for a wide range of applications in instrumentation, communications, avionics, and electronic countermeasures. The low-pass video amplifier is the simplest deflected-beam amplifier. The semiconductor target is fabricated in the class B configuration for operation directly into 50 Ω. Either cylindrical or sheet electron beam is used, depending on the power level and degree of amplifier linearity required. The very fast risetime is illustrated by a video amplifier developed by Tektronix for use as an oscilloscope vertical amplifier, which demonstrated 12 V output with a risetime of less than 0.22 nsec into a 100 Ω load, corresponding to a low pass bandwidth of 1.6 GHz.

High-power, broad-band bandpass amplifiers can provide cw output powers of hundreds of watts over bandwidths of several hundred megahertz. Thus, these amplifiers typically use much larger semiconductor targets than the video amplifiers, requiring a broad-band impedance transformation to match the diode terminal impedance to 50 Ω since the terminal impedance is roughly inversely proportional to output power or diode area. The impedance transformer configuration is selected to be the most applicable for the particular operating band. For example, broad-band transmission line transformer techniques are used to match impedance levels as low as 1 to 3 Ω at frequencies below 500 to 600 MHz (38). The lower transformation ratios that can be obtained using microstrip transformers for a broad-band amplifier at 1 GHz will limit the impedance to 5 to 10 Ω minimum.

The broad-band class B circuit shown in Fig. 31 presents a 3.1 Ω impedance across the diode terminals and the output is transformed to 50 Ω using cascaded 4 : 1 transmission line transformers. The three-element ladder networks consisting of C_1, L_2, provide a Chebyshev pass-band response. The specific class B circuit shown was invented (39) specifically for use in EBS amplifiers and has the advantage of minimizing the size of the bypass capacitors since it is not necessary to bypass the fundamental frequency. A semiconductor target shown in Fig. 27 incorporating this circuit has produced over 500 W of cw output power with a bandwidth of approximately 20 to 100 MHz. The overall amplifier efficiency was over 40% and the semiconductor target was dissipating over 750 W average with 20 mm² active diode area.

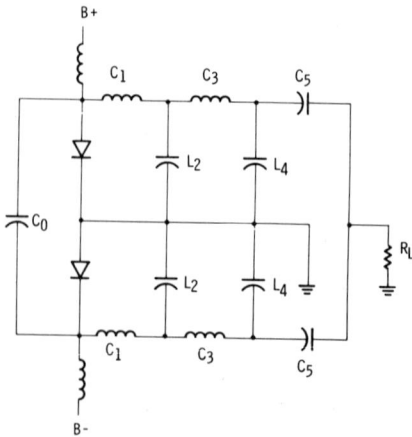

FIG. 31. Circuit providing class B operation with both good efficiency and low levels of harmonic distortion. The capacitor labeled C_0 bypasses the harmonics, while C_5 is a dc block. The other elements are used for impedance matching.

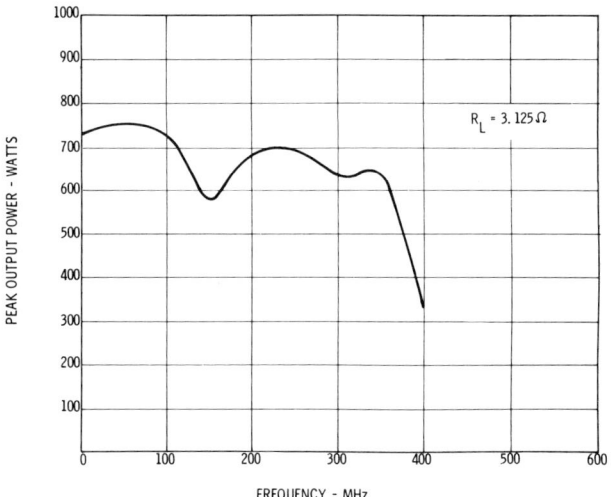

FIG. 32. Pulsed rf output power for deflected-beam amplifier with a three-section matching network and a six-diode target.

The optimized broad-band pulsed operation extending from 20 to 360 MHz is shown in Fig. 32, where an output ripple of ± 0.5 dB was obtained. Under these conditions the amplifier gain was approximately 25 dB. Due to the extremely flat response of the meander line input structure, the ripple in power gain is determined only by the ripple in the output-matching structure. Although this amplifier was not optimized for broad-band cw operation, over 200 W cw was obtained with a 3 dB bandwidth of approximately 350 MHz.

The distortion, which is a deviation from linearity, at high output levels can be assessed from Figs. 33 and 34. The upper curve in Fig. 33 shows the

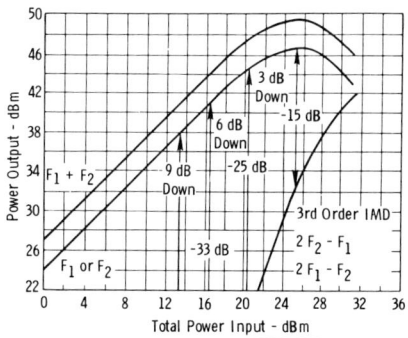

FIG. 33. Balanced two-tone operation of the deflected-beam class B rf amplifier showing low level of third-order intermodulation distortion.

single-signal transfer curve. The EBS has a linear small-signal region, with the gain constant to within ±0.5 dB over a 60 to 80 dB dynamic range. As the output power approaches saturation, the gain compression is typically 4.5–6.0 dB. The lower curves show the third-order intermodulation distortion (IMD) when the amplifier is driven with two equal amplitude signals slightly offset in frequency. Note that at saturation the third-order IMD is 16 dB below the carriers, and with the input signal reduced by 6 dB the IMD drops to −33 dB. Harmonic distortion is also quite low for these amplifiers. Due to the cancellation produced by class B operation the even harmonics are typically at least 50 dB below the carrier and the third harmonic, which has the highest level of any harmonic, is typically at least 30 dB below the carrier throughout the amplifier dynamic range. Figure 34

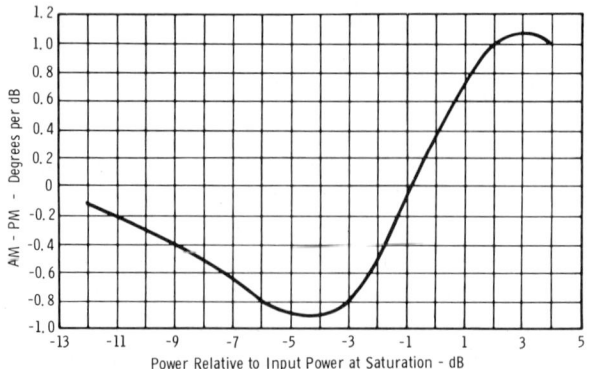

FIG. 34. AM/PM distortion for a class B deflected-beam rf power amplifier. The maximum phase deviation from small signal to saturation is only 6.3°, much lower than most other high-efficiency power amplifiers.

shows the change in output phase in the EBS output produced by changing the amplitude of the input signal. Note that over the entire amplifier dynamic range the AM/PM conversion is less than 1°/dB, which is an extremely low level for a high-efficiency amplifier.

The low distortion levels are possible because the diode is maintained in the high-field region through essentially the entire amplifier dynamic range. Saturation of the output power is produced geometrically by deflecting the electron beam off the diode rather than due to a large reduction in the carrier velocity, as occurs in most devices. This type of EBS amplifier is expected to play a significant role in future uhf and microwave communications systems.

C. Signal Processors

The inherent flexibility of the EBS, in particular the extremely rapid access of large-area diode arrays covering an electron beam, makes it possible to devise new signal-acquisition and/or signal-characterizing systems. Once the information has been written onto a target array, various methods can be used to store it and read it out. The simplest readout is an individual output from each diode; however, a more practical configuration holds the information on the array and then reads it out sequentially.

The fundamental limitations on speed and amount of information stored by an EBS signal processor are determined by the beam size and current, scan rate, and diode current gain, as well as linearity requirements. The beam size determines the maximum target density and this, together with beam scan rate, determines the sample time for an individual element. Maximum scan rate is determined by the beam current density.

1. *Transient Signal Samplers*

Typically, wide-bandwidth oscilloscopes have been used for the analysis of high-speed transient waveforms. The waveform is first photographed and then the information is transferred into a computer for further analysis. The development of a double-ended EBS scan converter, utilizing a silicon diode array, has eliminated this laborious process (*40*). The EBS captures and stores high-speed single transients, or low-repetition-rate signals, until they can be read out into conventional slow-speed processing or display circuitry.

The double-ended scan converter sketched in Fig. 35 contains separate writing and reading guns on opposite sides of the target, with the reading gun facing the junction side of the diode array. High-energy electrons for writing a signal on the target are supplied by a CRT-type gun operating at 10 kV. The reading gun supplies a low-energy beam that scans the target, neutralizing written charge images and providing a sequential readout signal

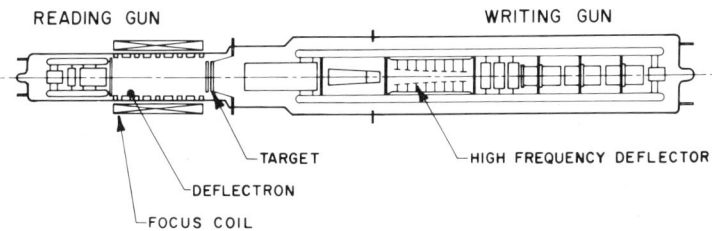

FIG. 35. Cross section of a dual-gun EBS scan converter where back-biased diodes are used instead of MOS capacitors found in conventional scan converters.

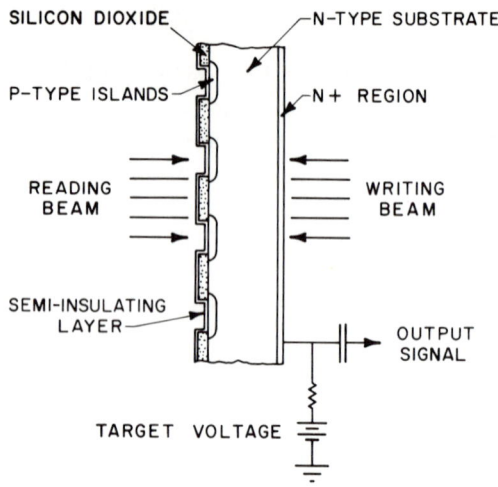

FIG. 36. Schematic diagram of a portion of the diode array target used in the EBS scan converter.

similar to that in a vidicon camera tube. Figure 36 shows the target in cross section, consisting of a planar array of p–n junctions formed on an n-type silicon wafer. Diodes are approximately 5 μm diameter on 12.5 μm centers and the active area of the target is 0.85 × 1.27 cm.

During operation the target substrate potential is held at about 10 V positive with respect to the reading gun cathode and the diodes are reverse biased. On scanning with the reading beam, the target is charged negatively to the cathode potential. During writing, 10 kV electrons bombard the target on the side opposite the diodes and create approximately 2000 electron–hole pairs near the entrance surface, charging the target negatively. The holes drift across the depletion region formed by the reverse biased diodes, discharging the target surface in the illuminated area. When the reading beam next scans this area, the diodes are recharged and the signal current is obtained in the target lead.

With the EBS scan converter, information writing speeds of 2×10^{12} tracewidths/sec have been demonstrated sufficiently to display a full-screen 2 GHz sine wave. To reliably digitize the readout for signal processing, a signal-to-noise ratio of about 20 : 1 is required, which the device will provide at speeds up to 5×10^{11} tracewidths/sec. Resolution at a 50% modulation level is 400 TV lines per horizontal scan.

The double-ended scan converter can be simplified by elimination of the reading gun, which requires restructuring the target to be read sequentially by other means. A charge-coupled scan converter not only reduces the size and complexity but also eliminates the inaccuracies associated with the

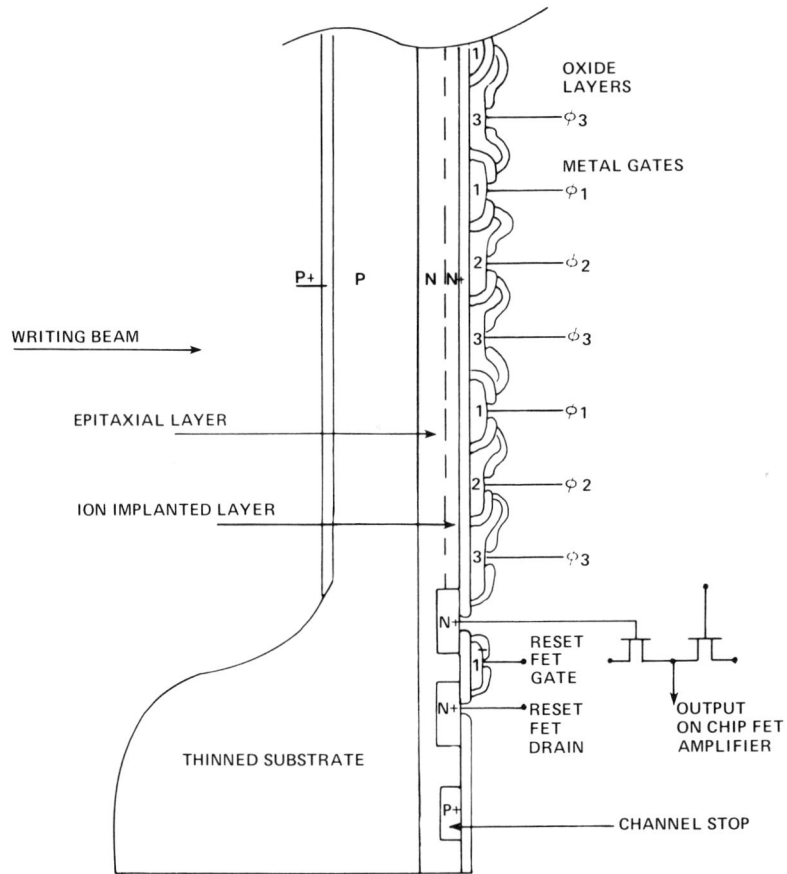

FIG. 37. Cross section of a charge-coupled semiconductor target that can store and sequentially output information written on it by an electron beam for use in a single-gun EBS scan converter.

analog reading beam (41). An electron-beam-addressed CCD (charge-coupled device) array is shown in Fig. 37. The charge that is written in the array is proportional to the input signal charge and it is this signal that is sequentially read out. At present, the charge-coupled scan converter is in the experimental development phase and devices are not yet commercially available.

2. Analog-to-Digital Converters

The scan converter is not well suited to the continuous acquisition and digitizing of an analog signal because of its characteristic of recording a single frame and then reading it out in serial form. An EBS analog-to-digital

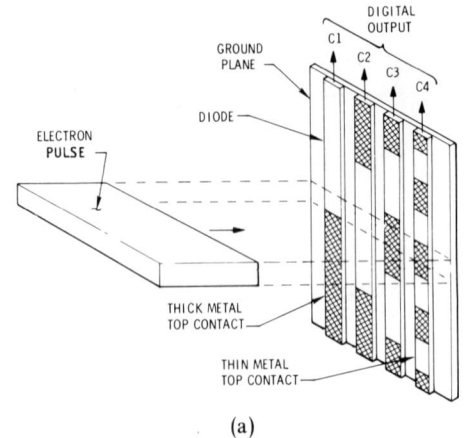

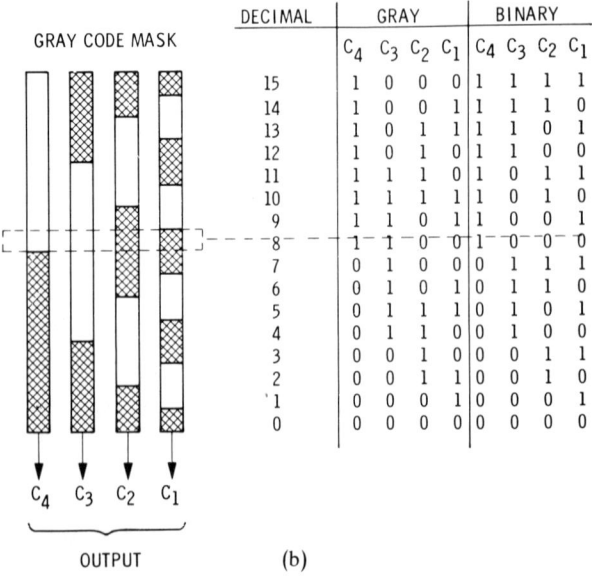

Fig. 38. (a) The electron beam in the A/D converter alone simultaneously illuminates all the columns in the semiconductor target. The position of the electron beam determines the output code. (b) The semiconductor target is designed to generate a Gray code output since the Gray code minimizes output errors due to small changes in electron beam position.

(A/D) converter is appropriate where long strings of data must be acquired and converted to digital format. One basic configuration of this device utilizes an electron beam and input sections that are almost identical to the deflected-beam EBS amplifier. The primary difference is in the target structure, shown in Fig. 38, which is designed to capture and digitize signals to 8-bit accuracy. A narrow-sheet beam is deflected at constant velocity along the array to provide an output that is digitized in 8-bit binary code proportional to the input signal. A Gray code (42) is used to minimize the number of simultaneous cross-overs as the signal changes from ON state to OFF state in each diode string. The accuracy is primarily limited by the ability to generate very fine electron beams that can be precisely aligned and swept over the target. An experimental EBS A/D converter uses 8 or 9 diode strings with an array length of approximately 0.25 in. Thus, for 8-bit accuracy, the smallest diode aperture is approximately 0.001 in., and the electron beam must also be approximately 0.001 in. by 0.050 in. The digitizing speed of such a device is limited by the required output circuit bandwidth, which is much greater than the frequency being digitized, and by the need for an external sampling system that can handle the data flow coming out of the A/D converter.

Since the limitation is typically in the output circuit that must sample the data flowing from the continuously addressed target, an improvement would be to include a strobing pulse within the EBS A/D converter. This can be done, as shown in Fig. 39. Using the concept of deflection blanking, the beam can be chopped into many short pulses at a predetermined clock frequency. This provides an integral sampling scheme and it is then only necessary to store the output bits long enough to dump them into a large-capacity memory. Simultaneously, the output circuit bandwidth required is greatly reduced.

As stated previously, there is no other technique that can compete with an electron beam in accessing extremely large areas, so that an EBS device can always outperform an all-solid-state device for speed of signal acquisition. The limitations become one of ability to handle the large masses of

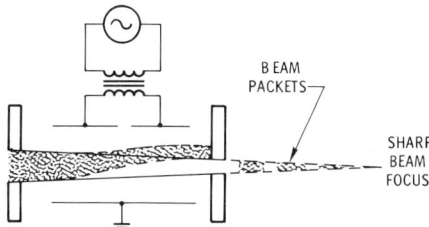

FIG. 39. Deflection blanking using a structure such as this can chop the electron beam into short packets while retaining sharp focus.

output data from the EBS. However, for low-speed or limited-data acquisition it is unlikely that the EBS technology will be competitive because of size, cost, and power supply complexity. EBS is best suited for those applications where high data rates, above 100 MHz, must be acquired and processed. The design limitations on accuracy are determined primarily by the ability to focus and deflect electron beams. Eight-bit accuracy is the current design standard, and 9 bits are achievable. Beyond that, the task gets very difficult.

The upper frequency limit for signal processors cannot, at the present time, be established. The electron-beam deflection technology can be extended to 10 GHz and electron beams can be designed to provide enough beam current to result in useful diode output at frequencies in this range. For EBS scan converters, there is no real difficulty in reading out the data; however, for the EBS A/D converter there is a severe problem as to what to do with the information coming out of the target array at this rate.

At the present time EBS processors are limited to those areas where there are rather specialized requirements for high speeds, single events, and/or long strings of data. As these devices mature and additional ones are developed, their cost will decrease and broader usage can be expected. An example of this may be in test and measurement instruments such as oscilloscopes. Current technology uses the CRT to display a waveform and either a storage CRT or camera film to store the information for further processing. In future oscilloscope designs the signal acquisition may be done by a device similar to the scan converter and then either be displayed on a CRT or stored electronically as desired by the user. As such applications develop, the EBS technology could become a key component for test and measurement applications.

VI. Life and Reliability

Demonstration of long life and high reliability of EBS devices has been a success story seldom matched by any other active device, solid state or otherwise. However, the circumstances under which this demonstration was made caused termination of most of the EBS activity in the United States. Under direction from an agency of the U.S. Department of Defense, government-funded EBS development work was stopped in early 1973 to improve reliability and demonstrate 2500 hours of life on cw rf amplifiers. A total of eight early Watkins-Johnson laboratory amplifiers were life tested in excess of 21,000 hours each at power dissipation levels up to twice the design levels, without a single failure. The net effect of this edict, in addition to demonstrating an EBS mean-time-to-failure (MTTF) of over 20 years and a diode MTTF of over 40 years, was to interrupt development of EBS devices, which was progressing in a satisfactory manner. Although the 2500 hours life was quickly demonstrated, development funding was diverted elsewhere and progress was set back by approximately two years.

The concern by DOD about device reliability was well founded because of early problems by many experimenters. The primary difficulty was with the specific problem of degradation of bare-junction, mesa diodes. However, at the time of the edict, third-generation, radiation-hardened, passivated planar diodes were being developed and were incorporated into some of the life test devices. Thus, the reliability improvements had already been made but not yet demonstrated by extended life tests. As a result of the impressive amount of life data accumulated, EBS life has been firmly established. Many of the life and reliability aspects of EBS devices have been previously described (43–45) and will not be repeated.

During the past two years improvements have been in the area of manufacturing technology, more specifically, in vacuum processing and in identifying and correcting some basic material problems. Work is expected to continue indefinitely on improving and upgrading the design and manufacturing technology to improve reliability, as well as performance. However, at the present time there are no major areas of weakness to prevent the achievement of an operating life expectancy of five to ten years, which is greater than the mean operating life of most systems that would utilize EBS devices.

VII. Future Potential

In the course of developing EBS devices to their present state, it was first necessary to establish a hybrid technology that would yield reliable, reproducible device characteristics. The next stage was to develop some prototype devices and upgrade their performance into standard products. The past several years have been directed toward these goals and the resulting technology base now makes it feasible to approach the predicted performance capabilities in practical EBS devices. Productized high-power video amplifiers and prototype rf amplifiers have already been developed and the potential exists for continued growth in performance and product types. EBS reliability has now been firmly established and performance superiority over competing devices from other technologies has been demonstrated in a number of applications.

Projections of improved performance levels and areas of application for new devices are appropriate. Projections which have been made in the past (46) were not unrealistic, but they failed to materialize in the time predicted because of the lower-than-expected support levels. Two handicaps have yet to be overcome in order for rapid growth to occur: unfamiliarity with the device by equipment designers and high cost due to the relative infancy of the technology. However, by far the dominant factor affecting the rate of growth of EBS is the dependence upon technology support from the U.S. Department of Defense. With the advent of a new national administration, an expected increase in support may fail to materialize, thereby affecting these present projections.

The high-peak-power EBS video amplifiers have already found application in electronic countermeasure systems that utilize their high-voltage, fast-pulse risetime capability. Continued growth in device performance can be expected, with output levels in excess of 2 kV at 10 A expected within one to two years, depending upon the demand for these higher output levels. A wide range of other video amplifier applications exists such as laboratory instrumentation, nondestructive testing, electrooptic communications, and radar modulators. At present the primary limitation to widespread usage, especially in the nonmilitary areas is the relatively high device price, rather than performance levels. When the price is brought down to $1000 or below, the market interest is expected to increase rapidly. Progress is being made in reducing costs and this trend is expected to continue so that devices of this type will be available for a few hundred dollars within five years.

Consideration has recently been given to using EBS video amplifiers in power converter applications where fast turn-on and turn-off could result in improving the size, weight, and efficiency of dc–dc converters. It is entirely feasible to develop devices operating at average output power levels of 10^4 to 10^5 W. Substantial development will be required to achieve these levels, but the resulting devices can be expected to become competitive with both power transistors and thyristors.

For assessing EBS capabilities, the rf spectrum can be conveniently separated into regions from dc to 1 GHz, 1 to 2 GHz, and above 2 GHz. Most new systems at frequencies below 1 GHz and with power levels below 1 kW are being designed with transistor amplifiers, which have successfully increased system reliability compared with vacuum tube amplifiers; however, the anticipated reduction in acquisition cost has proven illusory due to the continued high cost and limited output power capability of rf power transistors. The EBS is expected to be highly competitive in this frequency range in power amplifier applications because of the high output power, gain, and efficiency that can be provided. Single devices capable of 500 W cw output with 200 to 500 MHz bandwidth are expected within two years. These amplifiers, including the necessary power supplies, will provide superior electrical performance and comparable reliability to transistor amplifiers, requiring 10–20 transistors, at less than two-thirds the cost.

The region from 1 to 2 GHz is also expected to be a primary area for the application of EBS devices and, in fact, is an area where both conventional semiconductor and vacuum tubes have significant disadvantages as power amplifiers. Microwave tubes, such as klystrons and TWT's are large and expensive, while triodes are notoriously unreliable. Transistors suffer from the dual problem of low output power capability and low gain. For example, the highest peak output power provided by developmental transistors even at the lower end of this frequency range is only 300 W with gain less than 10 dB. Applications needing in excess of 1 kW peak power, such as IFF, TACAN, or radar, typically require an amplifier composed of 8 to 10 power transistors, matching networks, and combiners. Transistors are being used by default because of the serious disadvantages of the various vacuum tube types, despite the added complexity and cost. Now that a single EBS device has demonstrated 1500 W peak output with 20 dB gain, application of the EBS to new systems above 1 GHz is anticipated. An increase of peak power to over 3 kW and average power to greater than 200 W in this frequency range can be expected in two years.

In contrast to the lower frequency ranges, the future of EBS above 2 GHz is indeterminate at present. The EBS still provides a substantial margin of output power above transistors; however, microwave tubes are comparable in size to the EBS and can provide higher output power. The EBS is simpler, requiring fewer parts and ultimately EBS costs could be significantly lower. However, the use of the EBS at microwave frequencies is expected to result from a combination of performance capabilities rather than as a direct replacement for other types of medium power amplifier. For example, the EBS is capable of providing moderate output power (50-100 W) with high efficiency while maintaining low distortion levels. Furthermore, the EBS can be designed to provide a combination of functions within a single amplifier such as antenna switching or phase shift keying.

The development of special EBS devices for signal processing has seen a relatively low level of activity until recently, but this area can be expected to grow extremely rapidly in the next several years. EBS scan converters are currently in use and more sophisticated instruments built around improved converters that will provide improved resolution and higher data rates can be anticipated. High-speed A/D converters to operate at multigigabit data rates will also become available. As experience is gained in applying EBS techniques to signal processing, even more sophisticated devices can be envisioned, which will combine other technologies, such as surface acoustic waves, with EBS to incorporate scanned-beam and optical-convolution techniques within the same device. Devices of these types will augment current signal-processing techniques and in many cases will permit sophisticated analog signal-processing functions to be accomplished with far less hardware than would be required with conventional digital techniques.

Additional suppliers can be expected to enter the marketplace as EBS acceptance and volume improves, both to develop unique devices and to provide a second source. In fact, many equipment manufacturers will not design critical components into equipment expected to enter large volume production without a second source being available. Additional suppliers entering the scene will both signify success for EBS and provide healthy competition for the present two suppliers, who are not competing with each other. Both the technology and the users are expected to benefit.

ACKNOWLEDGMENTS

The authors wish to acknowledge the support of the Office of Naval Research, Army Electronics Command Evans Laboratory, Air Force Rome Air Development Center, Naval Electronics Systems Command, and the National Space and Aeronautics Administration.

We gratefully acknowledge the assistance given to the early Watkins-Johnson effort by Dr. C. B. Norris, Jr., while serving as a consultant to the Company, and by Dr. Aaron Ballonoff, who was responsible for the development and manufacture of EBS diodes supplied to the Watkins-Johnson Company by the Signetics Corporation, as well as by many of our colleagues, especially Drs. L. A. Roberts and R. L. Arnold, and Messrs. J. A. Long, D. H. Smith, and B. W. Bell.

REFERENCES

1. C. B. Norris, Jr., The capabilities of electron-beam semiconductor active devices. *Proc. IEEE Tube Tech. Conf.*, New York, 1968.
2. C. B. Norris, Jr., Optimum design of electron beam-semiconductor linear low-pass amplifiers—Part I: Bandwidth and rise time. *IEEE Trans. Electron Devices* **ED-20**(5), 447–455 (1973).
3. C. B. Norris, Jr., Optimum design of electron beam-semiconductor linear low-pass amplifiers—Part II: Output capabilities. *IEEE Trans. Electron Devices* **ED-20**(9), 827–839 (1973).
4. R. de L. Kronig, Change of conductance of selenium due to electronic bombardment. *Phys. Rev.* **24**, 377 (1924).
5. E. S. Rittner, Use of photo-conductive semiconductors as amplifiers. *Phys. Rev.* **73**, 1212 (1948).
6. E. S. Rittner, Electron device with semiconductive target. U.S. Patent 2,540,490 (1951).
7. E. S. Rittner, Electron device with semiconductor target. U.S. Patent 2,749,471 (1956).
8. E. S. Rittner, F. E. Grace, and G. A. Buetel, Electron switching tubes and circuits therefor. U.S. Patent 2,588,292 (1952).
9. E. S. Rittner, F. E. Grace, and G. A. Buetel, Electron switching device. U.S. Patent 2,803,779 (1957).
10. N. C. Jamison, Electrical device. U.S. Patent 2,831,149 (1958).
11. W. E. Kirkpatrick and R. W. Sears, Semiconductor signal translating device. U.S. Patent 2,589,704 (19   ).
12. W. Ehrenberg, C. Lang, and R. West, The electron voltaic effect. *Proc. Phys. Soc.*, **64**, 424 (1951).
13. A. R. Moore and F. Herman, Electron bombardment induced conductivity in germanium point contact rectifiers. *Phys. Rev.* **78**, 472–473 (1951).
14. A. R. Moore, Semiconductor translating device. U.S. Patent 2,600,373 (1952).
15. A. R. Moore and F. Herman, Semiconductor signal translating system. U.S. Patent 2,691,076 (1954).
16. A. V. Brown, Electron beam switched $p-n$ junctions. *IEEE Trans. Electron Devices* **ED-10**, 8–12 (1963).
17. A. V. Brown and P. F. Evans, A decoder-driver system for magnetic-core memories using a 64-position switch tube. *IEEE Trans. Electron Devices* **ED-13**, 713–719 (1966).

18. C. B. Norris, Jr., The capabilities of electron-beam semiconductor active devices. *Proc. IEEE Tube Tech. Conf.*, New York, 1968.
19. Electron beam semiconductor set for market. *Electronics* Oct. 26, p. 39 (1970).
20. M. Namordi, A hybrid electron beam-*pn* junction device. *Proc. IEEE Int. Solid State Circuits Conf.*, p. 96 (1970).
21. See Norris (*1*).
22. See Norris (*1*).
23. See Norris (*2*).
24. A. Silzars, D. J. Bates, and A. Ballonoff, Electron bombardment semiconductor devices. *Proc. IEEE* **62**(8), 1119–1158 (1974).
25. H. J. Wolkstein, Design considerations for grid-controlled electron guns for pulsed traveling wave tubes. *RCA Rev.* September, pp. 389–413 (1960).
26. H. E. Gallagher, Gridded electron guns for high average power. *IRE Trans. Electron Devices*. March, pp. 234–241 (1962).
27. A. Silzars and D. J. Bates, Laminar flow electron gun and method. U.S. Patent 3,740,607 (1973).
28. See Silzars *et al.* (*24*).
29. D. J. Bates, A. Silzars, and L. A. Roberts, Rectangular beam laminar flow electron gun. U.S. Patent 3,980,919 (1976).
30. E. O. Johnson, Physical limitations on frequency and power parameters of transistors. *IEEE Int. Conf. Rec.* pp. 27–34 (1965).
31. W. W. Siekanowicz, *et al.*, Current gain characteristics of Schottky-barrier and *p–n* junction electron beam semiconductor diodes. *IEEE Trans. Electron Devices* **ED-21**(11), 691 (1974).
32. D. P. Kennedy and R. R. O'Brien, *IBM J.* May (1966).
33. Semi-Metals East, Inc., 172 Spruce St., Westbury, N.Y. 11590.
34. K. R. Spangenberg, "Vacuum Tubes." McGraw-Hill, New York, 1948.
35. R. I. Knight and B. W. Bell, Electron bombarded semiconductors for fast rise time modulators. *Countermeasures* **2**(8) (1976).
36. G. L. Matthaei, L. Young, and E. M. T. Jones, "Microwave Filters, Impedance Matching Networks and Coupling Structure," Sect. 11.08–11.10, pp. 681–719. McGraw-Hill, New York, 1964.
37. See Semi-Metals East (*33*).
38. J. Johnson, ed., "Solid Circuits." CTC, 1973.
39. R. I. Knight and D. H. Smith, Electron bombarded semiconductor devices. U.S. Patent application.
40. R. Hayes, A silicon diode array scan converter for high-speed transient recording. *IEEE Trans. Electron Devices* **ED-22**,   (1975).
41. G. A. Antcliffe, *et al.*, Operation of CCD's with stationary and moving electron-beam input. *IEEE Trans. Electron Devices* **ED-22**, 857–861 (1975).
42. F. J. Hill and G. R. Peterson, "Switching Theory and Logical Design," pp. 162–163. Wiley, New York, 19 .
43. D. J. Bates, R. I. Knight, G. Taylor, and A. Silzars, Current reliability results on electron bombarded semiconductor power devices. *IEEE Trans. Electron Devices* **ED-21**(11), 734 (1974).
44. D. J. Bates, A. Silzars, and A. Ballonoff, Reliability improvements in electron bombarded semiconductor power devices. *Proc. Annu. Reliability Phys. Symp.*, IEEE, 11th 1973.
45. D. J. Bates, A. Silzars, and A. Ballonoff, Some reliability aspects of electron bombarded semiconductor power devices. *Solid State Technol.* **17**(7) (1974).
46. D. J. Bates, Semiconductors inside tubes make high performance RF amplifiers. *Electronics* July 25 (1974).

Basic Concepts of Minicomputers

LLOYD KUSAK

Hewlett-Packard Co.
Rolling Meadows, Illinois

I. Historical Development	283
II. Characteristics of Minicomputer CPUs	285
III. Instruction Repertoire	287
IV. Data Checking	293
V. Bus Architecture	295
VI. Power-Fail Protection	297
VII. Memory Protect	298
VIII. Memory Interleaving	298
IX. Cache Memory	299
X. Interrupt Structure	300
XI. Direct Memory Access	305
XII. Microprogrammable Systems	306
XIII. Nonminicomputer Systems	308
XIV. Peripheral Devices	310
XV. Computer Consoles	310
XVI. Printing Devices	312
XVII. Data Entry I/O	313
XVIII. Magnetic Tapes	314
XIX. Tape Cassettes	315
XX. Disk Systems	315
XXI. Floppy Disks	316
XXII. Process Control Peripherals	317
XXIII. IEEE Standard Digital Interface	319
XXIV. Data Set Interfaces or Modems	320
XXV. Software	321
XXVI. Disk Operating Systems	325
XXVII. Growth of Minicomputer Applications	328
Glossary	330
References	344

I. Historical Development

In order to gain an understanding of the place of minicomputers in the hierarchy of all computer systems, it is appropriate to review the history of the development of the general purpose computer. This development is credited to a project at the University of Pennsylvania that resulted in the

ENIAC, a vacuum tube machine that was the first to utilize the stored program concept. The developers of this computer formed a small company for the purpose of marketing a successor machine called Universal Automatic Calculator or UNIVAC. Inasmuch as this corporation was not a financial success, it was taken over by Sperry-Rand and has since become a division of that company. Shortly thereafter in 1953, IBM recognized the usefulness of the computer in business applications and undertook a computer project of its own. By the late 1950s, there were many large-scale computers in use in a variety of applications. Because IBM already had the vast majority of the punched-card business and because it had designed its computers for business applications, it very quickly became the predominant computer manufacturer.

By the late 1950s most large-scale computers were designed so that they had features suited either for business or for scientific applications. The machine more suited for scientific applications was characterized by a large word length and a scientific instruction mix, a fast compute speed, and a minimum input/output capability. The business machine was not required to perform any complex mathematical functions, but it was required to handle large streams of data. Therefore, a machine more suited for a business application would contain a large number of input/output channels, a capability for decimal arithmetic, and a large storage capability.

By the early 1960s IBM was marketing computers with three sets of architecture. These were exemplified by the 7090, the 1401 or 7080, and the 7070. None of the lines of machines was compatible with the others. Probably because of support problems, where individual groups of hardware maintenance people and software support people were required for each class of architecture, in April 1963 IBM announced the 360 series, which combined the features of both business and scientific classes of equipment. The invention of the transistor was now having its effect on the character of not only the IBM equipment but also of machines produced by other manufacturers. Computers at this time were approximately five times faster than their predecessor tube-type machines, were substantially more reliable, occupied less space, and required less power to operate.

The development of the minicomputer is the sequel to the development of small process control computers, which were also being developed in the early 1960s. As an example, IBM had developed around 1959 the Model 1620 computer, which was a transistor-type machine. This was a character-oriented machine, but it was equipped with an instruction set that permitted scientific calculations. Because of its relatively lower cost compared to the 709 and 7090 machines, there was a great deal of interest in using such a computer for controlling more accurately the operation of large processing units. The 1620 was modified to become the 1710 computer, which had a capability for being connected to data acquisition equipment. Because of the interest in computerized process control, several corporations developed special-purpose computers, which were readily interfaceable to data acquisition equipment that could be directly connected to industrial plants. The intent was to supply complete systems in petroleum refining, power plant, and general manufacturing applications. Their main motivation was the ability to install a complete system, with installation of only a computer being of secondary interest. It is noteworthy that every one of these companies turned out to be financially unsuccessful in the total systems business. As the total systems companies were fading away, companies such as Scientific Data Systems (SDS) and Digital Equipment Corporation (DEC) took advantage of the latest solid state technology that was available at that time and began to develop small, general-purpose, scientifically oriented computers, which were similar in architecture to the predecessor process control computers. One of the early introductions by DEC was the PDP-8, a computer priced sufficiently low that it could be included as part of sophisticated instrument systems.

The PDP-8 contains a 12-bit word size because, at the time of its introduction, most large computers had word sizes that were multiples of 6. After IBM introduced the 360 series of computers, the trend changed to designing computers with word sizes that were multiples of 8. This change came about with IBM's decision to use 8-bit EBCDIC codes and with the decision by the communications industry to use 8-bit ASCII codes in place of the earlier 5-bit Baudot code. Most presently produced minicomputers have a 16-bit architecture, which gives them the advantage of a larger instruction set and a broader memory addressing capability.

Along with the development of different word sizes and expanded instruction sets, hardware technology was moving toward the integrated circuit concept. This resulted in a dramatic reduction in the size of computers as well as reduced power requirements and heat generation. It was basically the same step that took place in the transition from tube- to transistor-type machines. Because of the much-reduced physical size and much lower cost, the usefulness and application of the minicomputer very suddenly began to grow at a phenomenal rate.

II. Characteristics of Minicomputer CPUs

The generally accepted definition of a minicomputer is that it has a central processing unit (CPU), a memory, and a scientifically oriented instruction set and operates on a stored program concept in exactly the same manner as does the larger EDP computer system. As will be seen when the minicomputers are described in more detail, the application of minis and their growth have literally been such that in many cases they are very similar in capability to the middle-sized EDP computers. The main distinguishing characteristics of minicomputers have been their relative lower cost, their applications, and the types of software packages that have been available on delivery with them.

More specifically, the classical definition of a minicomputer would include the following features:

1. A central processor with an instruction set most suitable for scientifically oriented problems.

2. A conventional memory usually composed of ferrite core, but more-recent versions of minicomputers are now equipped with solid-state memories.

3. A word length of 12 or 16 bits, although other sizes have been available.

4. Generally an excellent input–output architecture chiefly because its initial applications were in the direct data acquisition applications and in process control.

5. Generally quite small size as a result of the electronics technology that has been used in their construction.

6. A typical price range beginning at $5000 to $10,000 (which might include a few peripherals).

7. Scientifically oriented applications, so that the software is also scientifically oriented. Usually only assemblers and FORTRAN compilers are available.

Initial versions of the minicomputer generally had very limited instruction sets because, at the time of their introduction, the architecture had to be built up from discrete hardware components. Sophisticated instruction sets were not used because they were not economically competitive. An alternate route was chosen by Interdata, a minicomputer company founded in 1966. This company used the concept of microprogramming to emulate the instruction set of the IBM 360 computer system. The disadvantage of microprogramming at that time was that it resulted in a computer architecture that was substantially slower than architectures wherein the CPU functions were designed in discrete hardware. Recent developments in microprogramming technology have been such that microprogrammed computers operate at close to the speed of computers whose CPUs are composed of discrete components. However, the advantage of microprogramming is that many features ordinarily found in software, such as floating point arithmetic, can be inexpensively put into hardware. Some minicomputer systems offer specific software subroutine functions that will enhance the operation of the minicomputer system to an extent that it operates many times faster than without the microprogrammed firmware package.

The early minicomputers were usually limited to 8192 words of memory. As memory costs have declined in price and as the applications of minicomputer systems have increased, there has been a continuing requirement for minicomputers containing larger memories. Because of the addressing limitation of the 16-bit word, it has not been conveniently feasible to address memory larger than 65,536 words. In the last two or three years several minicomputer vendors have announced offerings wherein the "minicomputer" has a capability of containing 256,000 and even 512,000 words of memory. Where such offerings are made, the minicomputer system contains a separate special instruction set that permits addressing of more than the base number of words of memory. Use of this specific instruction set tends to slow down the execution time of programs in the minicomputer and it is particularly cumbersome when page boundaries delineating various 32,000 word segments are crossed. Because of this problem, several commercial minicomputers contain larger numbers of registers and registers containing word sizes greater than 16 bits in order to be able to address these large memories.

Examples of these larger minicomputers, which are sometimes called midicomputers or mega minicomputers, are the InterData 8/32, which advertises a 32-bit architecture, and the Systems Engineering Laboratories SEL 32/50 computer, which also advertises a 32-bit addressing capability. Other minicomputer systems that have the ability of addressing more than 32,000 words are the DEC PDP-11/45 and PDP-11/70, which have the capability of containing up to one million bytes of memory. The Hewlett-Packard HP 3000 and the 21MX computers are also able to contain up to 512,000 bytes of memory. The addressing schemes in the HP 21MX, the DEC computers, and the Data General Eclipse Series all use a form of memory mapping that goes through an additional instruction set to be able to address that range of memory.

III. Instruction Repertoire

Probably without exception, the instruction length of every computer is a multiple of the word length that is used in that computer. Within that instruction, whether it is 16 bits, 32 bits, or otherwise, a number of bits is allocated for the operation code and the remaining number of bits is allocated for addressing memory. As a consequence, if 5 bits of a 16-bit word are used to define an instruction, and this would define 32 unique instructions, this would leave only 11 bits to select a memory address. Using only 11 bits for addressing purposes allows the computer system to address only 2048 words of memory directly. Consequently, in 16-bit machines, various schemes are used in order to be able to address more than the typical number of words of memory now found with minicomputer systems.

In the classical addressing scheme shown in Fig. 1, a computer performs all of its functions with a number of registers. In other words, an instruction will cause a word to be moved from memory into a register of the machine; it will then cause a second word to be moved from memory and added into that register; then the third function will be to take the sum of those two items of information and store them in another location in memory. Usually at least two such registers are provided, wherein arithmetic and logical

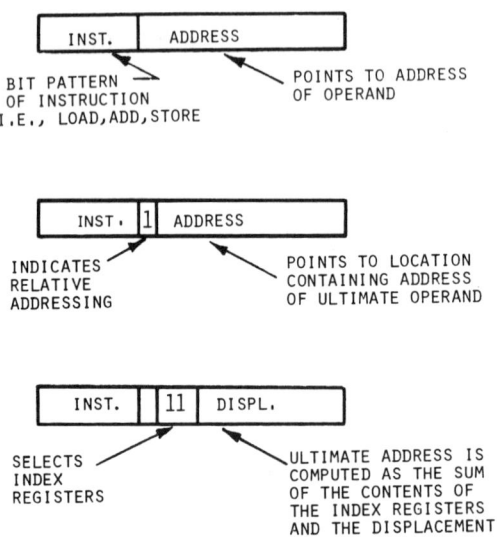

FIG. 1. Instruction formats.

functions are performed. The reason for two registers is that floating-point numbers are frequently represented as two successive 16-bit words and loaded up as a single 32-bit word. Then a second pair of 16-bit words is added as though it is a single 32-bit word to result in a 32-bit sum that may later be stored as a pair of 16-bit words. In some minicomputer systems, a larger number of registers is provided so that other functions may be in a partial state of execution along with the arithmetic operations. As an example, the Data General Nova Series uses two accumulators for arithmetic operations and two other registers for indexing purposes. The Hewlett-Packard 21MX Series also uses this architecture. The Digital Equipment computer system uses a total of eight registers for both indexing and arithmetic operations. The InterData series of computers, because it follows the IBM architecture, uses a set of 16 registers for addressing, indexing, and arithmetic functions.

In conventional addressing schemes, where a number of bits is designated for the operation and the remaining bits for the operand address, additional bits may be subtracted away from the operand address to indicate register addressing or indirect addressing. Usually if a machine has a small number of registers, namely four or less, it will combine these various methods within the same addressing format. Therefore, a computer such as the Data General Nova will set aside two bits to select the particular register that is being used. The selection of two bits gives it a possibility of any one of four registers, and it will set aside an additional bit to permit indirect addressing. Computer systems such as the InterData and the Digital Equipment PDP-11 series will often go to an instruction format whereby they address only registers. Consequently, one will see an instruction of the format where addition takes place from one register into another register. In this format, as in the InterData example, up to 16 registers can be specified within a 4-bit register addressing field. Because the Digital Equipment format in the PDP-11 requires only three bits for addressing, it has the ability to add an additional bit to indicate other functions. In the example shown in Fig. 2,

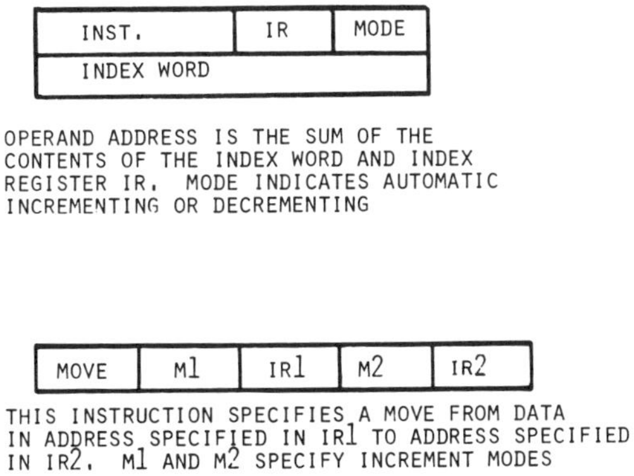

INST.	IR	MODE
INDEX WORD		

OPERAND ADDRESS IS THE SUM OF THE CONTENTS OF THE INDEX WORD AND INDEX REGISTER IR. MODE INDICATES AUTOMATIC INCREMENTING OR DECREMENTING

MOVE	M1	IR1	M2	IR2

THIS INSTRUCTION SPECIFIES A MOVE FROM DATA IN ADDRESS SPECIFIED IN IR1 TO ADDRESS SPECIFIED IN IR2. M1 AND M2 SPECIFY INCREMENT MODES

FIG. 2. More complex instruction formats.

where a number of sequential or repetitive functions can be performed, such as moving a number of words from one area of memory or to another, a mode bit may be set in addition to selecting three bits for a register. In that example, the mode bit can indicate automatic incrementing or automatic decrementing each time a register is used to perform a function.

The combination of auto increment and auto decrement permits a very comprehensive function to be performed each time that specific instruction is executed. When large arrays are to be moved from one area to another, the ability to use this type of instruction will permit the movement of data much more rapidly than if the conventional register-utilizing instructions are to be used. When a short number of words is to be used, however, the disadvantage is not as great because the registers must first be initialized to the proper addresses before the mass movement of information is begun. With short numbers of moves, it turns out that the initializing functions may take as much time as may be saved by the more comprehensive instruction. Furthermore, the execution of the very comprehensive move instruction does not take place in one or two cycles. Each execution of that instruction would require several cycles simply because several accesses to memory are required and the functioning is much more complex than the typical simple instruction execution.

Other modes of addressing are used but these are not as frequently found as those just described. These instructions are an immediate instruction and a relative instruction. For example, in a load immediate instruction, the data immediately following the memory location within which the instruction was located would be loaded into memory. The instruction counter or program counter would then skip the next memory location and go on to the following word as a source for its next instruction. In relative addressing the operand to be loaded into a register would be picked up as being relative to the location of the current instruction address. For example, an instruction located in memory address 1000, to load a value out of a location relative to itself by 100, would load out of memory location 1100. If that same instruction were located in memory location 5000, it would load the value from absolute memory location 5100.

Where more complex instruction formats are used, two- or three-word addressing is used. When a three-word instruction is picked up, the CPU recognizes it as a three-word instruction and increments the program counter to increment by 3 rather than by 1 as it would in the case of a typical one-word instruction.

Instruction sets of the earlier minicomputer systems such as the HP 2116 and the DEC PDP-8 computer divided memory up into a number of pages (Fig. 4). An instruction could address anywhere within the page within which that instruction was located and it could also address any location that was in the base page (or zero page as it was called). A bit would be set in the current instruction to indicate current or zero page addressing. The address range limitation then became one that was limited to the number of bits that were available. In the case of the PDP-8 computer it became 7 bits so that only 128 words could be addressed directly. In the case of the HP 2116 it became a 1024 word current page and a 1024 word base page. To cross page boundaries, execution of an indirect addressing instruction was required. When a programmer wrote a program in an absolute mode, he was required to keep track of exactly where his instructions were being executed, and he was obligated to generate the correct indirect addressing scheme

BASIC CONCEPTS OF MINICOMPUTERS

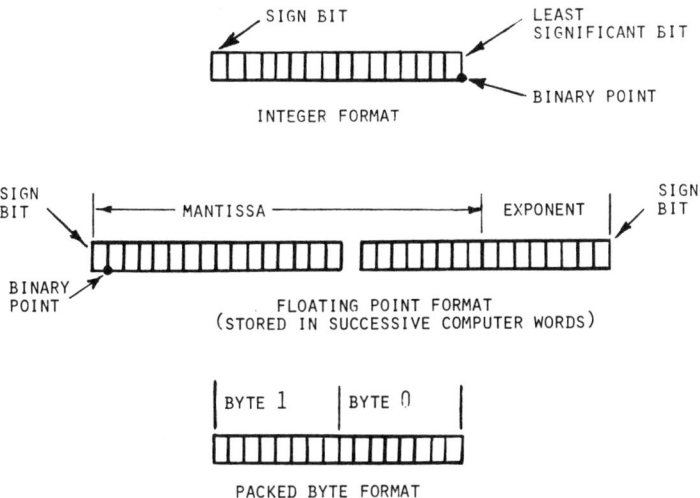

FIG. 3. Typical minicomputer data formats.

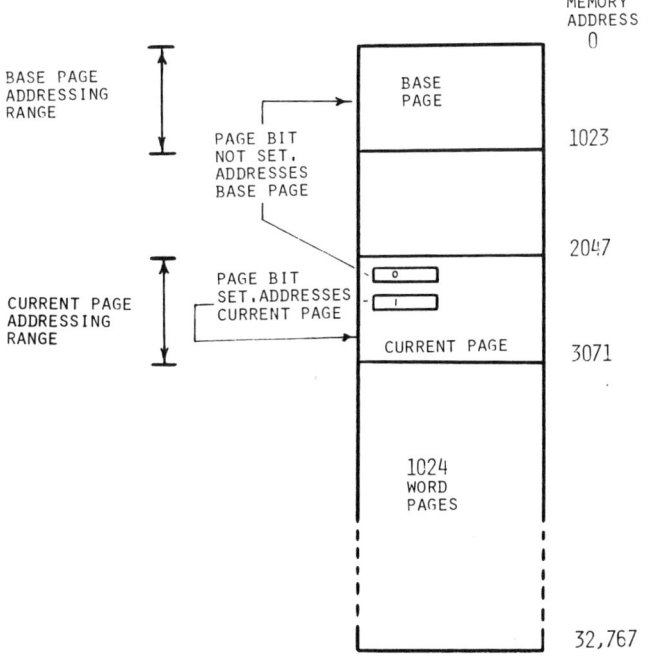

FIG. 4. Page addressing.

whenever it was required. When relocatable code was generated, the program loader was required to keep track of the current address and generate the indirect address instructions whenever they were required. Problems occasionally arise when a great number of these indirect instructions are required. In this case the amount of space allocated for indirect addressing may be used up, whereupon the loader will not be able to load the instructions required for the program even though there is apparently sufficient memory to store the instruction sequence required for the application program.

Because each method of memory addressing has its own advantages and disadvantages, it is very difficult to give a general answer as to which method is the best means for addressing. Certainly where small programs are being used and where the amount of crossing of page boundaries is minimal, then the paging method seems to be quite efficient and generally results in a smaller number of instructions being generated. Where very large programs are being generated, then some form of base register addressing or indexing scheme is probably more appropriate.

As the application of minicomputers has grown and as the costs of memory have dropped, several minicomputer systems have become available that contain more than 32,768 words of memory. In these large systems, the successor minicomputers maintain an instruction set that is compatible with the predecessor computers, which were not required to access this much memory. This requires maintaining the original addressing scheme, as well as the capability of addressing very large quantities of memory. Therefore, the large-memory addressing schemes are solved in a slightly more roundabout way than if a larger word were available for addressing.

The simplest method of addressing such large areas of memory is to use a set of bank registers whereby each bank addresses a segment of memory that may be 32,768 or 65,536 words in length. A special command is added to the computer system that permits the program counter to cross from one bank into another in just about the same way that current and zero page addressing is performed. The only exception is that this is a special instruction that is executed on special command and not as part of the regular instruction set. In this procedure, the understanding is that once a program crosses a bank boundary, the program is expected to stay in that bank for a substantial period of time or for a substantial number of functions. Otherwise, if crossing of bank boundaries were to happen frequently, this would greatly slow down the operation of the computer system.

A second scheme involves the use of memory mapping. In this procedure, a set of registers, which points to individual banks in memory, is added to the computer. Each register may contain 10 bits of information for addressing purposes. This information may be combined with the original 10 bits

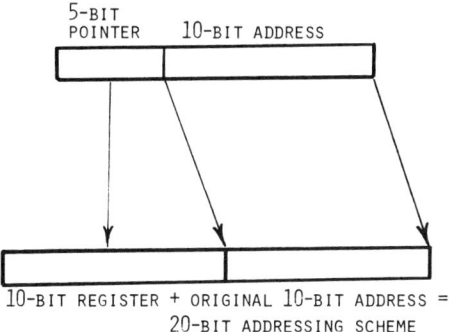

FIG. 5. A memory mapping method.

for addressing. The combined 20-bit addressing scheme permits directly addressing over one million words of memory on a quasi-direct basis. Modification of the 10-bit bank register permits selection of the bank that is desired. Modification of the lower 10 bits of addressing permits the user to transfer control to any word within the complete one million word range. This differs from the bank switching scheme from the standpoint that switching from one bank to another simply permits one to start at some predetermined location in the bank (Fig. 5).

Another addressing scheme requires the use of a relative address register. In this method, a register that has the ability to store more than 16 bits is preset at the starting location of a specific program. In a computer using 16-bit words, the addressing range again would be 65,536 words. More than one relative address or base register could be used such that one points to the instruction sequence and the other points to the stack of data that is being stored in memory. An advantage of a computer with an architecture of this type is that programs are very readily relocated and, provided that they are not modified during execution, their partially executed versions need not be stored on disk if preempted by a higher priority program. On restart, the original program is brought back from the disk and restarted from the point where it was previously interrupted.

IV. Data Checking

Some minicomputer systems do not offer any form of data checking on the basis that the hardware is substantially more reliable than conventional data processing hardware. However, most present-day minicomputers include either parity checking or some other more sophisticated data-checking procedure for checking the data in computer memory.

In the conventional parity-checking scheme, a check is made each time a computer reads a word from or stores it into memory. The number of bits in a word is determined and an additional bit is added in an additional bit position to make it an odd number if necessary. This, in effect, converts a conventional 16-bit computer into a 17-bit machine. The reason for odd parity is that there is a possibility of dropping or losing all the bits in a word of memory. The loss of all bits would be considered an even number and odd parity checking would catch this condition, whereas it would not be detectable if even parity checking were used.

When a parity error is detected it is possible to take some action to try to correct the problem. A minicomputer can be programmed to interrupt its normal processing and come to a halt or to go through some procedure to retry the instruction or to reread the data. If the error occurs in an instruction word, it is generally not a good idea to reattempt the instruction, because a new instruction may be generated that may have a very disruptive effect on the remainder of the program. If the parity error occurs in a data word, one may reattempt to read the data, or if the data are being stored in another location they can be reacquired and restored in that location with very little impact on the integrity of the system.

Among minicomputer systems with very much larger memories, there are at least two manufacturers who offer a more complex memory-checking method. This is a fault-correcting method that uses 5 additional bits, each of which is a parity check on a subgroup of the 16 data bits. The organization of the parity-checking bits is such that at least two parity-check bits are used to check each bit in the word (Fig. 6). In the error-correcting procedure, the

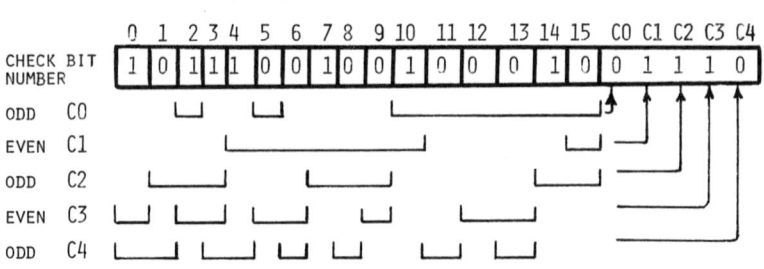

FIG. 6. Check scheme for memory fault correction.

logic assumes that the check bits are correct, and that the data bit has been dropped. This procedure permits the recalculation of any bit in the 16-bit word so long as only one bit has been lost. In the event that more than one bit has been lost, the fault-correcting memory method reverts to the same kind of parity-checking capability that ordinary parity checking would provide. In addition to the ability to correct for single bit faults, the fault-correcting feature may provide a capability to record all faults in some auxiliary memory. The record of errors can be invoked by the service person, who will then be able to determine where the errors occur and replace the faulty memory in a much shorter time.

V. Bus Architecture

There are two concepts used in minicomputer design for the transmission of data among the various functional components of minicomputers. In the conventional design, a fixed number of I/O channels is designed into the computer and one or more peripheral devices may be attached to each channel. Each channel contains logic to address and control each device attached to it. In conventional non-DMA accessing, the channel emits a request-to-send signal for each item of information to be transmitted. As each item is received by the channel, the CPU transfers it from the storage register on the channel to another location in memory. If the channel is equipped with DMA (Direct Memory Access), the transfer is accomplished without intervention by the CPU. In some computers, a limited number of channels is provided with DMA. With others, DMA can be assigned under program control to any channel.

In single-bus computer architecture (see Fig. 8), notably exemplified by the UNIBUS of the PDP-11 series of the Digital Equipment Corp., a single

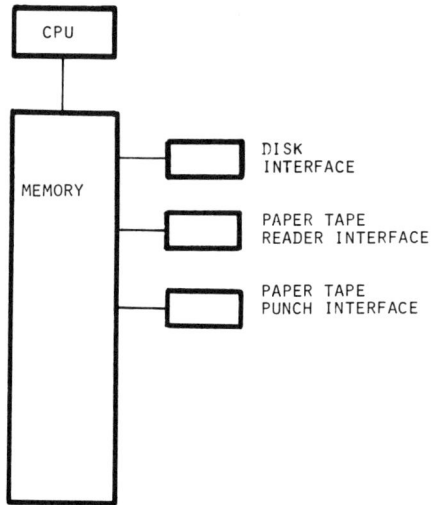

Fig. 7. Conventional minicomputer architecture.

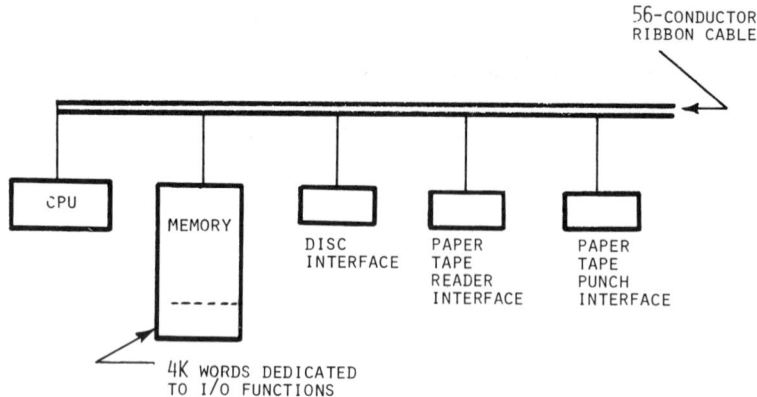

Fig. 8. Single bus architecture.

channel is used for the transmission of data among all the components of the minicomputer system. In the UNIBUS architecture, a 56-conductor ribbon cable is used to transmit control signals, status information, and data between components (Fig. 8). A master–slave function must be established between the two components that will exchange an item of data. The master (for example, the CPU) will interrogate a device, such as the memory in a request to send data. If the memory is not busy, the data will be sent. Other devices, such as a disk or paper tape reader, operate through controllers in the same way. The paper tape reader can request to send data to memory, and if the memory and the bus are not busy, the transfer is accomplished.

To enable the single bus system to function, the PDP-11 architecture allocates 4096 words of memory for communicating with the peripheral devices. In addition, each controller on the single bus is required to have appropriate decoding logic to function in a uniform manner. The resulting architecture is a system that provides a programmer with a convenient system for handling I/O functions.

The advantage of the single-bus concept is that there is no arbitrary limit on the number of channels into memory. The other advantage is that all devices operate asynchronously. A paper tape reader can transmit data to the paper tape punch without requiring control signals from the CPU.

Disadvantages of the single-bus architecture include its limitation of transmission speed. Even though the bus can transmit data at over a megahertz rate, when there are many high-speed devices connected to the system, the single bus may be a limitation on the I/O capability of the system. Higher-speed versions of the PDP-11 series have available, as an option, a second, higher-speed bus, called the FASTBUS.

Other disadvantages focus on the differences between the single bus and the multiple bus concept. Obviously, if there is a failure with any address decoding module on the single-bus system, causing a lockout, the complete system is inoperative. In the multiple-bus system, only the single channel and its associated device would be unavailable to the system.

IV. Power-Fail Protection

In a large number of applications, minicomputers are expected to run in a continuous and unattended mode. Often this is not possible because the power supply to the computer system is not consistent or there may be many power interruptions. If a computer does not have a feature to protect itself from such power interruptions, all or part of memory may be erased, causing the computer to function in an unpredictable manner, or it may lose its instruction set and come to a complete halt.

In order to prevent this problem, many minicomputers are equipped with a power-fail and automatic restart feature. Operation is such that when the power to the computer system falls below a safe working level, a very high-priority interrupt occurs, and control is given to a program that causes the computer to go through an orderly shutdown. The main function of this program is to save the contents of all the registers and the status of the operation of all I/O devices. When the computer restarts, it will restore all the registers and restart exactly at the position where it stopped. The mode of restarting can be at the user's discretion. He may choose to let the computer restart on a manual basis so he can keep a record of the power failure that took place. He may also restart the programs exactly where they stopped and continue as though no power interruption has taken place. Within the program, he may keep track of when power-fail occurred, and if the computer system also kept time, he would know how much time had been lost due to power losses.

In the past, most minicomputers had been equipped with ferrite core memories of a type that would retain their data during a power shutoff. Most of the recent offerings in minicomputer systems use solid-state memories that will lose their data when power is lost. Such minis have available as an option a battery power source that can maintain power on memory to prevent this data loss. This gives a solid-state memory computer the same advantage as a computer equipped with ferrite core memory. Such computers can be used in applications such as oilwell logging, automatic instrumentation systems, and process control systems where unattended, continuous operation is necessary.

VII. Memory Protect

Another feature commonly available in minicomputers is memory protect. Its purpose is to protect one area of memory from another. It is highly desirable when a computer system is doing more than one operation such as concurrent data acquisition and program development. It is therefore necessary to protect the well-debugged, continuously running program from one that is in a developmental stage. There are two techniques for providing memory protection. In architectures of earlier computers, a bit was added to each word in addition to the data bits to indicate that a word was to be protected. When this additional bit was set, a computer instruction could not write into that word unless it also had a similar bit set. A user who was in the developmental stage in his programming would not set protect bits in his instructions, which would not be permitted to write into protected areas of memory. In this organization, a computer that had parity checking as well as memory protect would in reality have an 18-bit internal word structure. A second technique of implementing memory protect is to use a bounds register or a fence register. The fence register is an additional register that is a part of the CPU structure. A special instruction is used to write the address of the boundary into the fence register. Regular user programs then cannot write below the address specified in that register.

If a memory protect area is violated, the computer will generate an interrupt that indicates that a memory protect error of some form has taken place. At this point, the user of the computer can direct it to take some alternate action. If memory protect violations are handled as part of a software executive, the executive usually takes the course of aborting or suspending the user program causing the violation.

VIII. Memory Interleaving

One of the common limitations of the execution speed of a computer is its memory speed. In the conventional execution of an instruction, in the first memory cycle a computer transfers the instruction from memory to the CPU; then in the next cycle, the operand is acted upon.

Although access to individual memories is obtained once per cycle, access to memories through separate ports can be initiated at time intervals of less than the cycle time for a single memory bank. In this organization, two or more banks can be accessed alternately to transmit data to the CPU at a rate that would be much greater than if only one bank were available. If the organization of the programs in the memory of the computer were such that all instructions were in one bank and all data in the other, the access rate could be doubled. Such a convenient arrangement of programs and data is not normally found, so that memories are normally divided into more than two banks in the expectation of a higher probability of accessing alternate banks during normal program execution.

If the memory is designed such that the 1st, 5th, 9th, etc., words are in Module I, the 2nd, 6th, 10th are in Module II, and the 3rd, 7th and 11th are in Module III, etc., then in a conventional sequential accessing procedure, four memory banks with an 800 nsec access time may operate in a fashion similar to one bank having an access time of 200 nsec (Fig. 9). A 16-way

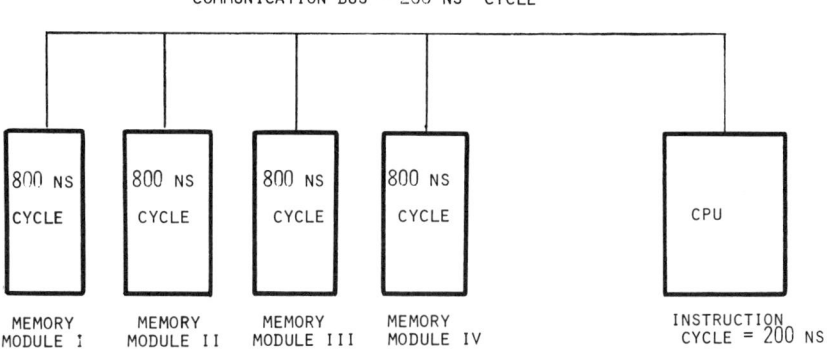

FIG. 9. Memory interleaving.

interleaving procedure of this type was used in the CDC 6600 computer, which was first delivered in 1963. It is very noteworthy that the method is now finding its way into minicomputers.

IX. CACHE MEMORY

Very high speed memories (300 nsec, bipolar) are available and can be used with minicomputers, but because of problems such as high power requirements, high heat dissipation, and high cost, they are not normally found as the only type of memory in a minicomputer. In some larger minicomputers, a section, or a cache, which may be as much as 2048 words in size, may be implemented in such a way that most of the execution takes place within that cache (Fig. 10). Instruction sequences will be moved in block mode into the cache from slower main memory and acted upon by the CPU, as though the entire memory were composed of the high-speed type. The small section of cache memory may be manageable in a minicomputer system from the standpoint of power supply and heat dissipation requirements, whereas it might not be manageable if the entire memory were composed of the bipolar type.

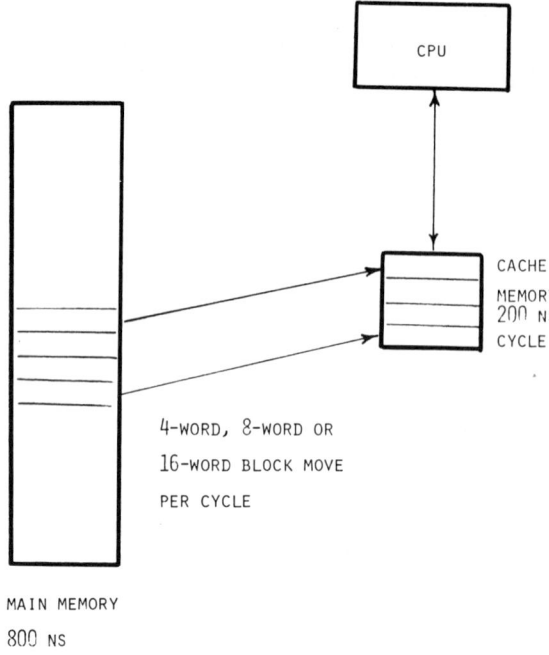

Fig. 10. Cache memory usage.

When the minicomputer includes this feature in its design, it will require an interleaving memory design in order to move data with sufficient speed to keep up with the cache memory, or it will require a wider bus, i.e., a capability to read several words in one cycle. In Fig. 10, the design indicates a four-word block move, which is achieved in one main memory cycle. The CPU accesses the instructions then stored in high-speed memory at a higher rate, i.e., 200 nsec. As it steps through the instruction sequence, new instructions are added in groups of four. Thus, this architecture can make a computer containing mostly 800 nsec memory act as though it were a computer containing mostly 200 nsec memory. The only catch is that programs frequently jump over many steps or jump back and forth among subroutines. In these cases, the execution speed then slows down to the execution speed limited by the main memory.

X. Interrupt Structure

The interrupt structure of minicomputers, as well as the input–output structure, is generally a much more sophisticated architecture than the input–output and interrupt structure of common business data processing computers. This interrupt and I/O structure is an outgrowth of the requirements of the early process control computers. Early applications of process control systems, where multilevel interrupt capability was desirable, were achieved by programming a quasi-interrupt scheme in which the program polled each device on a round robin basis to see if data were available.

Although this procedure can work almost as well as a hardware interrupt system, the program always runs the risk that an error will cause the program to be lost and information will no longer be collected. Furthermore, the programmer has to be extremely careful regarding the timings of this program in order to return to each device in time to collect information gathered by it before that information is lost.

The more common interrupt structure in minicomputers is that each input–output device has the capability of interrupting the computer directly. Permission for an interrupt to occur is usually handled under the control of the developer of the program. Where the computer has a multilevel priority system, the programmer may decide on a particular priority level for each device. He will then write an interrupt handling routine for each device, taking into consideration the characteristics of that device. Normally the program is designed so that the priority handler occupies as little time as possible and returns to the main program as quickly as it can. This means that the interrupt handling program simply accepts a character from the interrupting device and stores it in memory. At scheduled times, or when the complete message has been transmitted into memory, the application program that is to act upon it is rescheduled.

In some minicomputers, several devices are connected to the same I/O channel, and any one of these devices can cause an interrupt on that channel. In such a design, the computer programmer is obligated to poll the status of each device to determine which one caused the interrupt. This design is used when a number of low-speed devices such as teletypewriters are to be connected to the same I/O channel. In a situation such as this, it is highly unlikely that more than one teletypewriter would interrupt the computer within the same time cycle as the computer. Even if they were to interrupt within, say, the same microsecond, the computer would be able to poll each teletypewriter and accept a character from a very large number of them before the first teletypewriter would have time to send a second character through that I/O channel. Obviously, this is not the ideal design for ensuring that the computer collects all data from all peripheral devices. However, it is a design that is much more economical in terms of input–output channels and is usually found with very low cost computer systems.

A slight modification to the previous design is one that uses a special register to indicate the I/O status of each device on the I/O channel. In this architecture, the programmer is required to read the status register for that

I/O channel, and by checking each bit he will quickly determine which device caused the interrupt. Then he can accept the character from that channel and store it in memory in a buffer corresponding to the device on that channel. This permits the computer system to respond more quickly to the outside world and reduces the likelihood of losing data. Except for the fact that it is a quicker procedure to identify the interrupting device, this architecture is not greatly different from the architecture where the programmer has to poll the status of each device before he accepts the information from it.

The third form of input–output and interrupt architecture is an organization where each device is connected to a single channel. In this scheme, an interrupt on a particular channel gives control to a special location in memory, where a control word is assigned to process that interrupt. If, for example, a peripheral device were connected to I/O channel 15, the interrupt control word would be "location 15" in computer memory. When an interrupt occurs on that channel, the computer automatically takes its next instruction from the word in memory or the instruction in memory corresponding to that channel. The usual instruction stored in this word will be a jump to a subroutine (JSB) that will transfer control to an interrupt processing routine. When the interrupt routine has been completed, the return address set by the jump-to-subroutine instruction carries the program back to the original program that was operating before the interrupt occurred. This is the procedure used in the Hewlett-Packard 2100 series of minicomputers.

The Digital Equipment PDP-11 computers use somewhat the same scheme, although it is implemented slightly differently. In that computer system, a PC (program counter) and a PS (processor status) are two items of information maintained about every program that operates within the computer. The PC contains the address of the next instruction to be executed. The PS maintains information about the results of the last operation performed by the CPU. The PS register is subdivided into one-bit flags called condition codes. Each bit maintains unique information about a given operation. For example, the Z bit indicates that the last result of an operation was zero and the N bit indicates the last result was negative. The C and V bits maintain information dealing with carries and overflow, respectively, in arithmetic operations. These bits are sensed by other commands, which cause a change in the course of the program as a result of previous computations. For example, a branch on negative instruction would sense a bit to determine whether the previous result was a negative operation. Therefore, if an interrupt occurred in a program just after an operation had resulted in a negative result, it would obviously be necessary to save this information. When an interrupt occurs, the contents of the PC and PS are automatically pushed into a "stack," which is a set of memory words used for temporary storage. The number of addresses maintained on the stack is maintained by the stack pointer or register six of the computer.

To initiate operation of the interrupt responding program, the PC and PS are reset or loaded from two preassigned consecutive memory locations. The actual locations are chosen by the interface designer, and areas in low memory are usually selected. The first word that goes into the PC contains the starting address of the interrupt service routine and the second word contains the new PS. After the interrupt processing routine has been completed, a return from interrupt operation (RTI) is performed. The two top words of the stack are then placed into the PC and PS, which returns the interrupted program back to its original status and operation. The difference between the PDP-11 and the Hewlett-Packard 2100 is that in the former, the return information is in a specially designated area of memory; in the latter, any area of memory can be used by the program developer.

In both computer systems, a set of consecutive memory locations in low memory is used to store transfer instructions for handling responses to interrupts. This sequence of memory locations is called an interrupt vector and the means of response whereby each I/O channel gives control to a specific memory location is called a vectored interrupt (Fig. 11). The

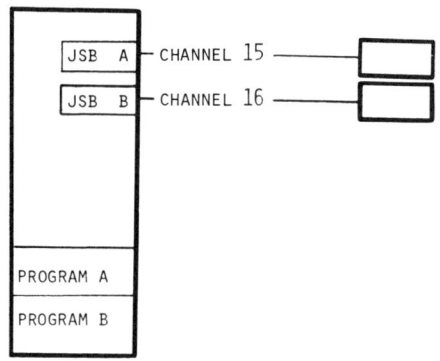

PROGRAMS A AND B HANDLE INTERRUPTS
FROM CHANNELS 15 AND 16, RESPECTIVELY

FIG. 11. Vectored interrupt handling.

vectored interrupt method has the advantage that the program does not have to poll or read the status of any device that may be on that input–output channel. Because each device has its unique I/O channel and a unique memory location to which control is given, the user automatically knows which device requires servicing. This reduces the amount of time required to service interrupts. As with all designs and procedures in computer systems, there is a cost to be paid for this advantage. The cost is paid in terms of a requirement to allocate a number of memory locations that are not usable in any other way, and a requirement for an I/O channel for each device.

To give the programmer additional flexibility on how his equipment responds to interrupts from specific I/O devices, in most computer systems he has control over which channels may interrupt on a priority basis over other channels. In one method, this is done by setting bits in a mask register. The presence of a bit allows an interrupt to take place on the channel corresponding to that bit position. If a program is not to be interrupted by a specific set of I/O channels, the first step in that program would be to clear the mask register bits for those channels. After that program has been completed, it is obligatory to reset the mask register to its normal status.

In all interrupt handling procedures, there is a point at which it is necessary to inhibit the computer from responding to an interrupt. During the inhibited period, status information regarding computational results as well as I/O functions is stored temporarily in memory. After the interrupt processing is completed, the reverse step takes place.

Interrupts are inhibited while the registers are restored to their preinterrupt status. In computers containing large numbers of registers, the recording and restoration intervals will take a relatively greater length of time. One computer, the IBM System 7, possesses a capability for automatically switching among any of four priority levels without having to store and restore registers. Each of the four priority levels has associated with it nine registers, which store the seven index registers, the accumulator, and the program counter. A switch from one priority state to another takes place in 800 nsec.

The use of multiple-register architecture as an aid for addressing and computational work as opposed to a simple address architecture that would aid in faster interrupt processing leads to differing design philosophies. The multiregister machines that permit direct addressing of large quantities of memory are more useful in problems where large arrays of data are to be processed. If the computer is to be connected to a large variety of peripheral devices that will be interrupting it in random manner and the response to each interrupt will require the saving of the data in all of the registers and saving of status information, the multiregister machine will be at a slight disadvantage.

Some care is required in interfacing the minicomputer to the various types of peripheral devices that it is required to operate. As a general rule, the faster the device the higher the priority that is given to it. The main concern is that the computer must be able to collect the data item from the device and store the item in memory before the device is ready with its next item.

XI. Direct Memory Access

Direct memory access (DMA), which is also called buffered input/output, is a means of transferring data directly to or from memory without requiring attention from the CPU. Before direct memory access can be initiated, a number of instructions need to be executed to initialize DMA and to permit the device to transmit data to the computer system. The initialization commands are required to give the DMA function the starting address of the buffer in memory into which data will be transferred and the number of data words to be transferred. Additional information given is an indication for the DMA controller to interrupt the computer when the DMA function has been completed. Interrupts may also be requested during an incomplete transfer of data. The user program or the operating executive would then be required to interrogate the status of the channel to determine the reason for the incomplete termination.

DMA operates on a cycle-stealing basis because there is usually only one access channel to memory. When the DMA channel needs access to memory, the CPU cannot simultaneously access it. The CPU will halt its normal functions for one cycle time to allow the DMA channel to read a data word from memory or to store it into memory. DMA rates are normally related to the cycle time of the computer system. If DMA has priority over the CPU, the DMA rate will usually be the inverse of the cycle time of the computer system. In other words, a computer with a cycle time of 2 μsec would be able to transfer data at the rate of 500,000 words per second. In machines where the CPU function has priority, the maximum transfer rate is one-half as great because the DMA channel has access on only every other cycle (see Fig. 12).

As the DMA transfer takes place, each time a data word is stored in memory a counter is incremented and compared against the upper limit counter, which has been preset by the programmer. The upper limit counter may be the last address to be stored or the total word count. After the last word has been stored, the DMA operation is terminated and an interrupt may be given to the computer if it has been requested by the program. The DMA channel functions rather independently of the main CPU and has a reasonable amount of logic associated with it. In some of the larger minicomputers, such as the SEL 32/55, the DMA channel is implemented with a microprocessor. The microprocessor can be programmed to do such functions as parity checking and data restoration, and it can retry the transfer for such faults as an intermittently faulty track on a disk. Such procedures are now programmed in software. The microprocessor-oriented system obviously will relieve the CPU of this burden as well as save the amount of memory required in the main processor to accommodate these functions.

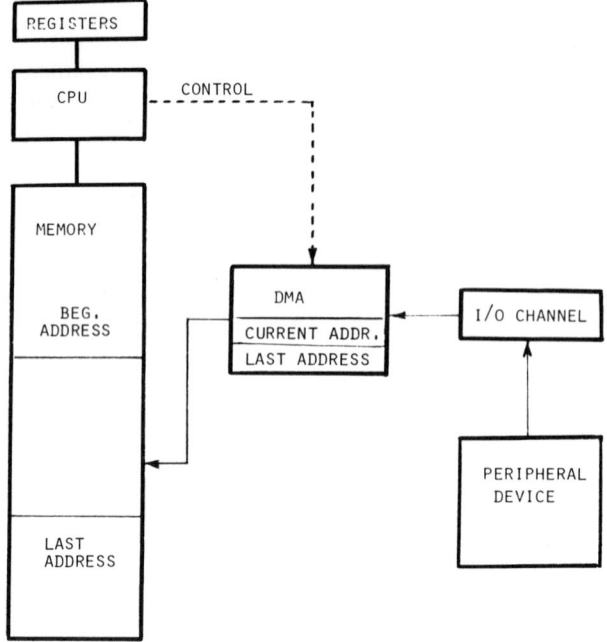

FIG. 12. DMA functions.

DMA is implemented in minicomputers on a "hardwired" as well as on a programmable basis. In the hardwired version, certain channels are designated as having a DMA capability. All high-speed devices must be attached to them. In the programmable designs, DMA can be allocated under program control to any I/O channel. Because DMA functions only for short periods of time (because of the limitation of computer memory), the ability to assign a DMA function to any channel, in effect, almost gives such a computer the same capability as one that has DMA hardwired to all channels.

XII. MICROPROGRAMMABLE SYSTEMS

Several minicomputers are now designed so that the instruction set is developed in microcode. The microcode is implemented in a read only memory (ROM), in which case it becomes a permanent part of the instruction set. Read only memory can be visualized as a semiconductor diode array whose bit patterns are read by the computer and interpreted as instructions. These bit patterns are written by a programmable ROM writer, which outputs voltage pulses into certain points in the array so as to fuse selected diodes.

This process generates the proper bit patterns to be interpreted by the computer. Additional forms of memory are read–write memories or random access memories (RAM), which are similar to conventional semiconductor memory. The RAM memory receives instructions from the main memory of the computer and executes these instructions as though it were a variable-instruction computer. The disadvantage of RAM memory is that because it is semiconductor in nature, the instruction set is lost if power is dropped from the minicomputer system. The advantage is in that a variable instruction set is available and a large number of firmware routines can be programmed that will execute substantially faster than if they were to be programmed using conventional computer instructions. If the minicomputer is of the microprogrammable type and even if the microprogramming instructions are not available to be implemented by the user, microprogrammability of the computer permits the hardware designer to give it a substantially larger variety of instructions than would be available if individual instructions were designed by individual hardware units. For that reason, the more recently introduced minicomputers possess instruction sets that are much richer in their repetoire than were the instruction sets for minicomputers that were in existence just a few years ago. Because most vendors wish to maintain a compatibility from their earlier computer editions through their present ones, they will normally microprogram the initial instruction set into their newer versions.

They will add instructions to generate a superset, which will permit all earlier programs to run, and at a faster speed. The earlier minicomputers did not have even a fixed-point multiply instruction and certainly did not have floating-point hardware. These instructions were handled by software subroutines. In addition to adding conventional instructions, it is possible in microcode to add the trigonometric functions, exponential and logarithmic functions, and frequently called FORTRAN subroutines. With these more sophisticated procedures programmed in microcode or in firmware, a minicomputer of basically the same cycle time will operate substantially faster when computationally oriented instructions are executed. The availability of microcoding is also why present computers possess move, byte manipulation, and string array comparison instructions. Such instructions require several cycles of machine time to execute but they are coded as a single assembly language instruction.

Microprogramming at the user level, even though it has been offered by a number of vendors, has not been spectacular in its use. The main areas of interest seem to be in numerical control and in special computational areas such as in programming a Fourier analysis system. In both examples, the purpose of the microcode is to execute the program much more rapidly than the equivalent software program. The other advantage is that when an instruction sequence is placed in microcode, it is much more difficult to copy the instruction set and use it on another computer system. The disadvantage of microcode is that the microcode is not consistent from the same vendor's successor minicomputer even though the conventional instruction set may be consistent. Consequently, microcoded sequences would have to be remicrocoded when a new version of that minicomputer is offered by the original vendor.

XIII. NONMINICOMPUTER SYSTEMS

The competition for applications of minicomputer systems is being challenged by a number of types of computing equipment that do not qualify within the strictest description of minicomputer system. At the low end is to be found the microcomputer, and at the upper end is found the very large minicomputer that is sometimes called a midicomputer or a megaminicomputer. Somewhere in the middle, we also find the sophisticated desk calculator, which often has a computational capability very comparable to a conventional minicomputer.

The concept of a microcomputer was first suggested at Intel Corporation in 1970. At that time, integrated circuit technology had progressed to a point where a sufficient number of components, such as flip–flops, shift registers, counters, and memory arrays, could be placed on a single chip. These components could be interconnected to produce the central processor of a simple computer. Rapid developments in miniaturization of integrated circuit technology had progressed to a point where it became practical to produce a modest computer on a single chip. The first microcomputer was introduced only in 1972.

Initial microcomputers possessed short word lengths and simple instruction sets. A typical microcomputer would have a 4-bit word and a maximum memory consisting of 4096 words. At the present time, the typical microcomputer has an 8-bit word and an addressing capability of 65,536 bytes. There are several microcomputers already in existence that have 16-bit words and addressing capabilities greater than 65,000 bytes. Speeds of the initial microcomputers were such that it took somewhere between 5 and 10 μsec to execute an addition instruction. More recent introductions of microcomputers indicate that speeds are very comparable to minicomputers and they possess very similar instruction architecture. The main disadvantage of microcomputers is that they have been developed by semiconductor manufacturers who have not spent an equivalent effort in developing software. Microcomputers are found in dedicated systems where programming is done in assembly language, with read-only memories so that the programs are indestructible and unchangeable. Interestingly enough, these are the same applications of the early minicomputers. Because of their lower cost and extremely small size, microcomputers can be installed as a hardware component on many sophisticated instruments. They can also be used as components of a larger minicomputer system. In many instances, the choice between a mini- and a microcomputer system will be made because of the availability of application software.

When the cost of programming and interfacing the microcomputer to the instrument can be amortized over a very large number of instruments, the microcomputer is certainly going to be the obvious choice. Where a computer is to be used in a process control application that is going to be a one-time study, the cost of programming can be considerable. A computer system that has a substantial software package associated with it will more than likely be the choice.

At the other end of the scale is the midicomputer or megamini. Briefly, the megamini is a computer containing 100,000 words of memory or more, a cycle time of under 1 μsec., at least a 16-bit word size, and relatively sophisticated software. Included in the software should be a multiprogramming system that is able to act in a real time data acquisition environment. In this class we find the SEL 32/55 computer, Data General's Eclipse 200 and 300, the Digital Equipment PDP-11/70, and the Hewlett-Packard Model 3000 Series II computer. These computers are usually provided with programming languages such as RPG, COBOL, ALGOL, and APL, in addition to the customary FORTRAN and assembly language offerings. These midicomputers are competitive with the lower-sized EDP computer systems usually found in business installations. The midicomputers do not appear to be replacing the typical IBM, Burroughs, UNIVAC, or Honeywell systems, but because of their greater acceptability, they are frequently considered as an alternate for new business applications.

The sophisticated desk calculator has also enjoyed a reasonable growth rate, although it has not been as spectacular as that of the minicomputer. In many instances it competes with the mini. Sophisticated desk calculators appeared in the late 1960s and have gradually grown in capability. The desk calculator is usually preprogrammed in some algebraic language or it may actually be programmed to accept BASIC or APL. Because the desk calculator is preprogrammed to operate in a higher level language, it is easier to use but it also operates at a much slower execution rate. The desk calculator finds greater application on a stand-alone basis where convenience is a requirement and where the speed of the minicomputer is not required. Early vendors of the desk calculator were Wang Laboratories, CompuCorp, and Hewlett-Packard. The latest entry is IBM's Model 5100 desk calculator.

Desk calculators have been interfaced to analog-to-digital converters, digital I/O devices, disks, and line printers, so that in their application they are indistinguishable from a minicomputer. The main difference between the mini and the desk calculator is that the sophisticated desk calculator is programmable in a single high level language. The minicomputer is programmable in a binary oriented language and the software system with which it is programmed is usually implemented through a process that translates the original high level language statements into the fundamental binary steps that the computer interprets.

It is also interesting to note that one of the greatest uses of the microcomputer is as the computing element within the desk calculator. The power of the desk calculator is reflected by the power of the microcomputer with which it does its computations. For that reason, as microcomputers have greater capabilities, succeeding models of desk calculators will also have greater capabilities.

XIV. Peripheral Devices

At the time when minicomputers were first introduced, the peripheral devices available, such as tape drives, disks, drums, and line printers, were the kinds of devices that were normally attached to large business center computers. These devices were generally designed for high performance and required the typical sterile atmosphere of the EDP computer room in which to operate. They were high-priced devices so that when one purchased a minicomputer in the $10,000 to $15,000 price range, it would be somewhat inappropriate to add a device such as a tape drive or a disk, which might cost another $25,000 to $50,000. During the development of the minicomputer, a large number of lower-performance, but much lower-cost devices, which have the ability to operate in a more hostile environment, have been developed specifically for the minicomputer industry. The characteristics of these peripheral devices to tend to be somewhat different; for example, one sees teletypewriters frequently used as consoles for minicomputers, while a higher speed typewriter is used as a console for the large business center computer. Card equipment is not very frequently used because of the expense in producing a good quality card punch. Instead, one will find paper tape reading and also paper tape punching equipment.

XV. Computer Consoles

At the present time, the ASR-33 teletypewriter is the most frequently used console with minicomputers. It has the advantage that it is a reasonably reliable piece of equipment and has the capability of producing hard copy, for punching paper tapes, and for reading paper tapes. It has the advantage, which is particularly desirable for a small minicomputer manufacturer, that there are a number of service organizations that will service the teletypewriter. It is also produced in very large quantities at a cost that is probably more competitive than could be produced by any other manufacturer who did not have the advantage of such volume. The most common models of teletypewriters are the ASR-33 and the ASR-35, which is the heavy-duty version. They operate at 10 characters per second. The Model 37 operates at 15 characters per second, has upper- and lowercase, and has the capability to print with different colored ribbons. A greater amount of flexibility has been designed into the carriage control so that symbols other than normal printing characters can be printed.

At the present time, there are available several CRT's that have a great deal more flexibility than can be found on a typical teletypewriter. At the low end of the capability scale, there are CRT's, which are basically teletypewriter replacements or "glass typewriters." The main advantage of such CRTs is their silent operation and their much higher transfer rate. Their main disadvantage is that they do not make permanent hard copy for record keeping purposes.

Several more-sophisticated CRTs have appeared and are being promoted as data entry stations. These CRTs are designed with a number of data checking features built within them. They are usually programmed and controlled by a microcomputer built within the device. Features offered include the ability to display a complete image of an invoice, and to designate certain fields as protected fields so that headings, titles, and instructions to the operator cannot be erased by data input and error checking. Programmable, or "soft keys" are available so that a predetermined character string can be transmitted with the depression of a single key. CRTs are also available with cartridges or cassette tape drive units for storage of large quantities of data. The disadvantage of not having hard copy with a CRT is met by designing them with interfaces to frequently available line printers or typewriters. The interface usually follows the RS232C convention and there are many hard copy devices available from independent peripherals manufacturers that have interfaces that adhere to that convention.

The Teletype Corporation has a similar offering with its Data Speed 40. At the present time, there is some controversy as to whether or not that company will be able to offer that equipment for direct connection to computer systems. The entire configuration certainly is available to connect into a data set or modem and can be used on a stand-alone basis to communicate with computers on a timeshare basis.

In addition to the character-generating CRTs, there are several found on a special application basis where they can be programmed to generate drawings, designs, and diagrams with the aid of a computer. Light pens have been used with these CRTs for motivating the program within the computer system, which modifies the images displayed upon them. Typical functions of the light pen are to expand a particular area of the diagram, to delete certain portions of a drawing, and to enter changes. A system of this type finds excellent application in automobile body styling. Another application is in circuit design, where the computer system can be programmed to generate a circuit diagram based upon the design information entered by the circuit design engineer.

XVI. Printing Devices

In applications such as process control or data acquisition, there is usually no requirement for printing large amounts of information. As minicomputers have become accepted in applications where computation and report generation is a major requirement, there has grown a need for medium-speed line printers. This need has been satisfied by the larger mini manufacturers with their own line of line-printing equipment. The smaller manufacturers rely upon independent peripherals manufacturers. It is, however, the usual practice of the mini manufacturers to provide service on the complete system that they provide. If a user were to purchase a minicomputer system from a small manufacturer, it might turn out that the line printer was bought from a second supplier and a disk from a third supplier. The original vendor usually accepts the responsibility to service the complete set of equipment that has been provided.

The line printers found with minicomputers are those that have less performance capability than those found in EDP centers. Their speeds range from 200 to 400 lines per minute, and in many instances they are limited to 80 columns of print characters. In heavily report-oriented installations, the typical EDP printer, having a capacity for 132 columns and printing at rates up to 1200 lines per minute, can be found.

The earlier line printers used either the drum or the print chain design. In this design, the drum or the print chain contains a string of characters, which rotate behind the paper on which the character is to be printed. When the character to be printed appears behind the correct print position, a hammer strikes, causing the character to be printed. Because of the mechanical precision required to manufacture them, the costs of these printers are high when considering the cost of minicomputers.

There are two other principles in character formation that have reduced the cost of building line printers. One method uses the ink-squirting principle, which draws the character on the paper. The print head moves across the paper and a number of ink-squirting jets issue ink dots, which merge into a reasonably well-formed character. The ink-squirting printers operate very quietly and can be programmed to write characters of varying sizes. Their main disadvantage is that they can make only a single copy per printing. Ink-squirting devices appeared to have a great deal of attention in the early 1970s, but it appears that problems in maintenance have caused interest in them to decline.

Several vendors have now introduced line printers that use a dot matrix printing head to produce the characters. Line printers using this principle have the advantage of a simpler mechanism and can be produced at lower cost. They can also make multiple copies and operate more quietly than typical impact printers. They are priced in a range between conventionally designed line printers and fast typewriters, and they have the advantage of printing at speeds comparable to the lower priced drum or chain printers. Their only disadvantage is that a dot matrix is used to form the character, which is not as artistically well formed as one that is engraved on a drum or on a print chain.

XVII. Data Entry I/O

Card equipment has been infrequently associated with minicomputers. The chief reason is the high cost of card readers and card punches. It still costs roughly $20,000 to $25,000 to produce a reliable card punch, and it is difficult to justify that price with a minicomputer system that may cost under $10,000. It is still reasonably common to find card readers that will read cards at approximately 200 to 300 cards per minute that are designed for the minicomputer industry. There are very many situations where cards are a very useful medium for entering information to a computer system. For example, in the electric and gas utilities, it is common to have a meter reader write information on cards and to read these cards by an optical sensing card reader. A typical card reader attached to a minicomputer system will sell in the neighborhood of $4000 to $5000 and it will read cards at 200 to 400 cards per minute.

Card punches are still very rarely to be found with minicomputer systems. In some instances, the minicomputer vendors have modified keypunch machines and used them as satellites to minicomputer systems. A keypunch operator can still generate cards in the normal fashion when the computer is not using it. Under minicomputer control, it will punch at the rate of 35 cards per minute. Such a solution permits the generation of punched cards along with the flexibility of having a keypunch system located within the data entry center.

If a large volume of card output is required, the normal procedure is to supply the minicomputer system with a magnetic tape drive, which will write data compatible with the larger EDP center. The EDP center will then read the information from its tape drive and generate the punched cards.

XVIII. Magnetic Tapes

As with line printers, magnetic tapes commonly found in business centers were first interfaced to minicomputers. As the demand and the minicomputer industry grew, many of the minicomputer vendors and peripherals suppliers introduced tape drives that were somewhat lower in performance but substantially lower in cost. With most minicomputer systems, one normally finds a tension arm drive rather than the higher cost vacuum-chamber device. The disadvantage of the tension arm drive is that transfer rates are limited to approximately 45 in./sec, while vacuum-chamber drives have tape transfer speeds of up to 150 in./sec.

Although there are substantial differences in transfer rates among tape drives, the formatting of the data on the magnetic tape has been well standardized because of the wide acceptance of tape drives in business centers. In many instances, minicomputers are used as data collection stations in place of the larger systems.

Early versions of recording formats store data at densities of 200, 556, or 800 frames per inch and in a seven-track recording format. The seven-bit recording format was used at the time when the six-bit BCD character was in current use. At the present time, recording formats of 800 and 1600 frames per inch are used and nine tracks of information are stored simultaneously. The nine-track format stores an eight-bit extended BCD character and a parity bit. The main difference between IBM's and most other vendors' formats is that ASCII formatting is the standard used by most minicomputer vendors. Because ASCII and EBCDIC formats are very similar, a simple translation routine is required to interpret EBCDIC format on a mini and vice versa.

Data are written on magnetic tape in physical record formats that are separated by an end-of-record gap that is 1/2-in. in length. The purpose of the gap is to give the tape drive unit time to decelerate, stop, and accelerate to the normal tape movement speed. The size of the physical record may vary; it can be as short as one character of information or as long as the total memory that is installed in the computer. A 32,768 word computer would have the ability to store 65,536 bytes of information in a single record. Because information is read from or written on a tape in record format, it is not possible to select information on a character basis within that record.

Another disadvantage of tape drives is that the data are in a sequential format, so that the tape drive must pass over all preceding records to get to records at any point. If information is to be added, the entire tape is usually rewritten. The update procedure is usually a long one and is not done on a frequent basis. The update information is usually stored on a separate tape and a sort program is used to insert all the new records and generate a new set of tapes, which contain the updated information.

XIX. Tape Cassettes

As an intermediate medium between the storage capability of the paper tape system and large magnetic tape drives, magnetic tape cassettes have appeared. Physically, the tape cassette is identical to the cassette found in many portable tape recorders. The differences are that the tape recording cassettes are checked out to a much higher quality level than those used for voice recorders.

The main disadvantage of cassette systems is that there has not been a uniform recording format across the industry. In a typical recording format, data are stored on two different tracks, in serial format. One format used in a commercially offered system, stores bits at 0.01456-in. intervals, separates 8-bit groups with 0.00182-in. intervals, and separates records of 8-bit groups with a record gap of 0.25 in.

Files are also terminated with end-of-file-mark characters. The point should be made that this is an example of a recording format that is not an industry standard. Unfortunately, each cassette tape manufacturer has a somewhat different version of a recording format. A typical quantity of storage on a cassette system is 180,000 characters per track or 360,000 characters per drive. A write–protect feature is implemented by breaking a small plastic tab at each end of the tape cassette.

Typical configurations of cassette systems utilize two or more drives on a functional unit. In such a configuration, one unit may serve as an input unit and the other may be an output system as in a paper tape reader/punch system.

While cassettes have been available for a substantial number of years, they have not achieved the popularity of floppy disk systems. One reason may be that there is no uniform recording format among manufacturers of cassette systems. Another reason may be that they are not as convenient as the floppy disks.

XX. Disk Systems

Early configurations of minicomputer systems implemented drums or head-per-track disk systems. A principal reason for the use of disks or drums was that in early process-control applications, these devices served as a lower-cost extension of computer memory. Moving-head devices were very slow, requiring as much as 1 sec to access data from tracks located at opposite sides of the storage medium. Where data storage was a prime requirement, the slow access rate was not such a problem.

In today's technology, drums and head-per-track disk devices are very infrequently used. Moving-head disks operate at 3600 rpm, providing access to every data word on a track within 1/60 sec. Additional delays are encountered as the read–write heads are positioned to other tracks. Positioning times range from 10 msec between adjacent tracks to about 75 msec from the innermost to the outermost track.

Currently, the most popular types of disk drives are those which, within a single drive mechanism, will contain a single fixed disk and a removable cartridge disk. The drive will contain four read–write heads, which read information from four tracks without requiring repositioning. In the more up-to-date versions, the data on a "cylinder" or four tracks can be continuously read in four revolutions of the disk, i.e., there is no delay as the control logic senses the end of one track to read from the next one.

Drive mechanisms capable of storing much larger quantities of information are also available for attachment to minicomputer systems. A typical large drive has as many as ten platters or 20 surfaces. In some instances, such drives are provided with two read–write mechanisms, permitting one to be reading or writing while the other one is seeking a new track position. Such drives have a capacity of up to 80 million bytes of information.

XXI. Floppy Disks

The floppy disk is a low-cost storage medium that made its appearance in the early 1970s. The recording medium is a circular sheet of mylar plastic coated with magnetic iron oxide. The read–write head rides in contact with the surface of the disk, but otherwise the mechanism is similar to a disk drive.

The major advantage of the floppy disk is that a de facto standard for recording formats has been set by IBM with its Model 3740 floppy unit. Minicomputer systems can collect and store information on such a drive and have it available for reading by the larger business data processing centers. In the IBM configuration, the diameter of the disk storage unit is 7.88 in. and the unit permanently enclosed in an 8×8 in. plastic or paper envelope. Access to the recording surface is through a read–write head, through a slot that extends radially from the center of the envelope to its edge. Present recording conventions use 77 tracks and store 40,000 bits per track. In the IBM formating convention, a track is divided into 26 addressable sectors, each of which contains 128 bytes. The number of tracks used is 73, which reduces the storage capacity to 424,944 data bytes. The diskette operates on 360 rpm, and a data transfer rate of 250,000 bits per second is achieved. There is a recording head positioning time of 35 msec.

A common objection to the floppy disk is that the read–write head rides on the surface of the disk, which can cause wear of the surface. Interest in floppy disk systems is motivated by its convenience as a recording medium, its ability to transfer information from one computer to another, and its low cost.

XXII. Process Control Peripherals

Analog-to-digital devices are offered to accept data at a high-level (0–10 V) or low-level (normally 0–100 mV) basis. The low-level converters operate with an amplifier within the device so that the actual digitizing function takes place at higher logic levels. These instruments frequently have automatic ranging capabilities and can transmit the appropriate bits to the minicomputer for correct data scaling. Some instruments automatically integrate the signal over a short period of time to filter out the effects induced by other electrical signals. A frequent practice is to integrate over a 1/60-sec period, which cancels out signals induced by 60-Hz power lines and their harmonics. Another technique is to use a flying capacitor. The capacitor tracks the external signal and, at sampling time, it is disconnected from the signal so that only a steady voltage is seen by the converter. Other techniques use limiting band-pass filters, which permit signals that change only within a given range to be accepted. Where these techniques are insufficient, the only route may be to supply amplifiers for each signal in the expectation that the induced signal may be a smaller proportion of the true signal (see Fig. 13).

Low-level signals are sampled at rates up to about 1000 points per

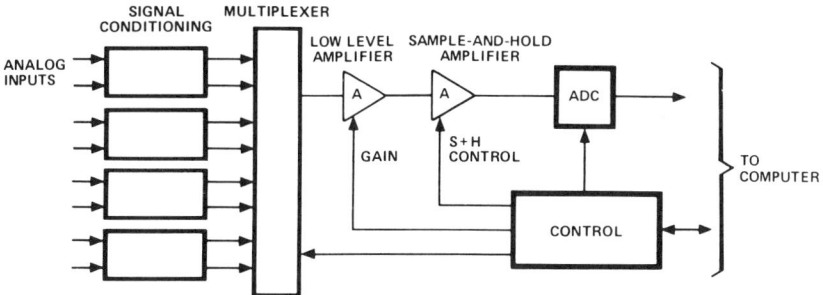

Fig. 13. Typical analog input circuit.

second. At high levels, instruments are available to scan at rates of up to 250,000 points per second. The resolution ranges from 14 bits at low speeds to 8 bits at high speeds. For simultaneous sampling, a multiplexer may be provided with individual sample-and-hold amplifiers at each channel. After the simultaneous multichannel sample, the multiplexer can sample each amplifier at its own sampling rate. These techniques are used frequently in vibration analysis studies.

Two types of analog-to-digital (A/D) converting elements are used to digitize the analog signal. In one, an analog signal increasing at a fixed rate is compared against the signal presented to the A/D converter. The length of time it takes to equate the internal signal to the external one is counted very precisely. The count is then presented to the computer as a measure of the amplitude of that signal. This is known as a ramp technique. A second type of converter is a successive approximation device (Fig. 14). In this design,

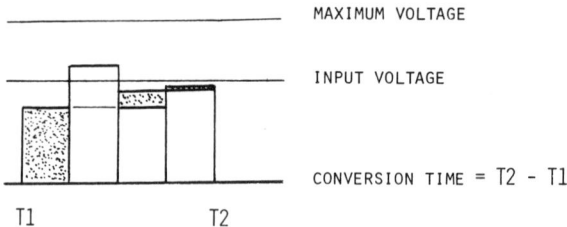

FIG. 14. Successive approximation A/D conversion.

the converter successively generates a signal approximately equal to the external signal by attempting to add voltages in binary fractions. Each step is an attempt to add an increment that is 50% of the difference between the external signal and the previously generated signal. If it is less, the first increment is retained and an attempt is made to add a second increment that is 50% of the remainder, or 25% of the maximum range. The next increment to be considered is half the previous one, which is $12\frac{1}{2}$% of the full range. The procedure continues until the resolution of the A/D system is exhausted.

Digital status is commonly sensed by contact closure sensing circuits. Typically, a 12-V signal is provided by the circuit and closure is sensed by the presence of the voltage at the input portion of the device. Contact closure sensing is used to sense the binary status of a device, such as whether a door is opened or closed or whether a valve is fully opened or fully closed. Digital inputs are readily read into the computer as full words, and the individual bits must be isolated by a program. The programming of digital outputs is handled in exactly the same manner in a computer program. A bit is set in the proper position in a computer word, which is then transmitted to the external device. In general, the function of a digital output is to close a contact or to complete a circuit that will control a peripheral device. Examples are turning on a light, an alarm, or a motor.

XXIII. IEEE Standard Digital Interface

In December 1974, the IEEE Standards Board accepted a standard for a common interfacing protocol among instruments, computers, calculators, and other accessories. The purpose of this protocol was to permit easy

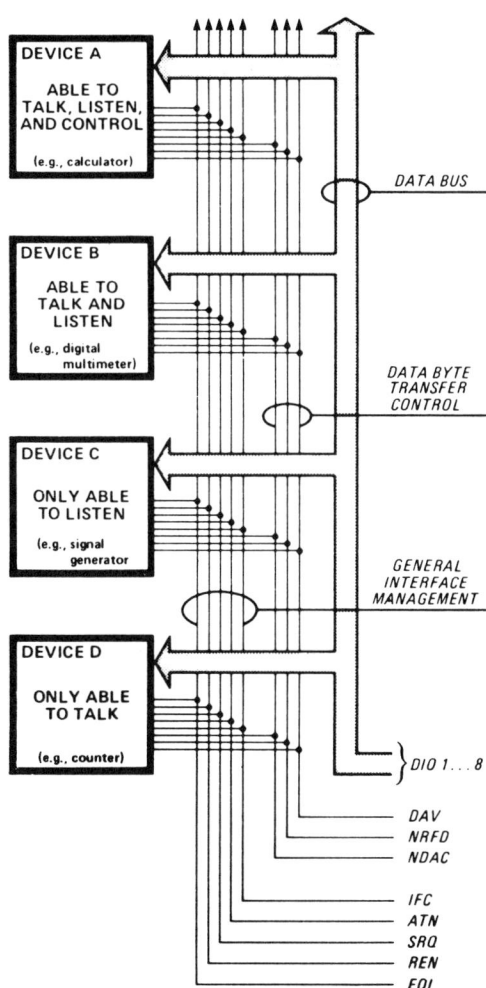

Fig. 15. IEEE standard interface.

interfacing among different manufacturers of computers, peripherals, and electronic instruments.

The standard defines a 16-conductor communications link, in which eight conductors are used for the transmission of data and the eight remaining conductors for signaling each device in accordance with its capabilities. Devices are defined as those able to issue command signals, i.e., controllers; those able to accept information, i.e., listeners; those able to transmit information, i.e., talkers (see Fig. 15). Specific designation can be given to a computer or a calculator on a system whereby it becomes the system controller. The IEEE standard is defined for a length of up to 20 m. Greater distances can be accommodated by use of a parallel-to-serial converter, which translates the eight-bit parallel signal into one that is bit-serial. This communications link is defined for up to 1000 m. The serial terminal unit can be interfaced to the telephone network and communications can be handled on a worldwide basis.

XXIV. Data Set Interfaces or Modems

Minicomputers have been well accepted in applications where they act as satellites to larger computers and also as small stand-alone timeshare systems. Most minicomputer vendors provide interface equipment that will connect to a conventional data set or acoustic coupler. The interface accepts a character in parallel format and translates it to bit serial format for the data set. The data set translates the bit serial stream to the analog mode of the telephone network. At the other end, the reverse translation takes place and the information is presented on a terminal such as a teletypewriter, or it may be accepted by another computer system. In the asynchronous mode of transmission, characters are sent as individual clusters of bits, with a bit added to indicate start-of-character and one or two added to indicate stop-of-character. For each 8-bit character transmitted, this translates to a 10- or an 11-bit format on the communications line.

The other transmission mode is synchronous. Characters are sent in groups and do not require the start or stop bits. This method of transmission is normally used between computers and high-speed terminals. At least two synchronous characters are first sent by the sending device for synchronization. The receiving device accepts the stream of bits and groups them, eight at a time, to reconstitute the message. Check characters, which are computed from the total number of bits in the previous characters, are usually sent at the end of the message. Synchronous communications are usually used between computers and higher-speed terminals such as CRTs.

XXV. Software

In the early 1960s there was a large number of small-scale computers used primarily for data acquisition and process control applications. These computers were the forerunners of the minicomputers as we now know them. They were characterized by scientific instruction sets, they embodied the latest technology in their construction and design, and they were provided with a minimum of software. In many instances the developers of programs for the applications of these computers were required to program them in machine binary programming code. This was extremely difficult, prone to error, and very costly. At the present time the requirement for software is such that no new minicomputer would be commercially successful without it. Because software represents a very large investment in a minicomputer system, each time a vendor introduces a new line of minicomputers, he maintains compatibility with the instruction set in the predecessor computer. This is the same procedure that has been followed by IBM in its 360 and 370 series of computers.

Within a minimum software package, a minicomputer system must include at least an assembler and one or more compilers. Assemblers are provided with a variety of features to aid the computer programmer. The simplest ones translate instructions into the bit pattern required by the computer on a one-for-one basis. Added features are control functions, such as selecting a new page for listing, generating a sequence of characters from a single line, and translating decimal to octal constants. More sophisticated assemblers have a macro capability that allows a user to label a sequence of instructions, and the assembler will insert that sequence each time the label is used. Assemblers generate code in absolute or relocatable format. In the former, a very simple loader may be used. In the latter, the loader must recompute all operand addresses. The advantage of the relocatable procedure is that a number of programs can be linked together at load time and need not be reassembled each time a modification is necessary in any one program.

The advantage of the assembler is that the computer programmer has full control over every individual attribute of the computer. A well-experienced programmer can usually generate code in less space and have it execute faster than code that is generated by a higher-level language compiler. The disadvantage is that he will be required to do much more coding than would be required with a higher-level language.

In broadening their marketing base, most minicomputer manufacturers offer several forms of higher-level compilers. The most common among these is FORTRAN. The initial acceptance of FORTRAN is based on the fact that the minicomputer was primarily used in the scientific community.

ALGOL is also used as a scientific language and is more frequently used in Europe. It has a great deal of similarity to FORTRAN and is somewhat more concise. However, so many programs have been written in FORTRAN and there is such a large reserve of FORTRAN programmers that it is doubtful if it will ever be supplanted as a commonly used scientific language. Both FORTRAN and ALGOL are compilers and they translate source programs into a machine-language program.

Most minicomputer systems are provided with a version of FORTRAN slightly different from the FORTRAN IV that has been developed by IBM. Usually they adhere to ANSI Standard FORTRAN IV, and in many cases

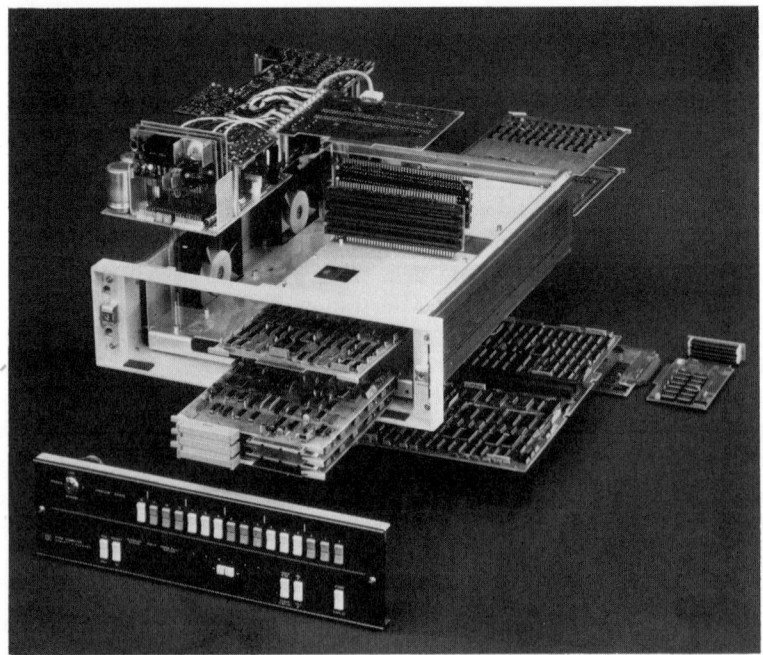

FIG. 16. Explosion photograph of Hewlett-Packard 2105 minicomputer, showing modularity of components. The power module (with optional power fail recovery system circuit boards) is at upper left. At right rear are two representative I/O interface boards, in place for insertion. The circuit board below the mainframe is the processor; optional firmware (ROM's) are to its right. At lower left is the control panel. Behind it, at lower right of the mainframe, are two modular memory boards and the memory controller. Above these, half inserted in the space reserved for it, is the dual-channel port controller (DMA) plug-in circuit board. In the center of the mainframe are the printed circuit boards, which serve the purpose of older-style wired backplanes.

they have some enhancements and some exceptions. These enhancements and exceptions are the sources of incompatibilities between compilers of various manufacturers. These variations are minor compared to the differences in the fundamental instruction sets of the computers provided by them.

As the cost of memories has declined, the use of FORTRAN and, to a lesser extent, ALGOL for process control and data acquisition applications has increased. Commands have been specified in FORTRAN for communicating with standard data processing peripherals such as line printers, tape drives, card readers, and consoles. The Instrument Society of America has sponsored a committee that has held regular meetings at Purdue University to develop a set of calling sequences for process control peripherals. The committee issued its report in 1972 as a standard for industrial computer system FORTRAN procedures. An example of a calling sequence for program functions is

$$\text{CALL START}(i, j, k, m)$$

This is the specified statement for beginning the execution of a program after a specified time delay. Another example is

$$\text{CALL TRNON}(i, j, m)$$

which causes a designated program (i) to be executed at a specific time of day (j). Another function is

$$\text{CALL WAIT}(j, k, m)$$

which specifies a time delay before a program continues execution.

The reasonable compatibility among FORTRAN compilers, as well as a programming standard for process control devices, gives the user more flexibility in his choice of minicomputer systems. In some instances, he may be required to buy additional memory in order to develop his program in FORTRAN. In the higher-level languages he retains the option of being able to change vendors readily, yet retain his investment in the programs that he has developed.

The BASIC interpreter is probably the next most popular higher-level language to be found with minicomputers. BASIC is implemented in two ways: as the major language used with timeshare systems and as a language used on a stand-alone basis. BASIC is an algebraic language as is FORTRAN, but in most implementations of BASIC the interpreter operates in conversational mode. It is very easy to learn, and because of its conversational implementation, errors can be corrected each time a statement is added to the program. In most implementations of BASIC, an area of memory is allocated to the interpreter and the remaining portion can be dedicated to individual user application programs. The main disadvantage of BASIC is that it operates more slowly than any machine language program. This is because BASIC, by convention, operates in interpreter mode, and each time it executes a statement it goes through the procedure of interpreting that statement, translating it to machine language, and executing that portion in machine language. Another disadvantage of BASIC is that it is in the memory of the computer at all times during execution and occupies a segment of that memory. All of the memory occupied by BASIC is not a complete loss to the user because a number of frequently used subroutines, such as trigonometric, exponential, and logarithmic functions, are included in this implementation. The outstanding advantage of BASIC is that it is very convenient to develop a useful program.

With the acceptance of large minicomputers, COBOL and RPG have also become readily available. COBOL initially was a requirement of the Department of Defense in the early 1960s and has now become probably the most popular business language. The specifications for COBOL are drafted by an American National Standard committee and specifications have been published in 1963, 1968, and 1974.

Because of IBM's dominance in the business data processing field, its version of COBOL and extensions is a de facto standard. The number of extensions or enhancements available in COBOL and the number of exceptions is probably greater in minicomputer COBOL than in minicomputer FORTRAN implementations. The important consideration here is the fact that COBOL compilers require much more memory than FORTRAN compilers. In order to implement them in smaller memory areas greater concessions have to be made in the implementation of the language.

A very favorable attribute of COBOL is that it is almost a self-documenting language. The implementation of the command structure requires much more description than the terse statements defining computational commands in FORTRAN. For example, the addition of two numbers to a sum is implemented in COBOL as follows:

ADD A TO B AND STORE IN C

COBOL possesses a much better capability than FORTRAN or BASIC in formatting reports. This is implemented in its picture command, which presents the location of data on a page, with description characters such as dollar signs and decimal points, which are automatically inserted at the correct locations.

The next most frequently used language in business applications is RPG (Report Program Generator). This language was originally developed by IBM and was in use substantially before minicomputers appeared. It was initially implemented on smaller computers and does not have the memory

FIG. 17. Typical family of minicomputers: HP 2113 at upper left; HP 2109 at right; and HP 2105 at lower left.

demands of COBOL. The main use of the language is to permit input and output of data and convenient data manipulation within memory. Several minicomputer vendors have implemented the RPG compiler on their larger line of minicomputers. These include Digital Equipment's PDP-11/45 and PDP 11/70, Data General's Eclipse computers, Hewlett-Packard's HP 3000.

XXVI. Disk Operating Systems

The availability of lower cost disks permitted the development of various forms of disk operating systems with minicomputers. Early versions of disk operating systems permitted libraries to be stored on the disk for rapid retrieval by the minicomputer. These disk storage files basically replaced paper tape or magnetic tape storage media. As the sophistication of disk operating systems increased, schedulers were added so that programs could automatically be retrieved and executed from the disk. The presence of disk systems has added a whole new dimension to the repertoire of software available with minis. Editors, compilers, and loaders have been modified to take advantage of the disk operating system environment. Typical disk operating systems now have the ability to take a source program, compile it, link the object code with required library subroutines already available on the disk, and produce a stand-alone usable program. Editors with DOS systems can be substantially more sophisticated than simple stand-alone editors. Where simple editors are limited to making additions or deletions on a character or line basis, disk operating system editors can do a complete file search to replace a given sequence of characters with another given sequence. The compilers installed in disk operating systems provide the capability to link into the input/output processing capability of the DOS system. This gives a software developer the privilege of spending more time on the solution of his problem and less time on the problem of input/output functions.

Most of the current minicomputer systems come supplied with a multi-programming operating system that has the capability of handling a very large number of jobs simultaneously. The operating system gives control to the program of the highest priority; when it goes into an input/output mode, which means the computer function for that program is suspended until the I/O function is completed, the multiprogramming system will automatically begin execution on the next highest priority program. This hierarchy continues until the CPU is fully occupied in execute mode and the peripheral devices are fully occupied in doing input/output functions. There is a problem in this arrangement in that certain peripherals may be used for short periods of time by programs of different priorities. This would result in a report that is interlaced with results or intermediate results from several programs. Provision is made to permit a program to lock onto a specific peripheral device in order to use it exclusively until that program has completed its execution. In the intervening time, programs that require the use of that device send their results to a file on the disk or magnetic tape. When the device, such as a line printer, is free again, these results are spooled out and listed. In this manner, a full uninterrupted report is presented as though that program had complete use of that particular peripheral device.

Most minicomputer systems also possess a real time capability in addition to a multiprogramming capability within their disk operating systems. The real time attribute permits external devices to randomly interrupt the computer, and permits the computer to exercise control of external processes while handling conventional batch processing jobs in a background mode. Examples of multiprogramming executive systems are Digital Equipment's RSX (for Real Time Systems Executive), Data General's MRDOS (for Multi-Programming Real Time Disc Operating System), and Hewlett-Packard's RTE (for Real Time Executive).

As an extension to their real time operating system, the more widely used minicomputers are provided with various forms of file management or data base management systems. File managers are used for convenient storage and rapid access of files of data, such as programs and computed results.

In several instances, a true data base management system is available that is not only a means of storing data in orderly fashion but also a means of rapidly retrieving related information from that data base. Examples of

FIG. 18. Mega-mini computer system with 512k (524,288) bytes of computer memory, three disk drives containing 47 megabytes each, three CRT displays, a 1250 lpm line printer, and a 500 cpm card reader. (Hewlett-Packard 3000 Series II, Model 9.)

data base management systems are TOTAL, currently available from a number of minicomputer vendors, INFOS, available from Data General, and IMAGE, available from Hewlett-Packard. The data base management system organizes the data in such a manner that related items may be retrieved rapidly without repeated searches. A common technique is to store data in a keyed format so that each record has a key that points to the next record with a similar attribute. In an information retrieval procedure, the computer system is not required to read every record to determine if a particular attribute is present. The inventory control manager therefore could retrieve all the parts that were required to be included in a given end product so long as all of the parts were keyed to that end product. By multiply keying these parts to different products, it becomes very convenient to look at alternate product mixes with a given set of parts.

Obviously, data bases can be structured about population trends, including such information as incomes, net worth, and credit ratings. Unfortunately such data bases have been viewed with a great deal of apprehension in the media; however, there is a great deal of concern to ensure that such data bases have adequate protection from unauthorized use. This is done by use of passwords and by access only to terminals that require a key to be enabled.

More-advanced versions of time-share systems have been installed as multiterminal operating systems, which provide more than a simple time-share function. These minicomputer systems provide the user access to other computers by transmitting files to them, which may enter their job streams. The job stream is acted upon by the large central computer in the same manner as if it had been entered by the local card reader. Because the job stream looks like any other file to the local time-share system, jobs can be constructed using any of the large computer languages (COBOL, RPG, PL/1 etc.), which are not available on the local system. In effect, the time-share system becomes a multiple RJE station in addition to having its own computational capability.

Other software commonly provided with minicomputers includes debugging aids, which consist of memory dumps, trace routines, and break-point routines. Trace routines follow the course of the computation and print the result for each step. Break-point routines halt the computer whenever it chooses to access an instruction or a data item in a predetermined word of memory. At that point, the user can start tracing the program or modify the instruction or data item. Most large minicomputer vendors maintain user libraries and user organizations. Both are beneficial to the user because they provide a source of programs that may be very similar to those he would otherwise have to develop himself. The user organization also provides a forum for the users of computing equipment to discuss similar problems and to exert substantial influence on the vendor to correct problem areas and to cause him to design new equipment that would be useful in a wider range of application.

Simulators, cross assemblers, and cross compilers are software offerings that are frequently provided with minicomputers. This is because large EDP centers include higher-speed line printers, tape drives, and card processing equipment, which are very useful for program development. The use of the large system avoids the expense of buying a program developmental system. On the other hand, the use of simulators is less efficient than using the original minicomputer in its native mode.

XXVII. Growth of Minicomputer Applications

Since the inception of minicomputers around 1965, the rate of minicomputer installations has grown at a pace of 25 to 30% per year. The value of these installations is still only about 15% of the larger EDP market, but the growth rate for minis is substantially greater. During this period, the average size per installation has gradually increased, reflecting the fact that larger quantities of peripherals are used with later minicomputer systems.

Minicomputers are sold on an outright purchase or on a fully paid out lease basis. Since minicomputers are sold rather than rented, there is a greater incentive for technological improvement. Because the mini manufacturer is not concerned about returns of equipment containing the previous year's technology, he is more interested in updating it to encourage customers to purchase the latest and most modern equipment. Although the manufacturers maintain a consistency between instruction sets and I/O conventions among the various editions of their product, substantial improvements such as memory capacities and speeds appear in two or three year intervals.

The concept of a family of minicomputers has been used in the minicomputer industry just as the concept of a family of computers has been used by the EDP industry. This permits the user to select the best match to his application in terms of cost and speed and permits easy upgrading because

FIG. 19. Solid-state 4096 bit RAM's are used to assemble a printed circuit memory board containing 65,536 bytes of memory. (From HP 3000 Series II computer.)

of consistent programming software and interfacing hardware over a relatively broad product line. Because minicomputers are sold rather than rented, they are built with a strong consideration for modularity. Minicomputer parts are designed in such a fashion that it is very easy to add components. Memories of up to 65,536 bytes, which are available on single 15 × 15 in. printed circuit boards, can be added by plugging them into a slot in the chassis of the computer. Additional peripheral devices can be added by plugging the interface controller, also now contained on a single printed circuit board, into the I/O chassis of the computer and connecting the peripheral device to the interface board by cable. Before the advent of LSI technology, as it is today, such controllers would often require a full cabinet themselves. Miniaturization allows maintenance to be achieved on a board exchange basis so that there is no need for customer engineers to decode the logic of an interface board and determine which component has failed. The faulty board is then sent back to the factory, where a computerized fault detection hardware and software system is used to determine where the fault exists on the board.

New peripheral devices that have made an appearance in the last two or three years are the floppy disk and the "smart" terminal. Both devices enhance the minicomputer's utility to act as a stand-alone data entry system, whereby the minicomputer can collect data in a form that is readily transferable to the corporate business data processing center. The smart terminal, being microprocessor driven, can be programmed to perform a very sophisticated sequence of checking functions before accepting the data that an operator wishes to transmit to a minicomputer system. The floppy disk has become a convenient medium for collecting information on a small system and transferring it to the large EDP center. This is because an industry-accepted recording format has been established by the leading EDP manufacturer (IBM) and has been accepted by the mini vendors.

The short-term outlook for minicomputers is very favorable, with a continuing growth in applications. The microcomputer is expected to absorb some of the applications now being served by the smaller minicomputer systems, but in most cases the microcomputer is finding application in completely new areas. The conventional minicomputer system has grown much larger and will become a more significant factor in the competition for smaller business data processing applications.

Glossary

Because of the unique terms used in the literature on minicomputers, a glossary is presented defining the most frequently used terms. Several of the references in the bibliography were used as a source for the terms selected, which are, in the author's opinion, those most commonly used.

Absolute: Pertaining to an address fully defined by a memory address number, or to a program that contains such addresses (as opposed to one containing symbolic addresses).

Access time: The time interval between the request for information and when it is available; usually the time required to transmit a word from computer memory into the CPU.

Accumulator: A register in which numbers are totaled, manipulated, or temporarily stored for transfers to and from memory or external devices.

Acoustic coupler: A modem designed for use with the telephone system whereby an acoustic connection is provided between the telephone handset and the data terminal. The coupler translates the tone signals used in the telephone system to the stream of pulses that are interpreted as characters by the terminal.

Address: A number (noun) that identifies one location in memory; also (verb) the process of directing the computer to read a specified memory location (synonymous with "reference").

Address modification: A programming technique of changing the address referred to by a memory reference instruction, so that each time that particular instruction is executed, it will affect a different memory location.

Address word: A computer word that contains only the address of a memory location.

Algorithm: A term used to describe the logic by which a given result is obtained.

Alphanumeric code: A code whose code set consists of letters, digits, and special characters.

Alter: A modification of the contents of an accumulator.

ALU: Arithmetic Logic Unit, a computational subsystem that performs the mathematical operations of a digital system.

Analog: Pertaining to information that can have continuously variable values, as opposed to digital information, which can be varied in degrees no smaller than the value of the least significant digit.

And: A logical operation in which the resultant quantity (or signal) is true if all of the input values are true, and is false if at least one of the input values is false.

ANSI: American National Standards Institute.

Arithmetic logic: The circuitry involved in manipulating the information contained in a computer's accumulators.

Arithmetic operation: A mathematical operation involving fundamental arithmetic (addition, subtraction, multiplication, division), specifically excluding logical and shifting operations.

ASCII: American National Standard Code for Information Interchange, which uses 7-bit characters (8 bits including parity check), for information interchange among data processing systems, communication systems, and associated equipment. The ASCII set consists of control characters and graphic characters. Also called USASCII.

ASR (Automatic Send Receive): A teletypewriter unit with keyboard, paper tape reader, and paper tape punch. Messages may be prepared off-line on paper tape for subsequent automatic transmission.

Assembler: A computer program that converts a program prepared in symbolic form (i.e., using defined symbols and mnemonics to represent instructions, addresses, etc.) to binary machine language.

Asynchronous operation: Usually refers to a computer system in which the speed of operation is not related to the CPU frequency or cycle time.

Asynchronous transmission: Transmission in which the time interval between each character or word of data can vary arbitrarily. Each character of data consists of information bits preceded by a start bit (zero condition) and followed by a stop bit (one condition).

Attenuation: The difference between transmitted and received power due to transmission loss through equipment, lines, or other communications devices.

Auto answer: Capability of the computer to respond automatically to an incoming call initiated by dial equipment.

Automatic calling unit (ACU): A dialing device supplied by the communication common carriers, which permits a business machine automatically to dial calls over the communication networks.

Background processing: Low-priority processing permitted to take place when no higher priority real-time entries are being handled by a system. A batch processing job, such as payroll, might be treated as background processing subject to interruption on receipt of an inquiry from a terminal.

Band: A well-defined range of wavelengths, frequencies, or energies of optical, electrical, or acoustical radiation.

Bandwidth: The difference, expressed in hertz, between the highest and lowest frequencies of a band.

Base: The quantity of different digits used in a particular numbering system. The base in the binary numbering system is 2; thus there are two digits (0 and 1). In the decimal system (base 10), there are ten digits (0 through 9).

Base page: The lowest numbered page of a computer's memory. It can be directly addressed from any other page.

Batch processing: A term usually used with office-type computers that process each job one at a time until it is completed. Contrasted with multiprogramming systems where a large number of jobs may be processed concurrently with the completion of each depending on its priority and availability of peripheral devices.

Baud (derivative of Baudot): A unit of signaling speed in data transmission. In an equal-length code, one baud corresponds to a rate of one signal element per second. Thus, with a duration of the signal element of 20 msec, the modulation rate is 50 bauds (per second). Baud should not be related to bits per second as a general rule; however, there are times when they are synonymous.

Baudot code: A code for the transmission of data in which five bits represent one character. Derived by M. Baudot in 1872, and still used internationally for five-channel paper tape code, where each telegraph signal consists of five pulses preceded by a start pulse and succeeded by a stop pulse. The name is usually applied to the code used in many teleprinter systems, which was first used by Murray, a contemporary of Baudot.

Benchmark: A program, usually written in a higher-level language such as FORTRAN, ALGOL, or COBOL, which is used to test the relative speeds of a number of minicomputers.

Binary: Denoting the numbering system based on the radix 2. Binary digits are restricted to the values 0 and 1. The relationship between binary and decimal numbers is illustrated as follows:

Decimal	Binary
0	0
1	1
2	10
3	11
4	100
5	101
6	110
7	111
8	1000
9	1001

BASIC CONCEPTS OF MINICOMPUTERS 333

Binary-coded decimal (bcd): A coding method for representing each decimal digit (0-9) by specific combinations of four bits. For example, the 8-4-2-1 bcd code commonly used with computers represents 1 as 0001 and 9 as 1001.

Binary point: The fractional dividing point of a binary numeral; equivalent to decimal point in the decimal numbering system.

Binary program: A program (or its recorded form) in which all information is in binary machine language.

Binary synchronous communications (BSC): A transmission line discipline developed by IBM. It uses a defined set of control characters and control character sequences, for synchronized transmission of binary coded data between stations in a data communications system.

Bistable: Pertaining to an electronic circuit having two stable states, controllable by external switching signals; analogous to an on–off switch.

Bit: Contraction of "binary digit," the smallest unit of information in a binary system. The digit can have one of only two values, 0 or 1.

Bit density: A physical specification referring to the number of bits that can be recorded per unit of length or area.

Bit rate: Speed at which bits are transmitted usually expressed in bits per second (bps), kilobits per second (kb/sec), or megabits per second (Mb/sec).

Bit-serial: One bit at a time, as opposed to bit-parallel, in which all bits of a character can be handled simultaneously.

Block: A group of data words, records, or characters handled as a single unit.

Block diagram: A diagram of a computer or a program in which the principal parts are represented by suitable geometrical figures to show the basic functions and the functional relationships among the parts.

Bootstrap: A technique or device designed to bring itself into a desired state by means of its own action, e.g., a machine routine whose first few instructions are sufficient to bring the rest of itself into the computer from an input device.

Branch: To transfer control of a program to another location in memory on the basis of a result or the value of an item of data.

Break: Telegraph term to denote interruption of a sender.

Breakpoint: A point in a program at which normal processing stops; usually used as an aid in determining intermediate results in program development.

Broadband: Communication channel having a bandwidth greater than a voice-grade channel, and therefore capable of higher-speed data transmission.

Buffer: A storage device used to compensate for a difference in data rate flow or time of occurrence of events when transmitting data from one device to another.

Bus: A major electrical path connecting two or more electrical circuits.

Byte: A sequence of adjacent binary digits operated upon as a unit and usually shorter than a word. Usually refers to eight adjacent bits that may represent an ASCII or EBCDIC character or two decimal digits.

Call: In computer programming, the operation of calling into action a subroutine, usually by supplying the required parameters and executing a jump to the entry point of the subroutine.

Carrier, communications common: A company that furnishes communications services to the general public, usually as a monopoly, and is regulated by appropriate local, state, or federal agencies.

Central processing unit (CPU): Part of a computer system that contains the main storage, arithmetic unit, and special register groups. It performs arithmetic operations, controls instruction processing, and provides timing signals and other housekeeping operations.

Central station: Term applied to a central computer in a data communications system, because of the function it performs as the main processor of information communicated over the system.

Channel: A path along which signals can be sent, e.g., data channel, input/output channel.

Character: An elementary symbol that may be either alphanumeric, a punctuation mark, or any other symbol that may be read, stored, or written. In Baudot code, a character consists of five information bits; in American Standard Code for Information Interchange (ASCII), a character consists of eight information bits.

Check character (or digit): One or more characters (or digits) carried in a symbol, word, or block encoded, depending on the remaining elements, in such a way that if an error occurs it will be detected (excluding compensating errors).

Checkerboard: An alternating pattern of zeros and ones stored in a computer for testing purposes.

Clock: A generator of pulses that controls the timing of switching circuits.

Code: A system of symbols that can be used by machines, such as a computer, and that in specific arrangements have a special external meaning.

Combinatorial logic system: Digital system *not* utilizing memory elements. Contrasts with sequential logic.

Communication system: A computer system having facilities for long-distance transfers of information between remote and central stations.

Compiler: A language translation program, used to transform symbols meaningful to a human operator to codes meaningful to a computer. More restrictively, a program that translates a machine-independent source language into the machine language of a specific computer, thus excluding assemblers.

Computer (digital): An electronic instrument capable of accepting, storing, and arithmetically manipulating information, which includes both data and the controlling program. The information is handled in the form of coded binary digits (0 and 1), represented by dual voltage levels, magnetic states, punched holes, etc.

Configuration: The arrangement of either hardware instruments or software routines when combined to operate as a system.

Configurator: A computer program whose purpose is to combine a number of program segments into an integrated whole, in a specific desired manner (configuration).

Contention: A method of line control in which the terminals request to transmit. If the requested channel is free, transmission goes ahead; if it is not free, the terminal waits until it becomes free. The queue of contention requests can be built up by the computer in either a prearranged sequence or in the sequence in which the requests are made.

Control block: The circuitry that performs the control functions of the CPU. It is responsible for decoding microprogrammed instructions and then generating the internal control signals that perform the operations requested.

Control program: The program responsible for handling input/output for both terminals and file storage, establishing processing priorities, maintaining waiting lists of work in process, activating operational programs, and performing other supervisory functions in a real-time system. Words sometimes used synonymously to designate such a program include driver, executive, monitor, and supervisor.

Conversational mode: A procedure for communication between a terminal and the computer in which each entry from the terminal elicits a response from the computer.

Converter: Device for changing one form of information to another so as to make the information compatible with a different machine (e.g., card to tape conversion, or Baudot to ASCII conversion).

Core: The smallest element of a core storage memory module. It is a ring of ferrite material that can be magnetized in clockwise or counterclockwise directions to represent the binary digits 0 and 1.

CROM (*Control read only memory*): This is a major component in the control block of some microprocessors. It is an ROM that has been microprogrammed to decode control logic.

Cross-assembler: When the program is assembled by the same computer that it will run on, the program that performs the assembly is referred to simply as an assembler. If the program is assembled by some other processor of a different architecture, the process is referred to as cross-assembly. Occasionally the phrase "native assembler" will be used to distinguish it from a cross-assembler.

Crosstalk: Information in one channel overlapping into another channel causing distortion or interference.

Current page: The memory page comprising all those locations that are on the same page as a given instruction.

Cycle stealing: A common type of memory cycle acquisition for I/O data transfers in which the central processor is stalled one or more cycle times, if necessary, by an I/O transfer to enable the data to be obtained from or placed into memory, thus "stealing" cycles from the central processor. This type of operation slows processor functions since the processor has fewer usable cycles per unit of time.

Cycle time: The time interval between the consecutive starts of successive accesses to a storage location. Contrast with access time. For example, if it takes 3 μsec to read a word out of core storage and 1 μsec more to rewrite the word before another read operation can be initiated, then the unit has an access time of 3 μsec and a cycle time of 4 μsec.

Cyclic redundancy check (*CRC*): An error detection scheme in which the check character is generated by taking the remainder after dividing all the serialized bits in a block of data by a predetermined binary number.

Data acquisition: The gathering, measuring, digitizing, and recording of continuous form (analog) information.

Data link: Consists of the communications lines, modems, and other communications equipment that permit the transmission of information in data format between two or more terminals.

Data origination: The earliest stage at which the source material is first put into machine readable form or directly into electrical signals.

Dataphone: A trademark as well as a service mark of the AT&T Company. As a trademark it identifies the data sets or modems manufactured and supplied by the Bell System for use in the transmission of data over the regular telephone network. As a service mark it identifies the transmission of data over the regular telephone network.

Data reduction: The transformation of raw information gathered by measuring or recording equipment into a more condensed, organized, or useful form.

Data set: A device that performs the modulation/demodulation and control functions necessary to provide compatibility between business machines and communications facilities (see Modem).

Data speed service: Data from teletypewriters or other business machines that produce punched paper tape are carried over the regular telephone network at 750, 1050, or 1200 bits per second.

Data word: A computer word consisting of a number, a fact, or other information that is to be processed by the computer.

Debug: Checking for and correcting errors in a program.

Decimal: Denoting the numbering system based on the radix 10.

Decrement: Changing the value of a number in the negative direction. If not otherwise stated, a decrement by one is usually assumed.

Dedicated lines: Private circuits between two stations, such as between a terminal and the communications equipment of a computer.

Demodulation: The process of retrieving an original signal from a modulated carrier wave. This technique is used in data sets to make communication signals compatible with business machine signals.

Device: An electronic or electromechanical instrument. Most commonly implies measuring, reading, or recording equipment.

Diagnostic (adj.): Relating to test programs for detection of errors in the functioning of hardware or software, or the messages resulting from such tests. Also (noun), the test program or message itself.

Digital voltmeter: An electronic voltage measuring device that provides a readout in digital form on the instrument panel and commonly (essential for computer purposes) also codes the measurement result in binary-coded decimal form as an electrical output.

Direct memory access (DMA): A means of transferring a block of information words directly between an external device and the computer's memory, bypassing the need for repeating a service routine for each word. This method greatly speeds the transfer process.

Disable: A signal condition that prohibits some specific event from proceeding.

Disk storage: A means of storing binary digits in the form of magnetized spots on a circular metal plate coated with a magnetic material. The information is stored and retrieved by read–write heads, which may be positioned over the surface of the disk either by moving the heads or the disk itself.

Documentation: Manuals and other printed materials (tables, listings, diagrams, etc.), which provide instructive information for usage and maintenance of a manufactured product, including both hardware and software.

Double-length word: A word that, due to its length, requires two computer words to represent it. Double-length words are normally stored in two adjacent memory locations.

Driver: An input/output routine to provide automatic operation of a specific device with the computer.

Dump: To record memory contents on an external medium (e.g., tape).

Duplex transmission: Simultaneous two-way independent transmission in both directions. Also called full-duplex transmission (see Half-duplex).

EBCDIC: Extended Binary-Coded-Decimal Interchange Code. An 8-bit character code used primarily in IBM equipment. The code provides for 256 different bit patterns.

Echo check: A check of accuracy of transmission in which the information that was transmitted is returned to the source and compared with the original information.

Edit: To modify the form or format of data, e.g., to insert or delete characters such as page numbers or decimal points.

Effective address: The address of a memory location ultimately affected by a memory reference instruction. It is possible for one instruction to go through several indirect addresses to reach the effective address.

Emulate: To imitate one system with another such that the imitating system accepts the same data, executes the same programs, and achieves the same results as the imitated system. The imitation is usually achieved by a separate hardware module that interprets the original instructions via microcode.

Enable: A signal condition that permits some specific event to proceed, whenever it is ready to do so.

Encode: To apply a set of unambiguous rules specifying the way in which data may be represented such that a subsequent decoding is possible. In hardware, the act of initiating the operation of an external device.

ENIAC: Electronic Numerical Integrater and Calculator.

Exclusive or: A logical operation in which the resultant quantity (or signal) is true if at least one (but not all) of the input values is true, and false if the input values are all true or all false.

Execute: To perform fully a specific operation, such as would be accomplished by an instruction or a program.

Execute phase: A predetermined state of the internal computer logic that causes the computer to interpret as data the information read out of memory during a memory cycle.

Exit sequence: A series of instructions to conclude operation in one area of a program and to move to another area.

Exponent: In a floating point representation, the numeral, of a pair of numerals representing a number, that indicates the power to which the base is raised.

Fetch phase: A predetermined state of the internal computer logic that causes the computer to interpret as an instruction the information read out of memory during a memory cycle.

Firmware: Software instructions that have been permanently frozen into an ROM.

Fixed point: A numerical notation in which the fractional point (whether decimal, octal, or binary) appears at a constant, predetermined position. Compare with floating point.

Fixed-point binary number: A binary number represented by a sign bit and one or more number bits, with a binary point fixed usually to the right of the least significant bit.

Flag bit: A signal, or the stored indication of this signal, that indicates the readiness of a peripheral device to transfer information to the computer.

Flip-flop: An electronic circuit having two stable states, and thus capable of storing a binary digit. Its states are controlled by signal levels at the circuit input and are sensed by signal levels at the circuit output.

Flip-flop (storage element): A circuit having two stable states and the capability of changing from one state to another with the application of a control signal and remaining in that state after removal of signals.

Floating point: A numerical notation in which the integer and the exponent of a number are separately represented (frequently by two computer words), so that the implied position of the fractional point (decimal, octal, or binary) can be freely varied with respect to the integer digits. Compare with fixed point.

Flowchart: A diagram representing the operation of a computer program.

Foreground processing: High-priority processing usually resulting from real-time entries and given precedence by means of interrupts over lower priority "background" processing.

Format: A predetermined arrangement of bits or characters.

FORTRAN: A programming language (or the compiler that translates this language) that permits programs to be written in a form resembling algebra, rather than in detailed instruction-by-instruction form (as for assemblers).

FORTRAN library: A collection of programs for the computer to provide the user with commonly used mathematical and formatting routines.

Four-wire: Four physical wires (two pairs) used to make up the electrical circuit to which a data set may be connected. The term four-wire does not necessarily imply full-duplex operation. It in no way defines the signal or signals on the circuit. The advantage of using four-wire half-duplex operation is that it eliminates the so-called turnaround (150 msec to 10 sec) of the data set and the circuit.

Gate: An electronic circuit capable of performing logical functions such as "and," "or," and "nor."

Hardware: Electronic or electromechanical components, instruments, or systems, contrasted with software.

Hardware diagnostics: A collection of programs for the computer, designed to assist in the identification of hardware malfunctions.

Half-duplex (HD, HDX) circuit: A circuit designed for transmission in either direction, but not both directions simultaneously.

Hand shaking: The exchange of predetermined signals for purposes of control when a connection is established between two data sets.

Hertz (Hz): A measure of frequency or bandwidth, the same as cycles per second.

Hexadecimal: A number system based on a radix of 16. Numbers 0 through 9 are represented as in the decimal system; 10 through 15 are represented by A through F.

High core: Core memory locations having high-numbered addresses.

High-level language: A problem-oriented programming language as distinguished from a machine-oriented programming language. The former's instruction approach is closer to the needs of the problems to be handled than the language of the machine on which they are to be implemented.

High speed: Wide-band capability. Above 9600 bits per second.

Immediate addressing: In this mode of addressing, the operand contains the value to be operated on, and no address reference is required.

Inclusive or: A logical operation in which the resultant quantity (or signal) is true if at least one of the input values is true, and false if the input values are all false.

Increment: To change the value of a number in the positive direction. If not otherwise stated, an increment by one is usually assumed.

Incremental magnetic tape: A form of magnetic tape recording in which the recording transport advances by small increments (e.g., 0.005 in.), stopping the tape advancement long enough to record one character at the spot located under the recording head.

Index register: This register is used for storing numeric values used for computing operand addresses. A frequent use is to add the contents of the index register to the operand address to obtain the ultimate address.

Indirect address: The address initially specified by an instruction when it is desired to use that location to redirect the computer to some other location to find the "effective address" for the instruction.

Indirect phase: A predetermined state of the internal computer logic that causes the computer to interpret as an address the information read out of memory during a memory cycle.

Information: A unit or set of knowledge represented in the form of discrete "words," consisting of an arrangement of symbols or (so far as the digital computer is concerned) binary digits.

Inhibit: To prevent a specific event from occurring.

Initialize: The procedure of setting various parts of a stored program to starting values, so that the program will behave the same way each time it is repeated. The procedures are included as part of the program itself.

Input: Information transferred from a peripheral device into the computer. Also can apply to the transfer process itself.

Input/output (I/O): Relating to the equipment or method used for transmitting information into or out of the computer.

Input/output channel: The complete input or output facility for one individual device or function, including its assigned position in the computer, the interface circuitry, and the external device.

Input/output system: The circuitry involved in transferring information between the computer and its peripheral devices.

Instruction: A written statement, or the equivalent computer-acceptable code, that tells the computer to execute a specified single operation.

Instruction code: The arrangement of binary digits that tell the computer to execute a particular instruction.

Instruction logic: The circuitry involved in moving binary information between registers, memory, and buffers in prescribed manners, according to instruction codes.

Instruction register: A register that indicates the location of the next computer instruction to be interpreted.

Instruction word: A computer word containing an instruction code. The code bits may occupy all or (as in the case of memory reference instruction words) only part of the word.

Interface: The connecting circuitry that links the central processor of a computer system to its peripheral devices.

Interpretive routine: A routine that performs the execution of a program by translating each source language instruction into a sequence of machine instructions and executing them before translating the next instruction. Thus, each instruction must be translated every time it is to be executed.

Interrupt: The process, initiated by an external device, that causes the computer to interrupt a program in progress, generally for the purpose of transferring information between that device and the computer.

Interrupt location: A memory location whose contents (always an instruction) are executed upon interrupt by a specific device.

Interrupt phase: A predetermined state of the internal computer logic that causes the computer to suspend operation of a program in progress and branch to a specific service routine.

Jump: An instruction that breaks the strict sequential location-by-location operation of a program and directs the computer to continue at another specified location anywhere in memory.

KSR (keyboard send/receive): A teletypewriter unit with a keyboard but no paper tape unit.

Label: Any arrangement of symbols, usually alphanumeric, used in place of an absolute memory address in computer programming.

Language: The set of symbols, rules, and conventions used to convey information, either at the human level or at the computer level. The term is used to refer to programming procedures such as FORTRAN, BASIC, or Assembler.

Library routine: A routine designed to accomplish some commonly used mathematical functions and kept permanently available on a library program tape (e.g., FORTRAN library).

Line protocol (also called *Line discipline*): A set of defined control sequences used to direct the flow of communications traffic, e.g., binary synchronous line protocol.

Load: Put information into (memory, a register, etc.). Also (e.g., loading tape), to put the information medium into the appropriate device.

Loader: A program designed to assist in transferring information from an external device into a computer's memory.

Location: A group of storage elements in the computer's memory that can store one computer word. Each such location is identified by a number ("address") to facilitate storage and retrieval of information in selectable locations.

Logic diagram: A diagram that represents the detailed internal functioning of electronic hardware, using binary logic symbols rather than electronic component symbols.

Logic equation: A written mathematical statement, using symbols and rules derived from Boolean algebra. Specifically (hardware design), a means of stating the conditions required to obtain a given signal.

Logic operation: A mathematical process based on the principles of truth tables, e.g., "and," "inclusive or," and "exclusive or" operations.

Longitudinal redundancy check (LRC): A system of error control based on transmission of a block check character based on present rules. The check formation rule is applied in the same manner to each character.

Loop: A sequence of instructions in which the last instruction is a jump back to the first instruction.

Low core: Core memory locations having low-numbered addresses.

Memory reference: The address of the memory location specified by a memory reference instruction, i.e., the location affected by the instruction.

Message switching: The switching technique of receiving a message, storing it until the proper outgoing circuit and station are available, and then retransmitting it to its destination.

Microinstruction: An instruction that forms part of a larger, composite instruction.

Microprogramming: Control technique used to implement the stored program control function. Typically the technique is to use a preprogrammed read-only memory chip to contain several control sequences that normally occur together.

Mnemonic: An abbreviation or arrangement of symbols used to assist human memory.

Modem: A contraction of modulator–demodulator. The term may be used when the modulator and demodulator are associated in the same signal-conversion equipment.

Modulation: The process by which some characteristics of one wave are varied in accordance with another wave. This technique is used in data sets to make business-machine signals compatible with communication facilities.

Modulation, amplitude: A method of transmission whereby the signal wave voltage is impressed upon a higher-frequency carrier wave such that the carrier amplitude varies with the signal wave.

Multilevel indirect: Indirect addressing using two or more indirect addresses in sequence to find the effective address for the current instruction.

Multiple-precision: Refers to arithmetic in which the computer, for greater accuracy, uses two or more words to represent one number.

Multiplexing: The division of a transmission facility into two or more channels.

Multipoint circuit: A circuit interconnecting several stations that must communicate on a time-share basis.

Multiprocessing: More than one computer is used in the processing of one transaction.

Multiprogramming: More than one transaction is in one computer, with the transactions being processed in parallel.

Nesting: Used in programming when several computational sequences are enclosed within each other, such as in a matrix inversion computation.

Nonreturn to zero (NRZ): A technique of magnetic tape recording in which the recording device does not turn off the magnetizing flux between recording of individual characters. The flux is always at saturation level during recording, and bits are indicated by reversals of flux polarity.

Nontransparent mode: Transmission of characters in a defined character format, e.g., ASCII or EBCDIC, in which all defined control characters and control character sequences are recognized and treated as such.

Object language: The language to which a language is translated, usually the binary code, which is directly interpreted by the computer.

Octal: A numbering system based on the radix 8. Octal digits are restricted to the values 0 through 7.

Octal point: The fractional dividing point of an octal numeral; equivalent to decimal point in the decimal numbering system.

Off line: The operation of peripheral equipment not under control of the computer.

One's complement: A number so modified that the addition to the modified number and its original value, plus one, will equal an even power of two. A one's complement number is obtained mathematically by subtracting the original value from a string of 1's, and electronically by inverting the states of all bits in the number.

On-line computer system: A system in which the input data enter the computer directly from the point of origin and/or output data are transmitted directly to where they are used.

Operand: The data item upon which an action is performed.

Operating system: A software system that controls the scheduling and execution of application and system programs.

Output: Information transferred from the computer to a peripheral device. Also can apply to the transfer process itself.

Output coupler: An instrument that provides the interconnecting circuitry between a measuring instrument and a recording instrument.

Overflow: The condition that exists when the result of an addition in an accumulation has exceeded the maximum possible signed value ($+32767$ or -32768, decimal) in a 16-bit computer. Also the one-bit register that indicates this condition.

Packed word: A computer word containing two or more independent units of information. This is done to conserve storage when information requires relatively few bits of the computer word.

Page: An artificial division of memory consisting of a fixed number of locations, dictated by the direct addressing range of memory reference instructions.

Page zero: The memory page that includes the lowest numbered memory addresses.

Parallel transmission: All bits of a character are sent simultaneously.

Parity bit: A supplementary bit added to an information word to make the total of one-bits always either odd or even. This permits checking the accuracy of information transfers.

Parity check, horizontal: A parity check applied to a group of certain bits from every character in a block (see *Longitudinal redundancy check*).

Parity check, vertical: A parity check applied to a group of bits making up one character.

Pass: The complete process of reading a set of recorded information (one tape, one set of cards, etc.) through an input device, from beginning to end.

Peripheral device: An instrument or machine electrically connected to the computer, but not part of the computer itself.

Phase: One of several specific states of the internal computer logic, usually set up by instructions being executed, to determine how the computer should interpret information read out of memory.

Plane: An arrangement of ferrite cores on a matrix of control and sensing wires. Several planes stacked together form a "memory module."

Power failure control: A means of sensing primary power failure so that a special routine may be executed in the finite period of time available before the regulated dc supplied discharges to unusable levels. The special routine may be used to preserve the state of a program in progress, or to shut down external processes.

Priority: The automatic regulation of events so that chosen actions will take precedence over others in cases of timing conflict.

Privileged instruction: A computer instruction that is not available for use in ordinary programs written by users; its use is restricted to certain privileged routines generally part of the operating system.

Process control: Automatic control of manufacturing processes by use of a computer.

Processor: The central unit of a computer system (i.e., the device that accomplishes the arithmetic manipulations), exclusive of peripheral devices. Frequently (when used as adjective) also excludes interface components, even though normally contained within the processor unit; thus "processor" options exclude interface ("input/output") options.

Program: A plan for the solution of a problem by a computer, consisting of sequence of computer instructions.

Program counter: One of the registers in the CPU that holds addresses necessary to step the machine through the program.

Program listing: A printed record (or equivalent binary-output program) of the instructions in a program.

Programmable read only memory (PROM): A fixed program, read only, semiconductor memory storage element that can be programmed after packaging. Once the program has been entered into the memory, it cannot be overwritten.

Programmer: A person who writes computer programs. Also (hardware), an interface card or instrument that sets up (or "programs") the various functions of one measuring instrument.

Programming: The process of creating a program.

Pseudoinstruction: A symbolic statement, similar to assembly language instructions in general form, but meaningful only to the program containing it, rather than to the computer as a machine instruction.

Punched tape: A strip of tape, usually paper, on which information is represented by coded patterns of holes punched in columns across the width of the tape. Commonly there are eight hole positions (channels) across the tape.

Queue: A line-up of programs waiting for processing by the computer.

Random access memory (RAM): A memory from which all information can be obtained at the output with approximately the same time delay by choosing an address randomly and without first searching through a vast amount of irrelevant data.

Read: The process of transferring information from an input device into the computer. Also, the process of taking information out of the computer's memory.

Read only memory (ROM): A fixed program semiconductor storage element that has been preprogrammed with a permanent program.

Real time: Time elapsed between events occurring externally to the computer. A computer that accepts and processes information from one such event and is ready to accept new information before the next event occurs is said to operate in a "real-time environment."

Recursive: A subroutine that is written in such a manner that it can call itself.

Redundancy: The portion of the total information contained in a message that can be eliminated without loss of essential information.

Reentrant: A routine that can be used by two or more independent programs at the same time. This means that the reentrant routine cannot modify the contents of any of its own locations, and that any required temporary storage must be supplied along with each program using the reentrant routine.

Register: An array of hardware binary circuits (flip–flops, switches, etc.) for temporary storage of information. Unlike mass storage devices such as memory cores, registers can be wired to permit flexible control of the contained information, for arithmetic operations, shifts, transfers, etc.

Relative address: The number that specifies the difference between the absolute address and the base address.

Relocatable: Programs whose instructions can be loaded into any stated area of memory.

Relocating loader: A computer program capable of loading and combining relocatable programs (i.e., programs having symbolic rather than absolute addresses).

Remote batch: A batch-processing system that can be operated from remote terminals.

Response time: The amount of time elapsed between generation of an inquiry at a data communications terminal and receipt of a response at that same terminal. Response time includes transmission time to the computer; processing time at the computer (including access time to obtain any file records needed to answer the inquiry); and transmission time back to the terminal.

Rotate: A positional shift of all bits in an accumulator (and possibly an extend bit as well), with those bits lost off one end of the accumulator "rotated" to enter vacated positions at the other end.

Routine: A program or program segment designed to accomplish a single function.

Scratch-pad memory: A small local memory utilized to facilitate local data handling on a temporary basis.

Simplex mode: Operation of channel in one direction only with no capability of reversing.

Simulator: A program that allows the execution of program written for a computer of one architecture to execute on a computer of another architecture. Simulators usually refer only to software systems as contrasted to emulators.

Software: Computer programs. Also, the tapes or cards on which the programs are recorded.

Software package: A complete collection of related programs, not necessarily combined as a single entity.

Source program: A program (or its recorded form) written in some programming language other than machine language and thus requiring translation. The translated form is the "object program."

Statement: An instruction in any computer-related language other than machine language.

Store: To put information into a memory location, register, or device capable of retaining the information for later access.

Subroutine: Sequence of instructions designed to perform a single task, with provision included to allow some other program to cause execution of the task sequence as if it were part of its own program.

Symbolic address: A label assigned in place of absolute numeric addresses, usually for purposes of relocation (see *Relocatable*).

Synchronous transmission: A transmission mode in which the data characters and bits are transmitted at a fixed rate with the transmitter and receiver synchronized. There are no start or stop bits as in asynchronous transmission. Transmission timing is divided through synchronizing characters at the beginning of each message or block of data.

System: An assembly of units (e.g., hardware instruments or software routines), combined to work as a larger integrated unit having the capabilities of all the separate units.

Two's complement: A number so modified that the addition of the modified number and its original value will equal an even power of two. Also, a kind of arithmetic that represents negative numbers in two's complement form so that all addition can be accomplished in only one direction (positive incrementation).

Utility routine: A standard routine to assist in the operation of the computer (e.g., device drivers, sorting routines) as opposed to mathematical ("library") routines.

Vectored interrupt: An architecture in which each interrupt line is processed by a command beginning at a location unique to that line.

Vertical redundancy check (VRC): A check or parity bit added to each character in a message such that the number of bits in each character, including the parity bit, is odd (odd parity) or even (even parity).

Volatile storage: A storage device in which stored data are lost when the applied power is removed.

Waiting loop: A sequence of instructions (frequently only two) that are repeated indefinitely until a desired external event occurs.

Word: A set of binary digits handled by the computer as a unit of information. Its length is determined by hardware design, e.g., the number of cores per location, and the number of flip-flops per register.

Write: The process of transferring information from the computer to an output device. Also, the process of storing (or restoring) information into the computer's memory.

Acknowledgments

The material in this publication has been adapted in part from publications of several minicomputer vendors including Data General Corporation, Digital Equipment Corporation, and Hewlett-Packard Corporation. The material so published herein is the sole responsibility of the author.

References

"A Microprogramming Guide for the Hewlett-Packard 2100 Computer" (1971). Hewlett-Packard Company, Cupertino, California.
"A Pocket Guide to Interfacing the HP 2100 Computer" (1973). Hewlett-Packard Corporation, Cupertino, California.
"A Pocket Guide to the 2100 Computer" (1972). Hewlett-Packard Corporation, Cupertino, California.
"An Introduction to HP Computers" (1970). Hewlett-Packard Corporation, Palo Alto, California.
"Application and Interface Guide for the Model 9300 Vacuum Column Synchronous Digital Magnetic Tape Recorder" (1974). Kennedy Company, Altadena, California.
"Auerbach Computer Technology Reports." Auerbach Publishers, Philadelphia, Pennsylvania. (A subscription service presenting technical information on EDP and minicomputer systems.)
"Common Business Oriented Language (COBOL)" (1975). IBM Corporation, Poughkeepsie, New York.
"Datapro Reports on Minicomputers." Datapro Research Corporation, Delran, New Jersey. (A subscription service presenting technical information on micro and minicomputers.)
"8080 Assembly Language Programming Manual" (1975). Intel Corporation, Santa Clara, California.
"8080 Microcomputer System Manual" (1975). Intel Corporation, Santa Clara, California.
"General Information—Binary Synchronous Communications Manual" (1970). IBM Corporation, Research Triangle Park, North Carolina.
"How to use the Nova Computers" (1970). Data General Corporation, Southboro, Massachusetts.
Husson, S. S. (1970). "Microprogramming: Principles and Practices." McGraw-Hill, New York.
"IEEE Standard Digital Interface for Programmable Instrumentation" (1975). IEEE, New York, New York. (A Standard.)
"IMP-16 Programming and Assembler Manual" (1974). National Semiconductor Corporation, Santa Clara, California.
"Industrial Computer System FORTRAN Procedures for Executive Functions and Process Input-Output" (1972). Instrument Society of America, 400 Stanwix Street, Pittsburgh, Pennsylvania. (A Standard.)
McCracken, D. D. (1962). "A Guide to Algol Programming." Wiley, New York.
McCracken, D. D. (1970). "A Guide to COBOL Programming." Wiley, New York.
McCracken, D. D. (1972). "A Guide to FORTRAN IV Programming." Wiley, New York.
"Minicomputers and Microcomputers" (1974). Quantum Science Corporation, New York, New York. (A Report.)
Morris, D. E., Christopher, C. J., Chance, G. W., and Barney, D. B. (1976). Third generation calculator has computer-like capabilities. *Hewlett-Packard J.* June.

"M6800 Microcomputer System Design Data" (1976). Motorola Semiconductor Products, Phoenix, Arizona.

"990 Computer Family Systems Handbook" (1975). Texas Instruments Company, Dallas, Texas.

Pakin, S. (1972). "APL/360 Reference Manual." Science Research Associates, Chicago, Illinois.

"PDP-11 Handbook" (1975). Digital Equipment Corporation, Maynard, Massachusetts.

"PDP-11 Software Handbook" (1975). Digital Equipment Corporation, Maynard, Massachusetts.

"Peripherals and Interfacing Handbook (1971). Digital Equipment Corporation, Maynard, Massachusetts.

"Preface to Programming" (1971). Hewlett-Packard Corporation, Cupertino, California.

Ricci, D. W., and Nelson, G. E. (1974). Standard instrument interface simplifies system design. *Electronics* **47** (23), 95.

Rothenbuecher, O. H. (1977). The top 50 companies in the data processing industry. *Datamation* **23** (6), 61.

"RPG/3000 Compiler, System and Application Manual" (1975). Hewlett-Packard Company, Cupertino, California.

"Software Design for Microprocessors" (1976). Texas Instruments Company, Dallas, Texas.

"The HP 3000 Computer System Reference Manual" (1972). Hewlett-Packard Company, Cupertino, California.

"The Microprocessor Handbook" (1974). Texas Instruments Company, Dallas, Texas.

Wegner, P. (1968). "Programming Languages, Information Structures, and Machine Organization." McGraw-Hill, New York.

AUTHOR INDEX

Numbers in parentheses are reference numbers and indicate that an author's work is referred to although his name is not cited in the text. Numbers in italics show the page on which the complete reference is listed.

A

Abbott, D. A., 203(2.70), *215*
Adair, J. E., 203(2.57b), 205(2.57b), *215*
Adams, R. L., 186(2.35), 187(2.35), 193(2.35), 203(2.35), *214*
Adelsberger, U., 48, 74, *96*
Adler, R., 88, *90*
Adler, R. B., 121, *136*
Agouridis, D., 158(1.71, 1.73), 169(1.71, 1.73), 170(1.73), *174*
Akhiezer, A. I., 103, 107, 111, 115, 116, *135*
Akhiezer, I. A., 107, 115, 116, *135*
Albrecht, T., 73, *90*
Allan, D. W., 52, 53, 55, 57, 59, 60, 61, 66, *90, 97*
Alley, O. C., 69, 72, 86, *90*
Amusia, M. Y., 12, 13, 14, 15, 16, 17, 18, *30*
Anderle, R. J., 76, *90*
Anderson, O. D., 60, *90*
Andeweg, J., 151(1.23), *173*
Angeli, M. T., 61, *90*
Angus, J. A., 187(2.36), 193(2.36), 203(2.36), *214*
Antcliffe, G. A., 273(41), *281*
Archer, J. A., 158(1.77), 169(1.77), 170(1.77), *174*
Archer, R. J., 193(2.51, 2.54), 202(2.54), *215*
Armstrong, L. A., 13, 14, *32*
Armstrong, L., Jr., 11, *32*
Arnold, R., 158(1.46), *173*
Asai, S., 193(2.47), 195(2.47), *215*
Ash, E. A., 101, 114, *138*
Ashby, N., 86, *90*
Askne, J. I. H., 122, *135, 137*
Assemat, J. L., 151(1.36), 153(1.36), 154(1.36), 158(1.36), *173*
Audoin, C., 71, 72, *90, 95, 96*

Auld, B. A., 108, 110, 111, 112, 124, 126, 129, 132, *135*
Azoubib, J., 66, *90*

B

Bachner, F. J., 201(2.52), *215*
Baechtold, W., 203(2.76), 209(2.76), 210(2.76), *215*
Bagley, A. S., 42, *90*
Bailey, R. L., 142(1.11), 158(1.11), *173*
Ballonoff, A., 227(24), 230(28), 277(44, 45), *281*
Barber, M. R., 104, *138*
Barber, R. E., *90*
Bardeen, J., 142, *172*
Barillet, R., 71, *95*
Barnes, D. B., *345*
Barnes, J. A., 47, 49, 51, 52, 53, 55, 59, 60, 61, 79, *90*
Barnoski, M. K., 151(1.34, 1.35), 153(1.34, 1.35), 154(1.34, 1.35), *173*
Barrera, J. S., 193(2.54), 202(2.54), *215*
Barrett, D. L., 193(2.46), 194(2.46), *215*
Baruch, P., 151(1.44), 155(1.44), *173*
Bar'yakhter, V. G., 103, 111, *135*
Barybin, A. A., 102, 108, 110, 112, 113, 114, 115, 118, 119, 120, 122, 123, 124, 125, 126, 127, 128, 129, 130, 131, 132, 133, *135, 136*
Bates, D. J., 227(24), 230(27, 28, 29), 277(43, 44, 45, 46), *281*
Baugh, R. A., 53, *91*
Bazin, M., 88, *90*
Bechtel, N. G., 175, 178(2.17), *214*
Becker, G., 61, 65, 67, 80, *91, 93*
Beehler, R. E., 43, 62, 63, *91*
Bell, B. W., 264(35), *281*
Bell, H. E., 44, 67, 71, *93*

Bellinson, H. R., 48, *97*
Belohoubek, E. F., 158(1.48), 159(1.48), *173*
Bender, P. L., 88, 89, *91*
Benjamin, J. A., 149(1.24), 151(1.24), 158(1.24), 165(1.24), *173*
Bennett, R. H., 187(2.36), 193(2.36), 203(2.36), *214*
Berg, H. C., 71, *93*
Berger, H., 122, *136*
Bergmann, F., 158(1.61), 165(1.61), 166(1.61), *174*
Berkowitz, J., 15, 16, 17, 18, *31*
Bernard, H., 158(1.59), *174*
Berz, F., 151(1.31), *173*
Bespyatykh, Y. I., 103, *136*
Bianco, B., 106, *136*
Black, J. R., 151(1.38, 1.39), 156(1.38, 1.39), 157(1.38, 1.39), *173*
Blackman, R. B., 52, *91*
Blair, B. E., 42, 77, 80, *91*
Blanchard, B., 151(1.44), 155(1.44), *173*
Blocker, T. G., 186(2.35), 187(2.35), 193(2.35), 203(2.35), *214*
Blötekjaer, K., 102, 108, 125, *136*
Bloom, S., 122, 124, *138*
Bobroff, D. L., 115, 117, 118, 119, 120, 121, 122, 124, *136*
Boileau, E., 51, 54, *91*
Boroffka, H., 193(2.51), 195(2.51), *215*
Borre, K., 61, *91*
Bosch, B. G., 105, *136*
Bowers, R., 99, *136*
Bowler, D. L., 158(1.70), 169(1.70), *174*
Box, G. E. P., 48, 60, *91*
Brattain, W. H., 142, *172*
Brenner, N., *91*
Brewer, G. R., 151(1.41), 157(1.41), *173*
Briatore, L., 87, *91*
Briggs, R. J., 107, 125, *136*
Brilman, M. E., 151(1.30), 153(1.30), *173*
Brown, A. V., 223, *280*
Brytov, I. A., 9, 10, 11, *32*
Buchanan, B., 203(2.72), 209(2.72), *215*
Buchsbaum, S. J., 99, *136*
Budenstein, P. P., 158(1.69), 168(1.69), *174*
Buetel, G. A., 223(8, 9), *280*
Bulman, P. J., 104, 105, *136*
Bura, P., 203(2.60), *215*
Burke, P. G., 12, 13, 14, *30*
Burrell, K. H., 101, *137*

Butlin, R. S., 187(2.36), 193(2.36), 203(2.36), *214*
Byron, E., 82, *91*

C

Camisa, R. L., 193(2.45), 203(2.45, 2.57a, 2.58, 2.59), 205(2.57a), 206(2.57a), 208(2.57a), 212(2.57a), *214, 215*
Campbell, N., 37, *91*
Cannon, W. H., 87, *91*
Carley, D. R., 148(1.19, 1.20, 1.21), 151(1.19, 1.20, 1.21), *173*
Carter, V. L., 1, 2, *31*
Castaing, C., 151(1.44), 155(1.44), *173*
Cateora, J. V., 82, *91*
Caudano, R., 29, *30*
Causse, J., 158(1.45), *173*
Chance, G. W., *345*
Chang, N. S., 103, *136, 137*
Chang, T. N., 2, 12, *30*
Chappey, M., 188, *214*
Ch'en, D. R., 193(2.49), 203(2.49), 209(2.49), *215*
Chen, J. T. C., 158(1.47, 1.55), 160(1.55), 161(1.55), 167(1.47), *173, 174*
Chen, P. T., 158(1.50), 159(1.50), *173*
Cherepkov, N. A., 12, 13, 14, 15, 16, 17, 18, *30*
Chernysheva, L. V., 12, 13, 14, 15, 16, 17, 18, *30*
Chi, A. R., 49, 53, 59, 82, 88, *90, 91*
Chiabrera, A., 106, *136*
Chie, C. M., 55, *93*
Chiu, T. L., 175, 178(2.16), *214*
Chodorow, M., 108, 117, *136*
Chovnyuk, Y. B., 103, *138*
Christensen, D. A., 115, *136*
Christopher, D. J., *345*
Chu, L. J., 121, *136*
Chupka, W. A., 15, 16, 17, 18, *31*
Chynoweth, A. G., 99, 104, *136, 137*
Clemence, G. M., 73, 87, *91, 97*
Cocke, W. J., 87, *92*
Codling, K., 11, 12, 13, 14, 15, 16, 17, 18, *30, 31, 32*
Collin, R. E., 108, 126, *136*
Collins, J. H., 103, *136, 137*
Combet Farnoux, F., 2, 3, 7, 8, 9, 15, *30, 31*

AUTHOR INDEX

Conwell, E. M., 117, *136*
Cooke, H. F., 142(1.12), 151(1.12), 153(1.12), 158(1.52), 169(1.12), *173*, 193(2.49), 203(2.49), 209(2.49), *215*
Cooper, J., 15, *31*
Cooper, J. W., 1, 2, 3, 4, 5, 6, 7, 8, 9, 11, 16, 17, 18, 20, 21, 22, 23, 26, 27, *31*, *32*
Costain, C. C., 64, 68, *94*
Counselman, C. C., 76, 89, *92*
Cowan, D., 203(2.60), *215*
Crampton, S. B., 71, *93*
Creer, K. M., *92*
Cremonese, M., 7, *31*
Crouchley, J., 80, *95*
Cutler, L. S., 49, 53, 54, 55, 59, 69, 72, 86, 90, *92*

D

Daams, H., 64, 68, *94*
Dacey, G. C., 174, 177, *214*
D'Alembert, J. L. R., 86, *92*
Danjon, A., 73, *92*
Da Rocha, P., 206(2.65), *215*
Das, M. B., 175, *214*
Dash, S., 151(1.32), *173*
Dauge, A. M., 188(2.42), *214*
Davenport, J. W., 30, *31*
David, P., 175, 178(2.14), *214*
Davis, D. D., 82, *91*
Davis, S., 69, 72, 86, *90*
Dean, R. H., 105, 106, *136*, 193(2.48), 195(2.48), *215*
Debrecht, R. E., 187(2.38), 193(2.38), 203(2.38, 2.69), 206(2.69), *214*, *215*
Dehmer, J. L., 15, 16, 17, 18, 19, 30, *31*
Delandre, M., 151(1.36), 153(1.36), 154(1.36), 158(1.36), *173*
Delmas, J., 188(2.41, 2.42), *214*
Denker, S. P., 122, *136*
Desaintfuscien, M., 71, *95*
Dhaka, V. A., 151(1.40), 157(1.40), *173*
Dhez, P., 7, *31*
Dill, D., 19, 27, 28, 29, 30, *31*, *32*
Dingle, H., 84, 86, *92*
Dixon, R. C., 79, *92*
Doherty, R., 80, *92*
Dolan, R., 203(2.72), 209(2.72), *215*
Drangeid, K. E., 186(2.33), *214*

Dreeben, A., 187(2.38), 193(2.38), 203(2.38), *214*
Dreeben, A. B., 105, 106, *136*, 203(2.69), 206(2.69), *215*
Driver, M. C., 193(2.46), 194(2.46), *215*
Drukier, I., 193(2.45), 203(2.45, 2.57a, 2.58, 2.59), 205(2.57a), 206(2.57a), 208(2.57a), 212(2.57a), *214*, *215*
Dumetz, A., 151(1.36), 153(1.36), 154(1.36), 158(1.36), *173*
Durand, P., 163(1.57), *174*, 188(2.42), 203(2.64), *214*, *215*
Durney, C. H., 115, 122, *136*, *137*

E

Early, J. M., 142, *172*
Easton, R., 72, 82, *92*
Ebers, J. J., 142, *172*
Ederer, D. L., 9, 10, 11, *31*, *32*
Edson, W. A., *92*
Efremow, N., 201(2.52), *215*
Ehrenberg, W., 223, *280*
Emeis, R., 142, 146(1.8), *172*
Emery, F. E., 158(1.52), *173*
Engelmann, R. W. H., 105, *136*
Enslin, H., *92*
Ero, J. W., 175, 180(2.22), 210(2.22), *214*
Essen, L., 41, 62, 63, 75, 83, *92*, *94*
Ettenberg, M., 101, 102, *136*
Evans, P. F., 223(17), *280*

F

Fair, R. B., 151(1.27), 153(1.27), *173*, 203(2.82), 211(2.82), *215*
Fano, R. M., 121, *136*
Fano, U., 2, 6, 8, 19, 27, *31*, *32*
Fay, B., 105, *138*
Felber, H.-J., 40, *92*
Feller, W., 37, *92*
Finch, H. F., 75, *92*
Fischer, B., 66, 80, *91*, *92*
Fisher, L. C., 82, *92*
Fleming, P. L., 106, *136*
Fletcher, N. H., 142, 146(1.7), 147(1.7), *172*
Fock, V., 84, 85, *92*
Fodor, G., 158(1.45), *173*

Folts, H. C., 89, *92*
Fosque, H. S., 88, *91*
Freeman, J. C., 102, *136*
Freire, G. F., 122, 124, *136*
Frey, J., 202(2.55), 203(2.80), 209(2.80), 210(2.80), *215*
Frey, W., 105, *136*
Fujisawa, K., 101, 122, *136*, *137*
Fukui, H., 158(1.75), 169(1.75), *174*
Fukuta, M., 151(1.22), *173*, 186(2.34, 2.37), 187(2.34, 2.37), 188(2.34), 193(2.34, 2.37), 203(2.34, 2.37, 2.61), *214*, *215*
Funck, R., 188(2.40), 193(2.40), 197(2.40), 203(2.40, 2.66), 206(2.66), *214*, *215*
Furumoto, A., 186(2.37), 187(2.37), 193(2.37), 203(2.37), *214*

G

Gabillet, P., 151(1.36), 153(1.36), 154(1.36), 158(1.36), *173*
Gallagher, H. E., 230(26), *281*
Gandy, J., 72, *95*
Gardner, J. L., 12, 13, 14, 28, 29, *32*
Gattis, T. H., 89, *96*
Gerber, E. A., 62, *92*
Gerstner, D., 158(1.61), 165(1.61), 166(1.61), *174*
Ghandhi, S. K., 175, 178, *214*
Ghosh, H. N., 175, 178(2.16), *214*
Gicquel, R., 174, 176, 180(2.9), 189(2.9), 210(2.9), *214*
Glaze, D. J., 60, 61, 66, 90, *97*
Glicksman, M., 99, *137*
Goel, J., 203(2.57a, 2.58, 2.59), 205(2.57a), 206(2.57a), 208(2.57a), 212(2.57a), *215*
Goetzberger, A., 147(1.17), 151(1.17), *173*
Goldberg, B., 78, *92*
Goodrich, L. C., 115, 122, *137*
Gover, A., 101, *137*
Grace, F. E., 223(8, 9), *280*
Graneaud, M., 66, *90*
Gray, J. E., 59, 60, 61, *90*
Greaves, W. M. H., 48, 52, *92*
Grebene, A. B., 175, 178, *214*
Grosvalet, J., 175, 178(2.10), *214*
Grow, R. W., 115, 122, *136*, *137*
Guinot, B., 44, 61, 66, 75, 90, *92*
Gulyaev, Y. V., 103, *137*

Gunshor, R. L., 102, 103, *136*, *137*
Gurevich, V. L., 103, *137*

H

Haensel, R., 7, *31*
Hafner, E., 62, 72, *93*
Hagon, P. J., 103, *137*
Hahn, S. L., 65, *93*
Hahn, W. C., 115, *137*
Halford, D., 38, 44, 67, *93*
Hall, H., 6, *31*
Hall, R. G., 41, 59, 61, 62, 63, *94*, *97*
Hammer, J. M., 101, *137*
Hanna, V. F., 101, *137*
Hanson, D. W., 82, *91*, *92*
Han-Tzong-Yuan, 151(1.42), 157(1.42), 158(1.42, 1.56), 160(1.56), 162(1.42, 1.56), *173*, *174*
Harada, K., 82, 95, *97*
Hargrave, J., 80, *95*
Harrison, R. I., 122, *136*
Hart, B. I., 48, *97*
Hartmann, K., 158(1.49), 159(1.49), *173*
Hartnagel, H. L., 99, 104, *137*
Hasegawa, A., 115, 122, *137*
Hatzakis, M., 151(1.40), 157(1.40), *173*
Haus, H. A., 115, 117, 121, 122, 124, *136*, *137*, *138*, 203(2.78, 2.79), 209(2.78, 2.79), 210(2.78, 2.79), 211(2.78, 2.79), *215*
Hauser, J. R., 175, 178(2.11), *214*
Hayes, R., 271(40), *281*
Healey, D. J., 49, 53, 59, *90*
Hefele, J. C., 86, *93*
Heffner, H., 108, *137*
Hefley, G., 80, *92*
Heil, O., 174, *213*
Heinzmann, U., 3, 19, *31*
Hellwig, H., 38, 44, 60, 61, 66, 67, 70, 71, 72, 82, 90, *92*, *93*, *97*
Heno, Y., 7, *31*
Herlett, A., 142(1.8), 146(1.8), *172*
Herman, F., 223(13, 15), *280*
Hetzel, P., 80, *91*
Heuer, H., 3, 19, *31*
Hiatt, C. F., 203(2.74, 2.75), 209(2.74, 2.75), 210(2.74, 2.75), *215*
Higashizaka, A., 203(2.68), *215*

Hill, F. J., 275(42), *281*
Hines, M. E., 101, *137*
Hirsch, N., 193(2.50), 195(2.50), *215*
Hobson, G. S., 104, 105, *136*
Houlgate, R. G., 12, 13, 14, 15, 16, 17, 18, *30*, *31*
Howe, D. A., 53, 54, 55, *93*
Hower, P. L., 158(1.68), 168(1.68), *174*, 175, 178(2.17), *214*
Huang, H. C., 193(2.45), 203(2.45, 2.57a, 2.58), 205(2.57a), 206(2.57a), 208(2.57a), 212(2.57a), *214*, *215*
Hudson, R. D., 1, 2, *31*
Huebner, U., 61, *93*
Hughes, J. J., 106, *136*, 187(2.38, 2.39), 193(2.38, 2.39), 203(2.38, 2.39), *214*
Huguenin, G. R., 89, *93*
Hunsperger, R. G., 193(2.50), 195(2.50), *215*
Husson, S. S., *344*
Hutson, A. R., 102, *137*

I

Ichikawa, H., 101, *137*
Inuishi, Y., 122, *137*
Ishii, T., 122, *137*
Ishikawa, H., 186(2.34), 187(2.34), 188(2.34), 193(2.34), 203(2.34, 2.61), *214*, *215*
Ivanov, V. K., 12, 13, 14, *30*

J

Jaeglé, P., 7, 8, *31*
Jaggi, R., 186(2.33), *214*
Jamison, N. C., 223(10), *280*
Jarvis, S., Jr., 44, 66, 67, *93*, *97*
Jechart, E., 72, *95*
Jenkins, G. M., 48, 49, 52, 55, 60, *91*, *93*
Jensen, O. G., 87, *91*
Jivery, W. T., 15, 16, 17, 18, *31*
Johnson, C. C., 108, 110, 121, 129, *137*
Johnson, E. O., 142(1.9), *172*, 175, *214*, 238(30), *281*
Johnson, J., 268(38), *281*
Johnson, K. H., 30, *31*
Johnston, K. J., 77, *93*
Jolly, S., 187(2.39), 193(2.39), 203(2.39), *214*

Jolly, S. T., 193(2.45), 203(2.45, 2.57a), 205(2.57a), 206(2.57a), 208(2.57a), 212(2.57a), *214*, *215*
Jones, E. M. T., 264(36), *281*
Josephs, H. C., 158(1.67), 168(1.67), *174*
Joshi, M. L., 151(1.32), *173*
Joy, R. C., 158(1.60), 165(1.60), *174*
Juleff, E. M., 158(1.64), 165(1.64), 167(1.64), *174*
Jutzi, W., 203(2.63), *215*

K

Kakihana, S., 158(1.47), 167(1.47), *173*
Kakkuri, J., 78, *93*
Kalliomäki, K., 77, 78, *93*
Kaminski, J. F., 105, *136*
Kanbe, H., 105, 106, *137*
Kaner, E. A., 99, *137*
Kant, I., 35, 38, 40, *93*
Kartaschoff, P., 71, *94*
Kawamura, N., 203(2.71), 209(2.71), *215*
Kawasaki, K., 103, *137*
Kawazura, K., 105, *137*
Kaye, D. N., *344*
Keating, R. E., 86, *93*
Kellner, W., 193(2.51), 195(2.51), *215*
Kelly, H. P., 11, *31*
Kennedy, D. J., 3, 7, 8, 9, 10, 11, 12, 13, 14, 15, 16, 17, 18, 20, 21, 28, *31*, *32*
Kennedy, D. P., 175, 178, *214*, *281*
Kent, R. H., 48, *97*
Kerr, J. A., 151(1.31), *173*
Kessler, J., 3, 19, *31*
Kildishev, B. N., 103, *138*
Kim, C. K., 175, 178(2.13), *214*
Kim, H. B., 193(2.46), 194(2.46), *215*
Kimura, K., 203(2.67), *215*
Kino, G. S., 105, 108, 110, 124, 129, 132, *135*, *137*, *138*
Kinoshita, H., 75, *93*
Kirkpatrick, W. E., 223(11), *280*
Kisaki, H., 151(1.22), *173*
Kleppner, D., 71, *93*
Klüver, J. W., 115, 117, 121, 122, 124, *136*, *137*
Kmita, A. M., 103, *137*
Kniepkamp, H., 193(2.51), 195(2.51), *215*

Knight, R. I., 264(35), 268(39), 277(43), *281*
Knott, K. F., 158(1.78), 169(1.78), *174*
Kodera, H., 193(2.47), 195(2.47), 203(2.67), *215*
Kolmann, K., 151(1.29), 153(1.29), *173*
Konagai, M., 151, *173*
Kotelyansky, I. M., 103, *137*
Kotyczka, W., 158(1.49), 159(1.49), *173*
Kovalevsky, J., 36, *93*
Koyama, J., 105, *137*
Kozu, H., 203(2.71), 209(2.71), *215*
Kramer, G., 55, *93*
Krause, M. O., 13, 14, 15, 26, 27, *31, 32*
Kronig, de L. R., 222, *280*
Kronquist, R. L., 151(1.30), 153(1.30), *173*
Kruger, J. B., 151(1.42), 157(1.42), 158(1.42, 1.56), 160(1.56), 162(1.42, 1.56), *173, 174*
Kugel, C. P., *96*
Kumabe, K., 105, 106, *137*
Kunz, C., 7, *31*

L

Lagorsse, J. M., 151(1.36), 153(1.36), 154(1.36), 158(1.36), *173*
Lakin, K. M., 103, *136, 137*
Landau, L. D., 122, *137*
Lang, C., 223(12), *280*
Laplanche, J., 163(1.57), *174*, 203(2.64), *215*
Larouche, R., 71, *96*
Lavanceau, J., 70, *95*
Laviron, M., 188(2.40), 193(2.40), 197(2.40), 203(2.40), *214*
Le Bail, J. L., 206(2.65), *215*
Lecoy, G., 203(2.77), 209(2.77), *215*
Lecrosnier, D. P., 191(2.44), 193(2.44), 203(2.44), 206(2.44), *214*
Lee, R. E., 158(1.62), 165(1.62), 166(1.62), *174*
Leeson, D. B., 49, 53, 59, *90*
Lefeuvre, S., 101, *137*
Lehovec, K., 175, 178(2.18), *214*
Le Mee, J., 175, *214*
Leschiutta, S., 87, *91*
Levine, M. W., 71, *96*
Levinshtein, M. E., 104, 105, *137*
Liechti, C. A., 203(2.62), *215*
Lieverman, T. N., 89, *96*
Lifschitz, E. M., 122, *137*

Lilienfeld, J. E., 141, *172*, 174, *213*
Lin, C. D., 12, 13, *32*
Lind, B. I., 122, *135, 137*
Lindholm, F. A., 158(1.70), 169(1.70), *174*
Lindsey, W. C., 55, *93*
Linfield, R., 80, *92*
Lipsky, L., 11, *32*
Loeb, H. W., 52, *93*
Lofer, D. D., 151(1.34, 1.35), 153(1.34, 1.35), 154(1.34, 1.35), *173*
Lopukhin, B. M., 121, *137*
Louisell, W. H., 108, 109, 117, 121, 125, 135, *137*
Lukirskii, A. P., 9, 10, 11, *32*
Lukomskii, V. P., 103, *137, 138*

M

McCarthy, D. D., 74, *93*
McCoubrey, A. O., 70, *93*
McCracken, D. D., *344*
McCumber, D. E., 104, *137*
McDonald, B. A., 158(1.76), 169(1.76), *174*
McDonald, G. J. F., 75, *94*
McFee, J. H., 102, 103, *137*
McGreough, P. L., 148(1.20), 151(1.20), *173*
McGuire, E. J., 11, *32*
McGunigal, T. E., 49, 53, 59, *90*
Macksey, H. M., 186(2.35), 187(2.35), 193(2.35), 203(2.35), *214*
Madden, R. P., 11, *32*
Maeda, M., 186(2.34), 187(2.34), 188(2.34), 193(2.34), 203(2.34, 2.67), *214, 215*
Maekawa, S., 151(1.22), *173*
Maloney, T. J., 202(2.55), 203(2.80), 209(2.80), 210(2.80), *215*
Manson, S. T., 1, 2, 3, 4, 5, 6, 7, 8, 9, 10, 11, 12, 13, 14, 15, 16, 17, 18, 20, 21, 22, 23, 24, 25, 26, 27, 28, 29, *31, 32*
Marcuvitz, N., 108, *137*
Markowitz, W., 41, 61, 62, 63, 73, 75, *93, 94*
Marr, G. V., 12, 16, 17, 18, *31*
Marrison, W. A., 42, *94*
Martin, A. V. J., 175, *214*
Marx, R. E., 104, *138*
Masuda, M., 103, *137*
Matarese, R. J., 106, *136*, 193(2.48), 195(2.48), *215*

Matsuo, Y., 103, *136, 137*
Matthaei, G. L., 264(36), *281*
Medved', A. V., 103, *137*
Meeks, M. L., 77, *94*
Meissl, P., 61, *91*
Melchior, P., 74, *94*
Menoud, C., 71, *94*
Meyer, M., 122, *138*
Middelhoek, S., 186(2.33), *214*
Mimura, T., 186(2.37), 187(2.37), 193(2.37), 203(2.37), *214*
Missoni, G., 7, 8, *31*
Mockler, R. C., 43, 63, 65, *90, 91, 94*
Mohr, T., 186(2.33), *214*
Moline, R. A., 151(1.37), 153(1.37), 154(1.37), 155(1.37), *173*
Moll, J. L., 142, *172*
Monnier, J., 151(1.44), 155(1.44), *173*
Moore, A. R., 223, *280*
Moran, J. M., *94*
Morgan, A. H., 42, *91*
Morris, D., 64, 68, 71, *94*
Morris, D. E., *345*
Morton, J., 20, 21, *32*
Moser, A., 186(2.33), *214*
Motsch, C., 175(2.10), 178(2.10), *214*
Msezane, A., 3, 4, *32*
Mueller, D. W., 158(1.54), 160(1.54), *174*
Mueller, I. I., 73, *94*
Mueller, L., 53, *96*
Mulholland, J. D., 36, *94*
Mullen, J. A., 49, 53, 59, *90*
Mullendore, J., 69, 72, 86, *90*
Muller, P. M., 75, *94*
Mungall, A. G., 61, 63, 64, 65, 68, *94*
Munk, W. H., 75, *94*
Murai, F., 193(2.47), 195(2.47), *215*
Murmuzhev, B. A., 103, *138*
Murray, J. A., Jr., 79, *94*

N

Nacci, J. M., 151(1.37), 153(1.37), 154(1.37), 155(1.37), *173*
Nadan, J. S., 101, 102, *136*
Nagasako, I., 203(2.71), 209(2.71), *215*
Nakagiri, K., 71, *94*
Nakayama, Y., 203(2.61), *215*
Namordi, M., 223(20), *281*

Napoli, L. S., 106, *136*, 187(2.38, 2.39), 193(2.38, 2.39), 203(2.38, 2.39, 2.69), 206(2.69), *214, 215*
Narayan, S. V., 104, *138*
Narayan, S. Y., 193(2.45), 203(2.45, 2.57a, 2.58), 205(2.57a), 206(2.57a), 208(2.57a), 212(2.57a), *214, 215*
Navon, D., 158(1.62, 1.63), 165(1.62, 1.63), 166(1.62), 167(1.63), *174*
Nelson, G. E., *345*
Newcomb, S., 36, 40, 41, *94*
Newhouse, V. L., 102, *136*
Newton, I., 35, *94*
Newton, R. R., 75, *94*
Nicholas, K. H., 151(1.25), 152(1.25), *173*
Nii, R., 105, *137*
Nilsson, O., 122, *135*
Nishizawa, J., 177, 180(2.30), 210(2.28), *214*
Norris, C. B., Jr., 222, 223(21), 224(22), 226, 232, 234, *280, 281*

O

O'Brien, J. F., 148(1.20), 151(1.20), *173*
O'Brien, R. R., 175, 178, *214, 281*
Ogawa, M., 203(2.71), 209(2.71), *215*
Ohara, S., 105, *137*
O'Hora, N. P. J., 75, *94*
Olson, K. H., 151(1.37), 153(1.37), 154(1.37), 155(1.37), *173*
Onori, G., 7, *31*
Orthras, G., 158(1.59), *174*

P

Pakin, S., *345*
Panella, G., 75, *94*
Pankratz, J. M., 158(1.56), 160(1.56), *174*
Papoulis, A., 49, *94*
Parekh, J. P., 103, *138*
Parekh, P. C., 151(1.29), 153(1.29), *173*
Parker, D., 187(2.36), 193(2.36), 203(2.36), *214*
Parkinson, B. W., 78, 83, *94*
Parry, J. V. L., 41, 62, 63, *92, 94*
Pauli, W., 84, *94*
Pavel, F., 74, *94*

P

Payne, R. S., 151(1.37), 153(1.37), 154(1.37), 155(1.37), *173*
Peletminskii, S. V., 103, 111, *135*
Pelous, G. P., 191(2.44), 193(2.44), 203(2.44), 206(2.44), *214*
Penfield. P. Jr., 121, *138*
Percival, D. B., 56, 57, 59, 61, 70, *95*, *97*
Perlman, B. S., 104, *138*
Peter, R., 100, *138*
Peters, H. E., 70, 71, *93*, *95*
Peterson, G. R., 275(42), *281*
Petit, P., 71, *95*
Picinbono, B., 51, 54, *91*
Pierce, J. R., 100, 101, 108, 114, *138*
Platzman, P. M., 99, *138*
Polovin, R. V., 107, 115, 116, *135*
Pontius, D. H., 158(1.69), 168(1.69), *174*
Poole, W. E., 158(1.51), 165(1.51), *173*
Popescu, C., 158(1.66), 165(1.66), 166(1.66), *174*
Potts, C. E., 78, *95*
Pozhela, Y. K., 104, 105, *137*
Prabhu, V. K., 78, *95*
Presser, A., 158(1.48), 159(1.48), *173*
Prim, R. C., 174, *214*
Pritchard, R. L., 142, 144(1.5), *172*
Proverbio, E., *92*
Pruniaux, B., 151(1.36), 153(1.36), 154(1.36), 158(1.36), *173*
Pucel, R. A., 203(2.78, 2.79), 209(2.78, 2.79), 210(2.78, 2.79), 211(2.78, 2.79), *215*
Pustovoit, V. I., 103, *137*, *138*
Putkovich, K., 61, *95*

Q

Quate, C. F., 102, 103, 108, 125, *136*
Quesada, V., *95*

R

Racine, J., 71, *94*
Radler, K., 7, *31*
Ramo, S., 108, 121, 126, *138*
Ramsey, N. F., 62, 63, 67, 69, 71, *95*
Ramsey, N. R., 71, *93*
Rao, A. R. P., 6, *32*
Rao, E. V. C., 151(1.26), 153(1.26), *173*
Rayner, J., 69, 72, 86, *90*
Reddi, V. G. K., 151(1.33), 153(1.33), 154(1.33), 155(1.33), 158(1.33, 1.68), 168(1.68), *173*, *174*
Reder, F. H., 80, *95*
Reeder, T. M., 108, *137*
Reichert, W. F., 187(2.38, 2.39), 193(2.38, 2.39), 203(2.38, 2.39, 2.69), 206(2.69), *214*, *215*
Rein, H. M., 142(1.14), *173*
Reindl, K., 151(1.28), 153(1.28), *173*
Reinhardt, V. S., 70, *95*
Reiser, M., 175, *214*
Reisse, R. A., 69, 72, 86, *90*
Renkowitz, D., 158(1.51, 1.53), 165(1.51), *173*
Ricci, D. W., *345*
Richardson, J. M., 43, 63, *91*
Ridella, S., 106, *136*
Rigaud, D., 203(2.77), 209(2.77), *215*
Ringer, D. E., 72, *95*
Ristow, D., 193(2.51), 195(2.51), *215*
Rittner, E. S., 222, 223, *280*
Roberts, L. A., 230(29), *281*
Robinson, B. B., 103, 106, *136*, *138*
Robson, P. N., 105, *138*
Rochester, M. G., 40, 75, *95*
Roosild, S., 203(2.72), 209(2.72), *215*
Rose, A., 175, *214*
Rosenberg, G. D., *95*
Ross, I. M., 174, 177, *214*
Rothenbriecher, O. H., *345*
Ruch, J. G., 203(2.53), *215*
Rueger, L., 82, *92*
Runcorn, S. K., 75, *95*
Russell, B., 35, *95*
Rutman, J., 53, 55, *95*

S

Saburi, Y., 82, *95*, *97*
Salecker, H., *96*
Saltzberg, G. R., 89, *96*
Samson, J. A. R., 9, 10, 11, 12, 13, 14, 28, 29, *32*
Sasaki, T., 7, *31*
Sasso, G., 186(2.33), *214*

Sayed, M. M., 158(1.47), 167(1.47), *173*
Scarlett, R. M., 147(1.17), 151(1.17), *173*
Scavuzzo, R. J., 151(1.37), 153(1.37), 154(1.37), 155(1.37), *173*
Schad, T., 142(1.14), *173*
Schafft, H. A., 158(1.65), 165(1.65), 166(1.65), *174*
Scheibe, A., 48, 74, *96*
Schiffer, M., 88, *90*
Schild, A., 83, *96*
Schlig, E. S., 158(1.60), 165(1.60), *174*
Schlömann, E., 103, *138*
Schmidt, P., 175, *214*
Schroeder, R., 63, *96*
Schroedinger, E., 37, 45, *96*
Schwinger, J., 108, *137*
Searle, C. L., 53, 54, 55, *92*
Sears, R. W., 223(11), *280*
Seaton, M. J., 2, *32*
Sechi, F. N., 158(1.58), 161(1.58), 169(1.58), *174*
Seeley, W. G., 175, 178(2.18), *214*
Sexl, R. U., 87, *96*
Shackle, P. W., 142(1.16), *173*
Shapiro, I. I., 83, *96*
Shapiro, R. K., 103, *138*
Shaw, H. J., 103, *136*
Shibata, J., 180(2.30), *214*
Shimizu, N., 106, *137*
Shin-ichi-Akai, 115, 122, *138*
Shirley, D. A., 29, *32*
Shockley, W., 142, *172*, 174, 180(2.31), *213*, *214*
Shur, M. S., 104, 105, *137*
Shyu, J. S., 7, 9, 25, 26, *32*
Siekanowicz, W. W., *138*, 251(31), *281*
Silverberg, E. C., *96*
Silzars, A., 227(24), 230(27, 28, 29), 277(43, 44, 45), *281*
Simons, R. L., 11, *31*
Sitenko, A. G., 107, 115, 116, *135*
Skobov, V. G., 99, *137*
Slater, J. C., 30, *32*
Smith, D. H., 268(39), *281*
Smith, H. I., 201(2.52), *215*
Smith, H. M., 48, 75, *96*
Smith, W. B., 158(1.69), 168(1.69), *174*
Smith, W. L., 49, 53, 59, *90*
Sodini, D., 203(2.77), 209(2.77), *215*
Solymar, L., 101, 114, *138*
Sommerhalder, R., 186(2.33), *214*
Sonntag, B., 7, *31*
Soula, J. P., 151(1.30), 153(1.30), *173*
Spangenberg, K. R., 260(34), *281*
Spector, H. N., 103, *138*
Spencer-Jones, Sir H., 75, *96*
Spenke, E., 142(1.8), 146(1.8), *172*
Stanley, J. T., 82, *96*
Starace, A. F., 8, 13, 14, 27, 28, *31*, *32*
Statz, H., 203(2.78, 2.79), 209(2.78, 2.79), 210(2.78, 2.79), 211(2.78, 2.79), *215*
Steele, M. C., 99, 101, 103, 104, 107, 111, 112, 115, 116, 122, *136*, *138*
Steggerda, C. A., 69, 72, 86, *90*
Stein, S. R., 73, *96*
Stepanov, K. N., 107, 115, 116, *135*
Stephenson, F. R., 75, *94*
Sterzer, F., 104, *138*
Stone, R. R., Jr., 89, *96*
Stover, H. A., 89, *96*
Stoyko, A., 40, *96*
Stoyko, M. N., 48, 75, *96*
Stoyko, N., *96*
Strutt, M. J. O., 158(1.49), 159(1.49), *173*
Sturrock, P. A., 107, *138*
Suhl, P., 100, *138*
Sumi, M., 101, 102, *138*
Susskind, C., 108, 117, *136*
Suyama, K., 186(2.34), 187(2.34), 188(2.34), 193(2.34), 203(2.34, 2.61), *214*, *215*
Suzuki, H., 203(2.61), *215*
Suzuki, T., 102, *138*
Swanenburg, T. J. B., 101, *138*
Swanson, E. R., *96*
Swanson, J. R., 11, *32*
Sydnor, R. L., 49, 53, 59, *90*
Sykes, R. A., 62, *92*
Symms, L. S. T., 48, 52, *92*
Synge, J. L., 39, *96*
Szebehely, V., 87, *91*
Szustakowski, M., 103, *138*

T

Tager, A. S., 101, *138*
Takagi, H., 103, *137*
Takahashi, K., 151, *173*
Takayama, A. Y., 203(2.68), *215*
Takenchi, 203(2.68), *215*

AUTHOR INDEX

Tango, H., 177, 210(2.28), *214*
Tatum, J. G., 147(1.18), 151(1.18), *173*
Taylor, B. C., 104, 105, *136*
Taylor, G., 277(43), *281*
Taylor, K. T., 12, 13, 14, *30*
Terasaki, T., 180(2.30), *214*
Ter-Martirosyan, L. T., 108, 109, 135, *136, 138*
Teszner, S., 163(1.57), *174*, 174, 175, 176, 177, 180(2.8, 2.9, 2.25), 186(2.7), 189(2.7, 2.9), 190(2.43), 193(2.43), 199(2.43), 200, 203(2.43), 206(2.43), 210(2.8, 2.9, 2.25), *214*
Thiennot, J., 101, 102, *138*
Thim, H. W., 104, *138*
Thornley, R. F. M., 151(2.40), 157(1.40), *173*
Thurston, M. O., 142(1.13), 158(1.13), *173*
Tien, P. K., 103, *138*
Tomboulian, D. H., 9, 10, *31*
Torop, L., 20, 21, *32*
Tremere, D. A., 142(1.10), 158(1.72), 169(1.72), *173*, *174*
Triano, A., 105, *136*, 187(2.38), 193(2.38), 203(2.38), *214*
Tribes, R., 175(2.10), 178(2.10), *214*
Tronc, P., 151(1.26), 153(1.26), *173*
Tsvirko, Y. A., 103, *137*, *138*
Tucker, R. H., 75, *96*
Tujimura, I., 186(2.37), 187(2.37), 193(2.37), 203(2.37), *214*
Tukey, J. W., 52, *91*
Turneaure, J. P., *96*
Turner, J. A., 187(2.36), 193(2.36), 203(2.36, 2.70), *214, 215*
Tursunov, S. S., 103, *137*

U

Uhink, W., 74, *94*
Umeno, M., 103, *137*
Upadhyayula, C. L., 104, *138*

V

Vainshtein, L. A., 108, 112, 114, 121, 126, 129, 132, *138*
van den Hurck, T. H. J., 151(1.23), *173*
Van der Ziel, A., 158(1.71, 1.73, 1.74), 169(1.71, 1.73, 1.74), 170(1.73), *174*, 175, 180(2.22), 203(2.73, 2.74), 209(2.73, 2.74), 210(2.22, 2.73, 2.74), *214, 215*
Van Duzer, T., 122, *138*
van Flandern, T., 39, 87, *97*
Vanier, J., 53, 71, 72, *90, 93, 96*
Van Vliet, K. M., 203(2.74, 2.75), 209(2.74, 2.75), 210(2.74, 2.75), *215*
Vashkovsky, A. V., 103, *136, 138*
Vedenov, A. A., 99, *138*
Verbist, J., 29, *30*
Vergnolle, C., 188(2.40), 193(2.40), 197(2.40), 203(2.40), *214*
Verma, K., 158(1.55), 160(1.55), 161(1.55), *174*
Vessot, R., 53, *96*
Vessot, R. F. C., 49, 53, 54, 59, 68, 71, 72, 73, *90, 93, 96*
Viennet, J., 71, *95*
Vikrilov, I. K., 101, *138*
Vladmirov, V. V., 99, *138*
Vogel, J. S., 203(2.81), 211(2.81), *215*
Voltmer, F. M., 103, *139*
von Neumann, J., 48, 53, *97*
Vural, B., 99, 101, 103, 104, 107, 111, 112, 115, 116, 122, 124, *138*

W

Wahl, A. J., 142(1.15), *173*
Wainwright, A. E., 62, 69, *97*
Wait, J. R., 78, *97*
Walker, T. E. H., 19, *32*
Walls, F. L., 62, 69, *97*
Watts, D. G., 49, 52, 55, *93*
Weber, J. T., 19, *32*
Wecki, B., 103, *138*
Wegner, P., *345*
Weinreich, G., 102, *138*
Wells, J. S., 38, *93*
Wells, J. W., 75, *97*
Wessel-Berg, T., 122, 130, *139*
Wesson, P. S., *97*
West, J. B., 12, 13, 14, 15, 16, 17, 18, 20, 21, *30, 31, 32*
West, R., 223(12), *280*
Whinnery, J. R., 108, 121, 126, *138*
White, D. L., 102, *137, 139*
White, M. H., 142(1.13), 151(1.13), 158(1.13), *173*

White, R. M., 103, *139*
Whittier, R. J., 142(1.10), 158(1.72), 169(1.72), *173, 174*
Wholey, J. N., 193(2.49), 203(2.49), 209(2.49), *215*
Wieder, B., 78, *95*
Wiener, N., 35, *97*
Wigner, E. P., *96*
Williams, R. E., 69, 72, 86, *90*
Wineland, D. J., 66, *97*
Winkler, G. M. R., 39, 45, 49, 53, 59, 61, 70, 78, 80, 87, *90, 95, 97*
Wolard, E. W., *97*
Wolf, P., 186(2.33), *214*
Wolff, P. A., 99, *138*
Wolkstein, H. J., 230(25), *281*
Woodruff, P. R., 13, 14, 15, *30*
Woolard, E. W., 73, 75, *97*
Wuilleumier, F., 13, 14, 15, 27, *32*

Y

Yakovenko, V. M., 99, *137*
Yamamoto, M., 82, *95, 97*
Yamamoto, R., 203(2.68), *215*
Yamanishi, M., 103, *139*
Yang, E. S., 175, 177(2.19), 178(2.13, 2.19), *214*
Yariv, A., 101, *137*
Yoshida, K., 103, *139*
Yoshimura, K., 61, *97*
Young, L., 264(36), *281*
You-Sun Wu, 151(1.42), 157(1.42), 158(1.42, 1.56), 160(1.56), 162(1.42, 1.56), *173, 174*
Yu, A. Y. C., 151(1.33), 153(1.33), 154(1.33), 155(1.33), 158(1.33), *173*
Yuan, H. T., 158(1.54), 160(1.54), *174*

Z

Zare, R. N., 15, *31*
Zetterquist, M., 147(1.17), 151(1.17), *173*
Zimkina, T. M., 9, 10, 11, *32*
Zoroglu, D. S., 158(1.46), *173*
Zubkov, V. I., 103, *136, 138*
Zühlke, R., 142(1.14), *173*
Zuleeg, R., 180(2.29), *214*
Zydney, H. M., 89, *96*

SUBJECT INDEX

A

Acoustic-wave amplifier, 102–103
Active polarized medium
 defined, 111
 normal-mode excitation in, 110–135
Allan variance, in clock performance, 52
Aluminum passivation, in bipolar transistor fabrication, 157
Ammonia maser, 72
Analog-to-digital converters
 desk calculators and, 310
 EBS devices as, 273–275
 for minicomputers, 317–319
Argon, photoionization cross section for, 11
Argon 3p, asymmetry parameter β for, 16–17
Argon 3s subshell photoionization cross section, 12
Astronomy, time scales in, 36
Atomic photoelectron spectroscopy, 1–30
 photoelectron angular distributions in, 15–29
 photoionization cross sections in, 1–15
Atomic clocks, 44
 in airplanes, 86
 atomic resonances in, 62
 clock errors in, 50
Autoionizing resonances, photoelectron angular distribution in neighborhood of, 28
Autoregressive moving average type filter, 48
Averaging time measurements, 78

B

BARITT diode, 171, 216
BASIC interpreter, for minicomputer, 323–324
BIH, see Bureau International de l'Heure
Bipolar transistor, 141–172
 advantages and disadvantes of, 171–172
 base layer width in, 150
 characteristics of, 170–172
 for class C amplifier operation, 159–161
 collector depletion layer width in, 150
 complementary process and mask creation for, 156–158
 cut-off frequency for, 143–144
 design and fabrication of, 151–158
 distortion and noise characteristics of, 168–170
 electrical characteristics and performance of, 156–170
 electron migration in, 156
 ion implantation for, 153–155
 multimetal systems for, 157
 proton-enhanced diffusion in, 155
 structure type geometries in, 147–150
 symbolic presentation of, 142
 thermal properties and reliability problems in, 165–168
Boundary value problems, normal-mode theory in, 108–109
Bulk spin wave amplification, in ferromagnetic semiconductors, 103
Bureau International de l'Heure, 40

C

Cache memory, in minicomputers, 299–300
Cathode ray tube, beam envelope of, 231
CCIR, see International Telecommunications Union Consultative Committee
Celestial navigation, time measurement in, 88
Central processing units, of minicomputers, 285–287
CERVIT cavities, hydrogen masers and, 71
Cesium, residual frequencies of, 56
Cesium beam atomic frequency standards, 62–70
Cesium clocks, 44
 clock errors in, 51
 commercial, 68–70
 long-term performance of, 57
 on satellites, 82
Charge-coupled semiconductor device, 273

Chlorine, photoionization cross section of, 13–14
Chronographs, compared with clocks, 43
Class C amplifiers, bipolar transistors for, 159–161
Clock(s)
  atomic vs. nonatomic, 44
  defined, 43 n.
  general principles of, 43–44
  performance of, 45–61
  quartz crystal, 42–44, 48
  "unusefulness" of, 48
Clock-carrying satellites, 82
Clock erratics, 52–57
Clock error
  measurement of, 46
  white frequency noise and, 51, 59
Clock hardware, 61–73
Clock modeling, 47–48, 57–60
Clock noise, 52–57
  gaussian white noise and, 59
Clock paradox, in relativity, 83–84
Clock performance, 45–61
  Kramer's curvature variance and, 55
COBOL language, for minicomputers, 324–325
Communications, time measurement in, 89
Complex-associated equipment, 126–129
CompuCorp desk calculator, 309
Computers, historical development of, 283–285
  see also Minicomputers
Continuum electron, spin-orbit effect on, 19
Cooper minima, in photoionization, 2, 9, 20
Coulomb phase shifts, 16, 24–25
Cut-off frequency
  for bipolar transistors, 144
  for field-effect transistors, 180

D

Data General Eclipse microcomputer, 309
Data General Eclipse minicomputers, 287–289
Data General Nova minicomputers, 288–289
Deflected-beam amplifiers, EBS devices as, 266–273
Desk calculators, 309
Digital Equipment Corporation, 295, 302

Direct memory access, in minicomputers, 305–306
Disk storage files, for minicomputers, 325–328
DMA, *see* Direct memory access
Drain current saturation process, in field effect transistors, 178–179

E

Earth rotation, in time determination, 74–75
EBS, *see* Electron-bombarded semiconductor
Eigenfunction, complex-associated, 126
Electromagnetic fields, internal and external sources of, 112–115
Electromagnetic waveguides, 113
Electron-bombarded semiconductor, 221–280
  as analog-to-digital converter, 273–276
  analytic foundation of, 232–238
  beam voltage in, 248
  as current source shunted by capacitance, 238–241
  defined, 221–222
  as deflected-beam amplifier, 266–270
  deflection-modulated amplifier using, 229
  device configurations and performance for, 260–276
  diode output capacity for, 236–238
  diode perimeter in, 241–242
  epitaxial thickness and effective radius of curvature in, 253
  excess capacitance ratios in, 255
  future potential of, 277–280
  general characteristics of, 227–232
  historical perspective on, 222–224
  life and reliability of, 276–277
  linear output voltage and current capabilities of, 239–245
  mean time to failure for, 276
  mounting substrate for, 258–259
  operating principle of, 224–227
  output current optimization in, 235–236
  physical limitations of, 238–246
  as planar-gridded amplifier, 260–266
  reverse-breakdown voltage in, 252
  as rf amplifiers, 264–266
  risetime optimization in, 234–235
  as scan converter, 272

SUBJECT INDEX

semiconductor diode in, 246–258
semiconductor materials for, 256–258
as signal processors, 271–276
target of, 232–235
target assembly in, 246–259
temperature rise in, 243–244
tip-contact metallization for, 251
as transient signal sampler, 271–273
as video amplifier, 261–264, 278
Electron-bombarded semiconductor amplifier, applications of, 160–273, 278
Electronic navigation, in rho-rho mode, 88
Electron migration, defined, 156
Emitter area, for power transistors, 146
Emitter lateral geometry, 147–148
Emitter periphery/base area ratio, for power transistors, 146–147
ENIAC computer, 284
Ephemeris time, 36, 41
dynamical time scale and, 87

F

FASTBUS architecture, 296
Ferromagnetic semiconductors, bulk spin wave amplification in, 103
FET, see Field-effect transistor
Field-effect transistor, 174–213
advantages and disadvantages of, 192
basic idea of, 174
channel gate equivalent circuit for, 180
cut-off frequency for, 180
design and fabrication of, 193–203
electrical characteristics and performance of, 203–212
frequency-power performance for, 203–208
general theory of, 174–184
Gridistor geometry for, 190
horizontal channel structures in, 185–188, 193–197
ion implantation in, 199–201
MESFET, see MESFET (mesa-structure field-effect transistor)
noise and distortion characteristics of, 209–212
overlaid gate-junction structure for, 193–197
semiconductor material choice for, 201–203

structure and circuit of, 176, 190, 193–197
structure-type geometries for, 185–196
thermal properties and reliability of, 208–209
vertical channel structures in, 197–201
Floppy disks, for minicomputers, 316–317
FORTRAN compilers, 286
for minicomputers, 286, 321–325
FORTRAN subroutines, 307
Four-dimensional space, 85
Frequency-power performance, of FETs, 203–210

G

General Conference for Weights and Measures, 41
Geodesy, time measurements in, 88
Global Positioning System, 78
GPS satellites, clocks on, 82

H

Hewlett-Packard 2105 minicomputer, 322
Hewlett-Packard 3000 minicomputer, 287, 309
HF results, in photoionization, 2
HF Standard Frequency and Time Signal, 77
HS cross section, in photoionization, 2
HS wave functions, in photoionization, 7
Hydrogen masers, as clock reference, 70–72

I

IMAGE data base management system, 327
IMPATT diode, 172, 213, 216–219
INFOS data base management system, 327
International Atomic Time, 38, 41
International Business Machines Corp., 284, 309, 330
International Congress of Chronometry, 43
International meridian, 40
International Telecommunications Union Consultative Committee, 42
Irrotational electronic beams, 121

K

Kinetic power theorem, 121
Kolmogorov structure functions, 55
Kramer curvature variance, 55
Krypton, photoionization in, 6
Krypton 3p subshell, asymmetry parameter for, 26

L

Laser pulse ranging, in time determination, 76
London, University of, 223
LORAN C, 78, 80
Lorentz transformation, 84–85
Lunar laser ranging, in time determination, 76

M

Magnetostatic amplification, 103–104
MBPT results, in photoionization, 2
Mean sun, 40
MEDC field-effect transistor, 188
Memory interleaving, in minicomputers, 298–299
Memory speed, in computers, 298
MESFET (mesa-structure field-effect transistor), 185–188
  design and fabrication of, 195–197
  frequency-power performance of, 207
  thermal properties and reliability of, 209
Metrology, time measurement in, 89
Microcomputers, 308–310
Microwave power semiconductor devices, 141–219
  see also Bipolar transistors; Field-effect transistor; Three-terminal devices
  present and future development in, 216–219
Minicomputers
  analog-to-digital converters and, 317–319
  applications of, 328–330
  basic concepts of, 283–330
  BASIC interpreter and, 323–324
  bus architecture for, 295–297
  cache memory in, 299–300
  central processing units for, 285–287
  COBOL language for, 324–325
  computer consoles for, 310–312
  data base management systems and, 326–327
  data checking in, 293–295
  data entry I/O for, 313–314
  data set interfaces or modems for, 320
  desk calculators and, 309–310
  direct memory access in, 295, 305–306
  disk systems for, 315–316, 325–328
  floppy disks for, 316–317
  FORTRAN compiler for, 286, 321–325
  FORTRAN subroutines and, 307
  historical development of, 283–285
  IEEE standard digital interface and, 319–320
  instruction repertoire of, 287–293
  interrupt structure in, 300–304
  larger size, 287
  magnetic tapes for, 314–315
  memory areas in, 292
  memory interleaving in, 298–299
  memory mapping in, 292–293
  memory protect feature in, 298
  microprogrammable systems and, 306–308
  page addressing in, 291
  peripheral devices and, 310
  power-fail protection in, 297
  printing devices and, 312–313
  process control peripherals and, 317–319
  program counter and processor status items in, 302–303
  read-only memory in, 306
  real time capability of, 326
  report program generation for, 324–325
  "smart" terminal in, 330
  software for, 321–325
  tape cassettes for, 315
  teletypewriters and, 311
Modems, for minicomputers, 320
Molecular fields, multicenter nature of, 30
Molecules, photoionization of, 30
Monochronism, 39
Mutual coupling, among modes, 130–135

N

National Aeronautics and Space Administration, 222

SUBJECT INDEX

National Bureau of Standards, 42
  crystal clocks of, 75
Negative differential mobility, 100, 104–106
Neon, photoionization cross section for, 10
Neon 2p, asymmetry parameter for, 15
Neon 2s subshell photoionization cross section, 14
Node orthogonality, reciprocity theorem and, 125–130
Nondegenerate solid-state plasmas
  Eulerian and polarization description of, 115–121
  small-signal power theorem for, 122
Normalization relations, reciprocity theorem and, 125–130
Normal-mode excitation
  by external sources, 132–135
  generalized theory of, 110–135
Normal-mode method, in boundary-value problems, 108
Normal-mode theory, in wave interaction analysis, 106–110
NPL cesium standard, 75

O

Original eigenfunction, 126–129

P

Phillips Laboratories, 222
Photoelectron angular distributions, 15–29
Photoelectron-ion angular momentum exchange, 27
Photographic zenith tube, 73
Photoionization
  Cooper minima in, 2, 9
  HF calculation in, 8, 11
  HS wave functions in, 7
  MBPT results in, 11
Photoionization cross sections, 1–15
Physikalisch Technische Bundesanstalt, 41
Planar-gridded amplifier, EBS as, 260–266
Polarized medium waveguides
  electromagnetic fields of, 113
  normal-mode excitation in, 110–135
Polarization small-signal velocity, 119
Polarization vectors, 110–111

Power transistors
  base layer method in, 150
  electrical characteristics and performance of, 203–212
  emitter area for, 146
Poynting's theorem, in microwave vacuum electronics, 121
PRN, see Pseudorandom noise modulation
Proton-enhanced diffusion, in bipolar transistors, 155
Pseudorandom noise modulation, in time measurements, 78–79
Pseudorandom noise systems, 88–89
PTB, see Physikalisch Technische Bundesanstalt
PZT, see Photographic zenith tube

Q

Quartz crystal clock, 42, 48
Quartz crystal oscillators, 61–62

R

Random access memories, in minicomputers, 307
RCA Princeton Laboratory, 223
Read-only memory, in minicomputers, 306
Reciprocity theorem, 125–130
Relativity, time and, 83–85
Remote time measurements, 77–89
  bandwidth in, 78
  geometric path delay in, 77–78
Report program generator, for minicomputers, 324–325
Rf amplifiers, EBS as, 264–266
Rubidium clock, 82
Rubidium vapor cells, as low-cost clock, 72

S

Satellite time measurements, 82–83
Semiconductor devices, electron-bombarded, see Electron-bombarded semiconductor
Smiconductor diode, in EBS, 246–258
Semiconductor electronics limit, 116
Signal processors, EBS as, 271–276

SUBJECT INDEX

Slow-wave structures, 100–101
Small-signal power theorem, 121–125
"Smart" terminal, in minicomputers, 330
Sodium atoms, photoionization cross section for, 1–2
Solid-state plasmas
  nondegenerate, 115
  small-power theorem and, 122
Solid-state traveling-wave tube, 100–110
Solar time, 39, 41
Space perception, levels of, 35
Space tracking, time measurement and, 88
Sperry-Rand Corporation, 284
Spin-orbit effect, photoelectron angular distribution and, 19
Spin-wave amplification, 103–104
Standard Frequency and Time Signal, 77
Stanford University, 223
STWT, see Solid-state traveling-wave tube
Substitution analysis, in wave analysis, 107
Sun, solar time and, 40
SWS, see Slow-wave structures
SYMPHONIE satellite, 82
Systems Engineering Laboratories, 287

T

TACAN application, EBS in, 279
TAI, see International Atomic Time
TED diode, 171–172, 213
Teletype Corporation, 311
Teletypewriters, microcomputers and, 311
TEO diode, 216, 218
TFSS, see Thin-film semiconductor structures
Thermal runaway, in bipolar transistors, 167
Thin-film semiconductor structures, 106–110
  wave interactions in, 99–135
Three-terminal devices, 141–172
Time
  clock and, 38
  equal processes in, 36, 39
  events in, 35
  flow of, 34
  mean solar, 39–40
  objective reality and, 34
  relativity theory and, 38, 83–85
  uniform, 35–36
Time code generator, 43

Time determination, 73–77
  see also Time measurement
  Doppler satellite system and, 76
  earth rotation and, 74–75
  lunar laser ranging and, 76
  new methods of, 75–77
  VLBI in, 76–77
Time dissemination capabilities, 81
Time intervals, equal, 36
Timekeeping, 33–89
  clocks and, 43–73
  reviews, literature, and conferences on, 42
Time measurements
  accuracy of, 37
  applications of, 88–89
  consistency principle in, 39
  functional approach to, 36
  LORAN C in, 78–80
  polar variation in, 40
  pseudorandom noise modulation in, 78–79
  remote, 77–89
  be satellite, 82–83
Time scales, 38–39
  in astronomy, 36
  clock time and, 60–61
Time standard, history of, 39–42
TOTAL data base management system, 327
Transient signal samplers, EBS as, 271–273
Transistor, bipolar, see Bipolar transistor
Transistor common-base equivalent noise circuit, 169–170
TRAPATT diodes, 172, 213, 216, 219
Traveling-wave amplifiers
  design and analysis of, 100–110
  with negative differential mobility semiconductors, 104–106
  wave interactions in, 99
Traveling-wave tube, 99
  electron-bombarded semiconductors and, 278
TWA, see Traveling-wave amplifiers
TWT, see Traveling-wave tube

U

UNIBUS architecture, 295–296
UNIVAC computer, 284, 309
Universal time, 40–41
  determination of, 73–74

SUBJECT INDEX

Universal time clock, 41
UT, *see* Universal time
UT1, UT2 time scales, 40–41
UTC time scale, 38, 41

V

Vacuum-electronics limit, 116
Very long baseline radio interferometry, 38
  in time determination, 76
Video amplifiers, EBS devices as, 261–264, 278
VLBI, *see* Very long baseline radio interferometry

W

Wang Laboratories, 309
Watkins-Johnson Company, 222–223

Wave analysis, substitution analysis in, 107
Waveguides, electromagnetic fields and, 112–113
Waveguide structure, polarized medium coupling with, 133
Wave interaction analysis, normal-mode theory in, 106–110
Wave interactions
  normal-mode excitation by external sources in, 130–135
  polarized medium coupling in, 133
  reciprocating theorem in, 125–130
  in thin-film semiconductor structures, 99–135

X

Xenon, photoionization cross section for, 10
Xenon 4d, asymmetry parameter β for, 20
Xenon 5p, asymmetry parameter β for, 18

U.C. BERKELEY
ENGINEERING LIBRARY

RETURN   ENGINEERING LIBRARY
                        642-3366